International Solutions to Sustainable Energy, Policies and Applications

International Solutions to Sustainable Energy, Policies and Applications

Stephen A. Roosa
Ph.D., CEM, CSDP, REP

First published by Fairmont Press in 2018.

Published 2020 by River Publishers
River Publishers
Alsbjergvej 10, 9260 Gistrup, Denmark
www.riverpublishers.com

Distributed exclusively by Routledge
605 Third Avenue, New York, NY 10017
4 Park Square, Milton Park, Abingdon, Oxon OX14 4RN

First issued in paperback 2023

Routledge is an imprint of the Taylor & Francis Group, an informa business

Library of Congress Cataloging-in-Publication Data

Names: Roosa, Stephen A., 1952- editor.
Title: International solutions to sustainable energy, policies and applications / [edited by] Dr. Stephen A. Roosa.
Description: Lilburn, GA : The Fairmont Press, Inc., [2018] | Inc ludes bibliographical references and index.
Identifiers: LCCN 2017039956| ISBN 0881737895 (alk. paper) | ISBN 9788770222211
(electronic) | ISBN 9780815381020 (Taylor & Francis distribution : alk. paper)
Subjects: LCSH: Renewable energy sources. | Energy development. | Energy policy. | Sustainable engineering.
Classification: LCC TJ808 .I634 2018 | DDC 333.79/4--dc23
LC record available at https://lccn.loc.gov/2017039956

International Solutions to Sustainable Energy, Policies and Applications/Stephen A. Roosa

ISBN 9780881737899 (The Fairmont Press, Inc.)
ISBN 9780815381020 (print)
ISBN 9788770222211 (online)
ISBN 9781003150978 (ebook master)

ISBN 13: 978-87-7022-945-6 (pbk)
ISBN 13: 978-0-8153-8102-0 (hbk)
ISBN 13: 978-1-003-15097-8 (ebk)

Dedication

This book is dedicated to my three wonderful children, Sarah, Daniel and Maryelle. They have grown into caring and responsible adults. I know that during their lifetimes, they and others like them will continue to help ensure the sustainability of our planet. After all, we must focus our efforts on creating a better world for future generations.

Table of Contents

Foreword

"Sustainable Development"… think about it.

Preface

In 2008, Dr. Stephen Roosa published *The Sustainable Development Handbook* which became the seminal work that provided the world's practitioners with a much-needed guide to understanding sustainability as a solution to the world's developmental problems. His new work, *International Solutions to Sustainable Energy, Policy and Applications* is a major contribution that could help save our civilization as we know it. Dr. Roosa argues that "A new era is upon us—one that focuses on energy technologies and how we transition our economies to reduce greenhouse gas emissions by both reducing the combustion of carbon-based fuels and preventing the release of their gases into the atmosphere... New strategies, programs and technologies at an international scale are needed to mitigate this threat. Constructive, synchronized, and comprehensive efforts must be engaged. Our success will be ensured with international cooperation, creating partnerships, and implementing solutions—failure will be assured without it."

This book is a must read for activists, designers, engineering professionals, elected leaders, and others interested in finding solutions to the climate chaos we are beginning to experience. One of the most important challenges facing us is climate change—a force capable of disrupting civilization. Earth is being impacted by more frequent and extreme weather occurrences that include droughts, fires, high temperatures, flooding and horrific storms. There are an increasing number of places on Earth that are now considered uninhabitable. Our planet is being adversely impacted by greenhouse gasses and environmental issues associated with fossil fuel combustion. Science has shown that mankind's activities have contributed to the vast majority of these problems.

Dr. Roosa's book is not intended to pour additional fear into a landscape already plagued by it. The purpose of his book is to present options for a future that is still in our hands. It offers proof of international efforts of how we can face these problems and work toward viable, practical and proven solutions.

One key to our future is adopting ways to reduce the use of fossil fuels to satisfy our energy needs. With only 5% of the world's population, the U.S. uses 25% of the world's energy and disproportionately contributes to greenhouse gasses emissions that are making our world more uninhabitable. This trend must be reversed. If we follow the historic Paris Agreement, our lifestyles must change. By reducing our carbon dependency by driving cars less, consuming less fossil fuel, living in smaller and more compact communities, reducing commutes, polluting less, and recycling our homes and buildings.

The future of our civilization is linked to greener cities that use renewable energy. The people in these communities bike to work, ride trams to the grocery store and walk to sporting events. Millennials are driving this trend, creating a brighter future by demanding cities that are more like San Francisco, Amsterdam, Bristol and Portland. They want cities that are safer, healthier, prosperous and sustainable. This book explores successful mitigation strategies that are being adopted to help save the Earth from environmental disruption. It provides ideas and solutions. proving that people around the world are taking action locally and regionally. Roosa's book is like no other. It provides an agenda of solutions of how we can create greener environments and reduce carbon releases from our businesses, industries and vehicles. This book discusses solutions to reduce our carbon footprint, create good jobs, and make our lives more sustainable. It offers our elected officials workable policy solutions and is among the best books in this area.

In 1960, social critic Paul Goodman wrote in *Growing Up Absurd*, "A man has only one life and if during it he has no great environment, no community, he has been irreparably robbed of a human right." Neither socialism nor free enterprise can provide a popular or workable analysis to addressing climate warming and at the same time create greater health, safety, prosperity and justice. This book addresses how people can work together to create positive changes to promote a more sustainable world.

John, I. Gilderbloom, Ph.D.

Dr. Gilderbloom is a Professor in the Graduate Planning, Public Administration, Public Health, and Urban Affairs program at the University of Louisville, where he directs the Center for Sustainable Urban Neighborhoods (http://sun.louisville.edu). He is a Fellow of the Scholars Strategy Network housed at Harvard University under the direction of Professor Theda Skocpol. An international poll of thousands of urbanists, planners and architects, ranked him as one of the "world's top 100 urban thinkers." His research in urban sustainability has appeared in eight co-authored or edited books or journals, and 55 scholarly peer-reviewed journals, with opinion pieces in notable publications including *The Wall Street Journal, Washington Post, Los Angeles Times, Chicago Sun-Times, San Francisco Chronicle, Courier-Journal*, and *USA Today Magazine*.

Acknowledgements

During the course of my international travels in over 80 countries, I learned to better understand the various perspectives of people from other countries about the nexus between sustainability and energy. My personal catharsis can be traced to Hungary in 1999 when I was asked to be the keynote speaker at a major European energy conference. After presenting at length about how to justify energy projects based on economics (which I will call the U.S. business school model), a few of the Europeans engaged me in a lively discussion that gravitated toward the idea that "here, we just do sustainability projects because it is the right thing to do" (or the model that seems to make sense and would be fun to do). There are times when these categories overlap, which is where I began to focus most of my professional efforts.

I didn't realize it then, but that conversation began a new interest and series of investigations that somehow, among those divergent paths of my life, ultimately led to the idea for this book. The seminal events included an AEE trade mission to South America, where we met with engineers and managers engaged in energy projects and activities. The workshops I presented in India and Hong Kong helped me better understand the energy challenges in some of the Asian countries. I recall a hot mid-summer afternoon sitting in a government office in New Deli when the power failed and the lights went out. India's then Under Secretary for Renewable Energy turned to me and said it happened almost every day at about that time, a local and certain method of electrical load shedding. I wonder even now if this would ever be handled so casually in the U.S. There were three more recent trips to Russia arranged by the U.S. State Department (thanks to Masha Lvova, Irena Chernushkina, and Wendy Kolls) during which I gave over 40 presentations on sustainable development and renewable energy.

I began to realize that there was more for us to learn about the linkages between energy and sustainability and we needed to identify more examples. In Cape Town, South Africa, I taught a renewable energy seminar held at Africa's first LEED Platinum hotel. While you don't have to go to Africa to learn how to build a sustainable hotel, it was a learning experience for me. More examples are out there, just waiting for us to explore, and offering hope that the solutions to our problems can be found.

During my travels, I witnessed firsthand that the impacts of climate change were already occurring. This created in me greater incentive to develop this book, a catalog of international research, studies and solutions.

I have been truly blessed to have had the opportunity to learn and share experiences with people who helped me understand the salient aspects of how important sustainable development is as over-arching policy principle. I had little desire to change the world, but knew that levers existed. This book is one among many, serving as a means to inform. It provides a set of stories, each a window into sustainable solutions on an international scale. My sincere thanks to all the contributing authors, too many to name but without them there would be little meaningful content in this book. Keep in mind that for many of these authors, English is a second or third language. This made corresponding with them and editing their contributions challenging at times. To them I give all the credit. It is my hope that my skills enhanced their dialogues, achieved their intents, and clarified their messages.

I have been moved by the investment my friends, family, and professional counterparts have made in me. I hope that this book will serve as a down payment on their efforts. Early in my career, there were those who helped shape and support my belief that somehow society was not on the right track when it came to energy and improvements were needed. A few helped me understand the long-term implications of decisions we make and the importance of contributing to the education of others. Sarah Tate, helped me understand how design principles in buildings help shape structures in ways to solve environmental needs. Richard Levine, a man with a dynamic yet alternative view on how sustainable development can be a more successful policy, has always been an inspiration. Thomas Lyons and Charles Ziegler both assisted me with many of the concepts and theories that this book touches at a time when most of the core concepts were in their infancy. Wayne Turner and Barry Capehart have always amazed me at their ability to produce quality results in print. John Gilderbloom has been a great friend and supporter.

A special thanks to Al Thumann for his insights and support. The folks at the Fairmont Press supported this project since the onset and believed that a book on this topic was both timely and worthwhile. Linda Hutchings is such a fine lady as is Joy Maugans worked long and patiently to give this book the look it needed.

There are innumerable friends, professors and associates who took the time to listen to my ideas about sustainability and offered innumerable insights into to the topic and offered their thoughts and support, many of them befriended during my wanderings. Those who have influenced me include Leonard Ravitz, Herb Burns, Steven Koven, Doug Kelbaugh, Jerry Taylor, John Whitney, Carl Salas, Eric Woodroof, MaryAnne Lauderdale, Ron Vogel, L. J. Grobler, Victor Ottaviano, Albin Zsebik, Laurie Wigand-Jackson, Steve Parker, Richard Costello, Nailya Kutzhanova, Lin Ye, Arnold Schwarzenegger, Tim Janos, Larry Good and Peter Myers—all wonderful people doing great things. My thanks to them for the inspiration they have offered and the gift of their time.

Stephen A. Roosa, Ph.D., CEM, CSDP, REP
Editor

Introduction

A new era is upon us—one that focuses on energy technologies and how we transition our economies to reduce greenhouse gas emissions by both reducing the combustion of carbon-based fuels and preventing the release of their gases into the atmosphere. Excessive generation of these man-made emissions is directly linked to global climate change. The fragile nature of the earth's ecosystems and the potential irreversibility of climate change are a global environmental threat which calls for concerted action. New strategies, programs and technologies at an international scale are needed to mitigate this threat. Constructive, synchronized, and comprehensive efforts must be engaged. Our success will be ensured with international cooperation, creating partnerships, and implementing solutions—failure will be assured without it.

The policies we adopt must enable deployment of workable technologies that have potential to achieve carbon reduction goals. Coherent public disclosure of mitigation strategies amplifies deployment effectiveness across sectors, industries and levels of government. In particular, coordinating mitigation, legislation and policies at the local, state, federal and international levels is necessary and must be encouraged. We need to strongly advocate for integrated, cohesive carbon reduction strategies and offer unyielding evidence to support the need for more progressive measures to stimulate intergovernmental cooperation. Should successful mitigation strategies and technologies not be implemented, we will soon find that the costs of adaptation will hinder economic growth.

Our options to react to climate change are limited by our infrastructure. Without concerted action on a global scale, reducing the impacts of climate change will be increasingly difficult. There is a need to revamp our governmental policies and institute a new range of programs. Simultaneously, we must develop and implement new energy technologies that enable permanent reductions in atmospheric greenhouse gases. The complexity of this problem is confounded by the world's energy requirements, developmental needs, growing populations, progressive social goals, market uncertainties, resource constraints and untested technologies. The importance of addressing increasing atmospheric carbon emissions cannot be understated—both the sustainability of the planet's ecosystems and the survivability of mankind are at risk.

Our local policy initiatives (e.g., community action, economic incentives, community action, and appropriate infrastructure design) must encourage energy-efficient lifestyles, renewable energy deployment and more sustainable developments. Successful policy intervention is possible.

The crossroad that is upon us creates a multitude of dilemmas but offers exciting opportunities. One dilemma is that our failure to take action today to reduce atmospheric carbon emissions exacerbates the difficulty of reducing carbon levels in the future. Inaction and delay ultimately increases mitigation costs. To resolve this, we must now change our agendas, reform our policies, institute new programs and implement new technological solutions. There is much work to be accomplished and we have only just begun to recognize solutions. Effective solutions must be deployed to achieve the goal of reducing the quantities of greenhouse gases being released into the atmosphere.

Like many other environmental issues, people are accepting the reality that the long-term costs of response, adaptation, remediation and mitigation associated with climate change will ultimately far exceed the costs of inaction. The Paris Agreement, signed by representatives of 193 countries, is one step toward policy improvement at an international scale. To be effective, all countries must cooperatively participate. At the Marrakech Conference of Parties 22, countries agreed to "move forward purposefully to reduce greenhouse gas emissions and to foster adaptation efforts, thereby benefiting and supporting the 2030 Agenda for Sustainable Development and its Sustainable Development Goals."

Supporting continued development of renewable energy systems is a key part of the solution. Renewable energy technologies are proven job creators. It is one of the solutions that enables us to improve the economy, improve the environment, and encourage leading technological research. Renewable energy creates new opportunities for an alternative future, one that is far less dependent on fossil fuels. It is a future that envisions less greenhouse gas emissions, cleaner air and water resources and more healthy environments. To achieve this, we must deploy more renewable energy systems. Solutions on an international scale are needed. This book considers these solutions.

International Solutions to Sustainable Energy, Policies and Applications provides an in-depth examination of alternative energy sources, applications, technologies and policies as a means of meeting international sustainability goals. Key themes include policy analysis, program assessment, energy efficiency, clean energy and carbon reduction. This book includes a compiled set of chapters discussing the various forms of strategies being planned and implemented to reduce energy use and carbon emissions and the solutions offered by alternative energy. Taking an international perspective, contributors from countries including the U.S., Canada, Peru, Hungary, Iran, Ukraine, Jordan, the UAE, South Africa, Korea, India and China among others offer their perspectives on the issues and solutions they are focused upon. Economic issues and financial methodologies are also considered.

Chapter 1, authored by Dr. Stephen A. Roosa, considers sustainable development and its linkages to energy. Sustainability, in terms of energy use, can be considered to be: 1) providing energy that is environmentally benign; 2) not using energy in an overly consumptive manner; 3) providing options for renewable energy; 4) reducing the need for energy from non-renewable energy sources; and 5) distributing energy resources equitably. To explore more fully the relationship of energy to sustainability, it is important to consider how this relationship evolved. Sustainable solutions are the thrust of this book.

Chapter 2 offers an analysis of energy use policy criteria for developing nations. Energy plays an important role in the development of societies. Policy makers enact policies intended to increase the security of energy supply and consumption. However, in most cases the ethical issues of the policies are not considered when policies are created and implemented. This chapter considers energy policies and their ethical notions in developing countries. The focus is to determine and prioritize energy consumption policy criteria in order to achieve intergenerational justice in Iran. The results show that renewable energy provides the best alternatives for solutions when ethical notions are considered. It is authored by Ali Asghar Pourezzat, Ali Asghar Sadabadi, Mohammad Koohi Khor, Narges Salehi, Shahrabi, Alireza Aslani and Mohammad Mahdi Zolfagharzadeh.

Chapter 3 considers facility scale energy storage and its applications, technologies and barriers. Many facilities in the U.S. are deploying energy storage technologies for electric demand response programs. Technologies developed for energy storage show promise for managing short-term electrical demand peaks and longer-period demand response events. This chapter's author is Jesse Remillard, a project engineer who has investigated facility/campus-scale energy storage for efficiency program administrators and who recently completed a storage technology research report for an international consortium of utilities. These systems can be responsive to utility demand programs and time-of-use rates to reduce electrical peak demand costs. This chapter considers the technical properties of storage systems, including flywheel, compressed air, and various battery technologies. The technical and market barriers associated with distributed storage, along with proposed paths for resolving these barriers, are also discussed.

Chapter 4, authored by Dr. Md. Faruque Hossain, a former director of technical services for the City of New York, considers clean energy development from cyanobacterial biochemical products. It discusses an improved process technology to produce hydrogen biologically. A photobioreactor is a bioreactor that utilizes a light source to cultivate phototrophic microorganisms. The photochemical reaction among photons, ultraviolet (UV) light, and cyanobacterial biomaterials in photobioreactors offers a unique methodology for producing hydrogen energy. Using this technology, hydrogen production is significantly higher than from any other technology that has ever been used. In this chapter, we discover that producing hydrogen using cyanobacteria could be a method of meeting future global energy requirements.

Chapter 5 evaluates energy planning approaches for Canadian First Nation communities. There is a close link between energy security, economic prosperity, sustainability and sovereignty for indigenous communities in Canada. Geographically remote locations, absence of all-season roads, off-grid status, diesel fuel dependency and lack of alternative energy access causes energy insecurities along with serious economic, social, and local environmental concerns in northwestern Ontario's First Nation communities. Freedom from diesel dependency and scoping sustainable energy solutions are immediate priorities and motivate effective community energy planning. The chapter's overarching purpose is to bridge knowledge gaps regarding socio-cultural requirements, discuss the social costs in energy planning, and advance academic literature on indigenous perspectives on energy planning. It details the disconnects between theory and practice in energy planning for these communities. This chapter is authored by Roopa Rakshit, Chander Shahi, M.A. Smith and Adam Cornwell.

Chapter 6 offers a hybrid electrical generation solution for Peru's grid independent oil and gas well fields, an option that can be used elsewhere. In Perú today, fossil fuels including petroleum, diesel fuel and natural gas are used to produce electrical energy for oilfield extraction purposes. Due to their remote geographic locations, oil and gas fields are often not linked to interconnected electrical grids. High investment costs are required to produce electricity in these remote locations. Wind resources can produce electricity using wind turbine generators. Evidence of this includes Peru's 164 MW of installed capacity, distributed among four operating wind farms. This chapter presents research that assessed the implementation of a hybrid wind-thermal natural gas system to provide electrical power for an oilfield and estimates the reductions in fuel consumption and CO_2 gas emissions. It is authored by Francisco Porles, a senior mechanical engineer who focuses on energy infrastructure projects.

Chapter 7 assesses barriers to renewable energy development. The Republic of Trinidad and Tobago is located in the Caribbean, just off the coast of Venezuela. The economy is hydrocarbon-based with the energy industry contributing approximately 45% of the country's gross domestic product. This tropical country faces challenges that include suppressed oil and natural gas prices, falling hydrocarbon production, and gas curtailment in its petrochemical sector. The latest hydrocarbon assessments indicate that the country has proven natural gas reserves to last to only 2019. It is imperative that the economy becomes more diversified including its power sector. This chapter discusses the primary barriers to renewable energy development and implementation in Trinidad and Tobago. It also identifies the hurdles the country must overcome to formulate a sustainable energy strategy in the context of renewable energy, energy efficiency and greenhouse gas emissions reductions. Dr. Zaffar Khan and Atiyyah Khan authored this chapter.

Chapter 8 rigorously examines the connection between nuclear energy consumption and economic growth in Spain, an oil-importing country. This study by Francisco Porles, a senior mechanical engineer, empirically observed the nexus between nuclear energy consumption and economic growth in the country of Spain. Unit root and stationarity tests, Johansen cointegration tests, vector error correction models, and Granger causality tests were applied to annual data for the years 1968 to 2014. Empirical results confirmed the existence of a long-term equilibrium relationship between variables and showed that nuclear energy consumption using Granger causality tests causes economic growth in Spain, which leads to important regional policy implications. As environmental concerns increase, policy makers might consider increasing the share of nuclear power in Spain's energy portfolio since it provides energy with no carbon emissions.

Chapter 9 assesses renewable energy potentials in the regions of Hungary. The national energy mapping project created an interactive energy map for Hungary to serve as the basis for financial resource use planning and implementation from 2014 to 2020. The map incorporated all available regional energy related data in sectoral breakdown, using statistical databases and graphical information system methodologies. This chapter, authored by Dr. Bálint Hartmann, Dr. Attila Talamon, and Viktória Sugár, introduces the methodology and results of the evaluation process. The assessment considers hydropower, wind energy, solar, geothermal and biomass energy potentials. It discusses how the regional potential of each energy source was determined, provides an overview of the results, and considers recent renewable energy policy changes in Hungary.

Chapter 10 details statistical approaches to forecasting domestic energy consumption in Iceland, Norway, Denmark, Sweden and Finland. Most domestic energy consumption in the Nordic countries is used to provide space heating and hot water. This chapter presents a study that forecasts the annual energy consumption of the Nordic residential sectors by 2020 as a function of socio-economic and environmental factors, and offers a framework for the predictors in each country. The research models domestic energy use by applying the multiple linear regression, multivariate adaptive regression splines, and the artificial neural network (ANN) analysis methodologies, creating distinct models for each Nordic country. Using these models, Nordic countries domestic energy use by 2020 is forecasted and the causal links between energy consumption and the investigated predictors are assessed. The results showed that the ANN models have a superior capability of forecasting domestic energy use and specifying the importance of predictors compared to the regression models. The models revealed that changes in population, unemployment rate, work force, urban population, and the amount of CO_2 emissions from the residential sectors can cause significant variations in Nordic domestic sector energy use. Samad Ranjbar Ardakani, Seyed Mohsen Hossein, and Alireza Aslani, are the authors of this chapter.

Chapter 11 asks the question: Is saving energy more related to politics or business? Dr. Volodymyr Mamalyga considers how this question applies in the Ukraine. Recent history shows that the greatest successes in the field of energy have been achieved by the wealthy countries in Western Europe, North America and Japan. Greater attention to energy saving and environmental protection is occurring in wealthy countries of the Middle East and the developing countries of China and India. Less wealthy countries often lack the financial resources to implement widespread energy conservation and environmental protection improvements. Are optimal decisions regarding the use of energy efficient technologies based solely on economics? Or, are the reasons related to political circumstances? This chapter presents a pragmatic approach to feasibility assessments to achieve reductions in energy usage and generate energy cost savings. Case assessments including lamp replacements, wind turbine generators, frequency converters, throttling, and electric motors show that using energy efficient equipment is not always feasible. Their implementation is often largely explained by political influences in the project implementation decision-making process.

Chapter 12 looks at energy efficiency solutions in the United Arab Emirates (UAE). Energy efficiency is a strategy with three essential elements—focus, measurements and accountability. Buildings contribute to our energy dilemma both globally and locally in the UAE. Buildings lead electric consumption in the emirates of Abu Dhabi and Dubai, accounting for over 80% of the total annual electricity consumption. Maen Al-Nemrawi, the chapter's author, states its purpose as discussing how monitoring based commissioning is a viable methodology for achieving results and verifying energy savings. Commissioning provides a solution. With commissioning, risks from other energy efficiency initiatives can be reduced. The market analysis shows that savings can be achieved by targeting energy efficiency measures that offer a quick return on investment.

Chapter 13, authored by Samer Zawaydeh, focuses on the economic, environmental and social impacts of developing energy from sustainable resources in Jordan. The development of renewable energies from solar, wind, geothermal, hydro and tides improves energy security for oil importing countries and promotes sustainability. Renewables also create employment in industries that educate, design, finance, procure, manufacture, manage, construct, commission, and maintain the new clean energy technologies. This chapter details the status of the development of renewable energy sources in the Middle East and North Africa concentrating on the successful efforts in Jordan. It discusses the development of sustainable energies which are naturally available, socially acceptable and economically feasible in the Middle Eastern country of Jordan.

Chapter 14 advocates harnessing innovative solutions for petroleum refining in Nigeria. Nigeria, a major petroleum producer and exporter, suffers from lack of refining capabilities, surprisingly creating supply shortages. To satisfy demand, petroleum products must be imported from other countries. The situation is most dire in areas of the Niger Delta. This has contributed to the development of artisanal refining of petroleum—a homegrown solution. Several studies have considered various aspects of this locally innovative strategy, often focusing on its environmental problems. Little attention has been given to how local refining can be harnessed to ameliorate the problems associated with petroleum shortages in Nigeria. This chapter examines the benefits of local, innovative refineries and argues that they be legitimized and regulated. The author, Dr. Nathaniel Umukoro, provides a colorful and thought-provoking view of artesian solutions.

Chapter 15 explores options for domestic water heating in South Africa. Electric water heating (EWH) is the most common technology used to heat water in residential and commercial buildings in the country of South Africa. Water heating accounts for roughly 40% of the energy consumption of a typical South African home. The high capital costs of solar thermal water heating (SWH) systems in South Africa have made implementation of this technology very expensive. With reductions in costs, solar photovoltaic (PV) systems are economically competitive with EWH systems. The relatively steady annual electrical generation output of the PV systems when compared with the seasonal performance of the conventional SWH has created opportunities for solar PV to capture a larger share of the local market for water heating. This chapter reviews and assesses the three primary water heating technologies currently used in South Africa and provides an energy and financial analysis of each technology. It is authored by Ognyan Dintchev, G.S. Donev, J.L. Munda, O.M. Popoola, M.M. Wesigye and S. Worthmann.

Chapter 16 provides a strategic analysis of Iran's energy system. Due to the importance of energy to societies, an analysis of local, regional and national energy systems is salient for researchers and policy makers. Iran is one of the world's fastest growing energy consumers and its economy is highly dependent on energy exports. The patterns of Iran's present energy system impact its economy and development. This chapter considers Iran's extraction, processing, conversion, transmission, distribution, and consumption of energy. This research conducted by Alireza Heidari, Alireza Aslani, Ahmad Hajinezhad, Seyed Hassan Tayyar from the University of Tehran, assesses the energy system in Iran by identifying and analyzing its strengths, weaknesses, opportunities and threats. The chapter considers these from internal and external perspectives and concludes that energy has an important role in Iran's policy development.

Chapter 17 uses emissions accounting for ports in India as a means of defining sustainable programs. Seaports are global hubs for the transportation of goods. They have important roles in today's global societies and are critical nodes in transportation networks. Sustainable energy use impacts people, the world's environment, and is relevant to the operation and maintenance of ports. In this chapter, an inventory of greenhouse gas (GHG) emissions in the port of Chennai is made by accounting for the various port facilities, the housing areas, and the fishing harbor. GHG emissions are quantified by following international guidelines. The detailed estimation of energy consumption and emissions generated by the individual systems are useful for energy engineers when implementing energy conservation measures and renewable energy technologies. Implementation of GHG mitigation strategies for all port-related activities achieves significant GHG reductions, reducing the adverse impacts of global climate change. The authors of this chapter were Atulya Misra, Karthik Panchabikesan, Dr. Ayyasamy Elayaperumal and Dr. Ramalingam Velraj.

Chapter 18 deals with the issue of renewable energy based microgrids for the development of green port infrastructure. Ports as an industry account for 3% of the global greenhouse gas emissions. Sustainable initiatives and zero net energy goals are driving the use of renewable resources including solar photovoltaic and wind energy. The objective of this chapter is to discuss reducing greenhouse gas emissions inside ports by integrating suitable renewable energy technologies in an efficient manner using micro grids. Clean energy-based, direct current microgrids are considered as a revolutionary power solution. This chapter highlights and details the benefits of implementing renewable-based microgrids with energy storage technologies. This chapter was authored by Atulya Misra, Gayathri Venkataramani, Senthilkumar Gowrishankar, Dr. Ayyasam Elayaperumal and Dr. Ramalingam Velraj.

Chapter 19 considers carbon capture and geologic storage applications in India. India has ambitious developmental goals. Future development is expected to increase energy demands and subsequently GHG emissions. Carbon capture and storage (CCS) is one of the mitigation strategies that might be adopted in this context. This chapter summarizes the scope of deployment of CCS in India's power sector. It offers perspectives regarding CO_2 capture technologies vis-à-vis Indian power plants. The potential geologic CO_2 storage sites and their resulting storage capabilities are discussed with references to the research being performed. Suggestions include progressive ideas on the technologies and policies that can advance CCS in India. The authors, Udayan Singh and Garima Singh, conclude that CCS is an important transition technology to minimize GHG emissions while renewable energy resources are deployed.

Chapter 20 considers the barriers to wind energy development in India. Among the renewable energy sources, wind energy, biomass energy systems and solar PV have capacity and proven technology to provide continuous and reliable energy. Wind energy, the fastest developing energy source, is renewable and environment friendly and the systems that convert wind energy to electricity have developed rapidly. The authors, Sanjeev H. Kulkarni and Dr. T.R. Anil, provide a thought-provoking assessment which aims at identifying and ranking the barriers preventing the diffusion of renewable energy generation using wind energy. Some of the barriers identified are the lack of knowledge about energy sources, higher investment cost, preference for grid extension projects, lack of arrangements for the long-term operation and maintenance, the absence of certification systems, and the lack of financial instruments for renewable energy implementation. Five main barrier groups are considered and their dimensions are identified. They are then ranked based on a multicriteria decision making process. The results provide evidence of how consumers receive wind energy information and make decisions using their analytical capabilities.

Chapter 21 presents a study of industrial efficiency and environmental governance efficiency in five urban areas in China. Industrial efficiency is important for the development of regional economic policies. Based on a network data envelopment analysis methodology considering undesirable outputs and links between sub-processes, the authors studied the overall industrial efficiency, pollution governance efficiency and industrial production efficiency of China's largest five urban agglomerations from 2000 to 2014. Results indicate that poor efficiency of environmental pollution governance is the key factor that restrains industrial efficiency. Increasing the efficiency and technical levels of industrial pollution treatment is an important way to improve the ecological environment of urban areas and their overall industry efficiency. This will ultimately promote more sustainable urban economic and environmental development. The authors are Sufeng Wang, Ran Li, Jia Liu, Zhanglin Peng, Yu Bai, Sufeng Wang, Ran Li, Jia Liu, Zhanglin Peng and Yu.

Chapter 22 considers the impact of the Guangdong emissions trading scheme (ETS) pilot project on the costs and profits of coal-fired thermal power plants. A pilot carbon program has been launched for three years in Guangdong province of China, with the power industry contributing nearly 66% of the covered CO_2 emissions. This chapter reviews the policy design of the program and determines that the percentage of paid allowance is the only factor reflecting the carbon cost for a generator with an average efficiency. The impact of carbon cost on the overall cost of 300 MW, 600 MW and 1,000 MW plants are analyzed, and the results show that the ratio of carbon cost to total cost is about 0.5% for the power plants in the program. This has little influence on the plant operation. The impacts of the carbon cost on the cash flow of the three kinds of plants are performed, and the assessments show the benefit of the plant at specific paid allowances and carbon prices. This information can be used by the governments to improve policy design and by enterprises to manage their carbon assets. This unique assessment is authored by Yuejun Luo, Wenjun Wang, Xueyan Li and Daiqing Zhao.

Chapter 23 assesses networking through international energy organizations, making comparisons by trade type. Authored by Sang Yoon Shin and Jae-Kung Kim, this study investigates differences regarding international energy cooperation among countries. It classifies 49 major oil and natural gas trading countries into exporting, importing, and balanced countries, and compares two network characteristics of each group within a network formed through participation in 28 international energy organizations. The study's purpose was to compare the network characteristics of energy importing and exporting countries in their networks formed through participation in these organizations. The analysis results confirm that both the importing and balanced groups have higher values for both degree and power centrality indices compared to the exporting group. The results also indicate that countries having considerable imports occupy more central positions with more partners, which increases their network influence.

The chapters in this book show how the crossroad that is upon us creates a multitude of dilemmas that offer exciting opportunities. This book is about solutions, innovation and taking concerted action. One dilemma is that failure to take action today to reduce carbon emissions exacerbates the difficulty of reducing carbon levels in the future. To solve this problem, we must change our agendas, rewrite our policies, institute new programs and implement new technological solutions. This book offers evidence that fruitful efforts are being taken to these ends on an international scale.

Our mission awaits. There remains much work to be accomplished. In fact, we have only just begun to recognize solutions. *International Solutions to Sustainable Energy, Policies and Applications* deeply examines the solutions that are necessary to achieve the goals of instituting changes to energy policies, implementing sustainable energy solutions, applying renewable technologies, and reducing atmospheric carbon emissions.

Chapter 1

Linking Sustainable Development and Energy

Stephen A. Roosa, Ph.D., CEM, CSDP

"The human appetite for energy appears to be insatiable. There are good reasons for the continued growth of energy consumption in the future: the survival needs of the underprivileged billions, increased adult life expectations stemming from the population boon... and the desirable goal of improved quality of life for everyone. But there are equally good reasons for carefully examining what the consequences of energy growth will be after present consumption rates have quadrupled. When do we melt the polar icepacks? How much land can we afford to set aside for energy plants? How much photosynthetic smog can we tolerate?"

Alfred M. Worden
Proceedings of the 1st World Energy Engineering Congress, 1978

The evidence is undisputable—sustainable development is generating attention throughout the world. Evolving from the environmental movement that swept across the U.S. and Europe in the mid-1970s, sustainability is both a buzzword and an ideal and that is defining a new era. It is an important concept, worthy of the attention it is generating. Understanding what sustainability means in energy practice is important and has a wide range of implications. Sustainability manifests itself as a set of policies, programs, initiatives, and technologies, each with its own applications. Does sustainable development represent a new vision for our future? If so, how did this concept develop and how is it linked to energy? What is a sustainable energy solution?

This chapter considers the history, definition and applications of sustainable development, explains its origins, reviews the pertinent academic information available concerning the concept, and explores the view that sustainability may indeed offer an enlightened vision for the future. Yet another goal of this chapter is to study the intellectual terrain regarding sustainable development and to frame sustainability as an overarching strategy for developing social, political, economic, and technological policy frameworks. While there are a number of factors and events driving the interest in sustainability (e.g., population growth, development practices, etc.), this chapter focuses primarily on energy and environmental considerations. The broader policy issues of the agenda, such as patterns of energy use and mitigation efforts for climate change will be discussed.

INTRODUCTION

Sustainable development is a theoretical construct. Within the time frame of the human experience on Earth, it may be difficult if not impossible to achieve. Like most such constructs, its limitations include an inability to be a solution for all global problems. According to Harken, "Most global problems cannot be solved globally because they are global symptoms of local problems with roots in reductionist thinking that goes back to the scientific revolution and the beginnings of industrialism" [1].

This reality does nothing to diminish the importance of sustainable development. The concept fosters hope… a hope for a better future. Many aspects of the concept of sustainability have problem-solving implications for industry, institutions, corporations and governments. Most often, sustainability is defined more by the policies that are deployed and the agendas that these policies establish. To complicate the difficulties in implementing the agenda, understand that sustainability has far-reaching and universal implications. Sustainability is about resources, resilience, management policies, energy, social concerns, planning, economics, environmental impacts, construction practices and much more. Responding to its agenda has caused companies to rethink basic processes, institutions to upgrade infrastructure and governments to refocus their resources—yielding fresh and creative solutions to current problems.

Understanding sustainability is a complex task. Berke believed that sustainability was the paradigm or "framework to dramatically shift the practice of local participation from dominance by narrow special interests toward a more holistic and inclusive view" [2]. While achieving this, sustainability becomes an overarching principle that has a corresponding set of guidelines. One strength of the agenda is the recognition that the decisions and investments we make today have serious implications for our future and the future generations to come. The long-term impact of our present decisions demands serious consideration, meaning we must temper present actions with caution.

Sustainability clothes itself in a systems analysis approach that considers how processes are redesigned and managed, with the hope of yielding better long-term outcomes. More favorable outcomes are those that best meet the goals of the agenda after trade-offs are considered. While sustainability hopefully occurs when the agenda's guidelines are successfully implemented, sustainable development can be thought of as physical outcomes that occur when the guidelines are followed.

Is sustainable development a policy, a set of agendas, a management philosophy, a new set of solutions, or perhaps all of these? How did the agenda evolve and what are its components? How can sustainability, an abstraction, be defined? What solutions, if any, does it offer? Before such questions are addressed, let's consider what drives a sustainability agenda.

HOW SUSTAINABLE DEVELOPMENT EVOLVED

There are many forces responsible for the concept of sustainable development. These include social issues, economic concerns, resource allocation, environmental damage, population growth, access to potable water, health, and energy usage among others. Several of these will be discussed to demonstrate how sustainability evolved. There are a number of causes and effects that seem destined to make sustainable development an important priority of a new world agenda. Underlying causes of specific global problems (such as urban development, population growth, and energy use) and their effects (such as pollution, changes in infrastructure and climate change) drove the evolution. Specific policies supporting sustainable development helped refine the nature of its scope.

Sustainable development evolved like a tidal wave building beyond the horizon. Sustainability as a policy is growing in strength, changing how we think, changing our agendas, changing how we design buildings and infrastructure, changing the processes we use and changing the technological solutions we implement. It has redefined our views on the environment. The process is rippling through our corporations, governments and institutions, changing our patterns of energy consumption. The wave will soon come ashore and be upon us. As it washes over, it forces us to redesign our future and the future of our descendants.

Sustainability Emerges as a New Social Force

The theory of sustainability is rooted in the 1960s environmental movement when "problems such as overpopulation, resource depletion, decreasing water supplies, air pollution and the spread of chemicals and heavy metals in nature came into focus" [3]. Warning of the dire consequences of pesticide use, Rachel Carson's descriptions of vanishing species of birds in *Silent Spring*, when fears about overpopulation leading to resource disruptions, and potential food shortages, summarized the concerns of the day [4,5]. These events contributed to the theory that there might be *Limits to Growth* [6].

Sustainability rapidly evolved to become a buzzword for the dawning of the new century. Being sustainable has become the socially preferable approach to almost everything. There have been references to sustainable policies, sustainable communities, sustainable agriculture, sustainable horticulture, sustainable use of the oceans, sustainable ecosystems, sustainable housing, not to mention sustainable businesses, sustainable practices, sustainable business practices and sustainable ad nauseam. Sustainable development as currently defined traces its origins to the 1987 Bruntland Report of the World Commission on Development and the Environment. Expressed simply, sustainable development is commonly considered to be "development that meets the needs of the present without compromising the ability of future generations to meet their needs" [7]. Sustainability is an integrative concept.

Ultimately, the coining of sustainability has caused the term to become nearly ubiquitous. At times, it seems purposefully politicized, such as the use of sustainability in the U.S. Federal Agriculture Improvement and Reform Act (1996), which defined sustainability as a way "to continue primary emphasis on large-scale industrial agriculture for competitive production in a global export economy" [8]. However, few understand what is meant by sustainable development and fewer still have a clear idea of how its promise might be fulfilled.

The concepts of sustainability and sustainable development evoke a broad range of questions. What is sustainability and how does it relate to policies? What aspects of the sustainable development agendas have the greatest implications? What are sustainable development policies and are they worth pursing? To comprehend answers to such questions, one must consider how the concept of sustainable development came about.

Sustainability evolved from multiple sources. Among these are theories from the sciences that suggested that many problems and potential solutions, first thought to be individually manageable and resolvable, might be interrelated across natural and physical systems. Einstein was among those who led the way by suggesting that all "phenomenon of nature, all the laws of nature, are the same for all systems that move uniformly relative to one another" [9:46]. He also believed that everything indicated that the universe was ultimately progressing to "darkness and decay," suggesting that disorder and entropy would eventually result [9:105]. The idea that models of various systems may have common, interrelated characteristics was proposed by Bertalanffy in his *Theory of General Systems* [10]. Later, the broader implications and relevance his theories were applied across sciences previously considered to be unrelated [11].

Theories of ordering systems into hierarchies soon evolved [12]. The "systems idea" was very quickly adapted to the planning and design processes for large-scale systems such as cities. Miller's inspired work, *Living Systems*, resulted in the merger of hierarchy theory and systems theory, yielding broader applications and providing an understanding that organizations, cities, and transportation networks had structural similarities to living systems [13].

The resolution of commonalities among subsystems provided a theoretical basis for the interrelated and complex nature of processes inherent in living systems [13]. According to Clayton and Radcliffe, "the size and complexity of the earth system indicates that there could be, at any one time, a very large number of development paths and possible outcomes, a smaller subset of which would be relatively sustainable for the human species" [14]. There was a growing awareness that man's activities might unbalance natural ecosystems and create dysfunction within and among them. Some believed that this was leading to unpredictable ecosystem problems that were entropic and potentially pathological. Sustainable development advocates were increasingly concerned with managing current events and their possible outcomes in a manner that increased the probability of more favorable outcomes, ultimately offering greater potential for mankind's sustained existence.

National governments began to respond to a ground swell of public opinion that actions to mitigate environmental problems needed to be undertaken as the linkages between the environment and sustainability became focused. In 1969, the U.S. Congress passed the National Environmental Policy Act (NEPA), responding to the influences of "population growth, high-density urbanization, industrial expansion, resource exploitation" and declaring it a policy of the federal government "to create and maintain conditions under which man and nature can exist in productive harmony, and fulfill social, economic, and other requirements of present and future generations of Americans." NEPA introduced the concepts of both environmental harmony and intergenerational equity. The law was enacted in 1970 and represented the expansion of environmental governance at the federal level in the U.S. Other provisions of the NEPA included enhancing the quality of renewable resources, recycling depletable resources, and maintaining an environment that supports diversity.

The theory of sustainable development ultimately evolved to provide an even broader vision, addressing an even larger range of concerns. Yet its origins remain uncertain. Stephen Wheeler [15] observed that:

> "It is far from clear who was the first to use the term 'sustainable development' in its current sense. Rather, it seems one of those inevitable expressions—that so neatly express what many people are thinking—that once the words are mentioned they quickly become ubiquitous. The birth of the sustainability concept in the 1970s can be seen as the logical outgrowth of a new consciousness about global problems related to environment and development..."

Wheeler believes that catalysts for the change in "consciousness" included events such as the rise of ecological problems and "the 1973 oil embargo during which millions of people suddenly realized that their fossil fuel use could not continue to expand forever" [15].

As the stage was being set to address sustainable development during the 1992 United Nations Conference on Environment and Development, Gro Harlem Brundtland, then Prime Minister of Norway, asserted that "we should not be surprised that developing nations are approaching the Rio Summit with open economic demands... for them, it is essentially a conference about development and justice" [16]. On the other hand, there was a fear among the developed nations that they might be called upon to bear the primary financial burden of protecting the earth's biodiversity [16].

Redefining Sustainable Development

Sustainable development is often defined as the ability to meet the needs of the present without hampering the ability of future generations to meet their needs. For our purposes, sustainable development is defined as: The ability of physical development and environmental impacts to sustain long term habitation on the planet Earth by human and other indigenous species while providing:

1) An opportunity for environmentally safe, ecologically appropriate physical development;
2) Efficient use of natural resources;
3) A framework which allows improvement of the human condition and equal opportunity for current and future generations; and
4) Manageable growth.

Regardless of definition, non-sustainable development is the antithesis of sustainable development. It implies persistently unmanageable growth that is environmentally unsafe, that consumes resources such as energy ineffectively while degrading the human condition. Extreme cases of non-sustainable development are characterized by ecosystem disruption, paralyzed communications, dysfunctional transportation systems, persistent lack of resources and materials, pervasive environmental mismanagement, prolonged poverty and lack of medical care, perpetual destruction of infrastructure, and recurring military conflict. Such conditions, either individually or in combination, can be disruptive to environments, yielding non-sustainable results.

This new definition incorporates social concerns, such as those held by Perloff, who noted that, "A price must be paid for material progress in the destruction of irreplaceable natural resources, in air, water and noise pollution; and in millions of disadvantaged individuals and families who are left behind; and in cities that are also left behind with large sections in decay and poverty" [17]. Al-Homound believes "Usable resources are made available to mankind to be utilized for their benefit and well-being. Every individual bears the responsibility of not wasting or misusing usable resources. All moral codes are against such actions. Therefore, every individual should be educated and trained to become part of the management of resources for the benefit of generations to come" [18]. Having a moral belief that resource efficiency is beneficial certainly has global social implications. In addition to moral beliefs, there are other direct linkages between sustainability and resource management.

LINKING SUSTAINABILITY WITH ENERGY

Next, we explore the idea that energy is a key component of sustainable development. Can sustainability be implemented in the face of continually increasing energy consumption? The linkages among energy, sustainability and their relationships are an important component of achieving sustainability goals. According to Saha, "The main challenge to energy policy makers in the 21st century is how to develop and manage adequate, affordable and reliable energy services in a sustainable manner to fuel social and economic development" [19].

Sustainability, in terms of energy use, can be considered to: 1) be the ability to provide energy that is environmentally benign; 2) not use energy in an overly consumptive manner; 3) provide options for renewable energy; 4) reduce the need for energy from non-renewable energy sources; and 5) distribute energy resources equitably. To fully understand the relationship of energy to sustainability, it is important to consider its economic, physical and conceptual links.

Economic Linkages

Forms of renewable energy (solar, wind, hydropower, bio-energy and geothermal energy) reduce dependency on fossil fuels (coal, oil and natural gas) and yield improved environmental impacts. The action plan of the World Summit on Sustainable Development included a proposal that the use of renewable energy technologies be increased to 15% of worldwide energy production by 2010 [20]. For the U.S., this would have generated employment at a time of employment losses during a recession and reduced oil imports during a period of negative trade imbalances. Delegates from the United States (then the world's largest oil importer), Saudi Arabia (the world's largest oil exporter) and other states "were lobbying to eliminate the provision and set no specific goals" [20]. It is doubtful that the nature of the alliance in opposition to the proposal was lost on the delegates. The U.S. "solution" took the form of a counter proposal. It included a series of hastily organized industry and foundation partnerships with an ambiguous agenda, involving $600 million per year in expenditures with a commitment of not more than four years. This figure dwarfed the hundreds of billions of dollars expended annually by the U.S. on foreign oil alone. Regardless, this example shows that the world's governments had accepted (and valued) the idea that sustainability and energy were linked. It highlights the importance of delving into the linkages.

Physical Linkages

Energy can be defined as the "capacity of doing work and overcoming resistance" and as "strength or power efficiently exerted" [21]. Energy can also be defined as "the work that a physical system is capable of doing in changing from its actual state to a specified reference state" [22]. The term conservation is the "act or practice of conserving; protection from loss, waste, etc.; preservation; official care and protection of natural resources" [21]. In apolitical terms, consuming natural resources in a strategically appropriate manner without adverse environmental impact provides economic benefits. The phrase *conservation of energy* in a physical sense refers to the "principle that energy is never consumed but only changes form and that total energy in a physical system, such as a universe, cannot be increased or diminished" [21].

The Second Law of Thermodynamics (a.k.a., the Law of Increased Entropy) validates that the fundamental processes in nature are essentially irreversible. While the quantity of energy remains constant, the quality deteriorates. Debating the idea of running out of energy is fruitless and counter-productive. There will always be energy available.

Suggesting that there may not be adequate and useful forms of energy or the means of transforming the available energy into usable product is far more rational. This is not about energy resources but the services available from their use. For example, burning oil for heating is not the end product or service. The end product might be usable heated space, heated water or process heat. These products are actually services that can be provided by multiple means, not necessarily provided by combusting oil.

Conceptual Linkages

Consumer behavior creates patterns of consumption (the depletion of goods or services either by consumers or in the production of other goods) that can economically provide the means to satisfy socially beneficial needs [21]. *Consumptive* behavior (consuming or tending to consume; destructive; wasteful) can increase product costs, cause shortages, equity imbalances, or contribute to economic externalities [21].

Conceptual thinking regards decision-making processes and management options as steering functions, while development and energy are considered production forces [23]. As a result, the pervasive concerns regarding energy use cause energy consumption to have broader global implications. Interestingly, energy usage is often highly decentralized while energy generation and production tends to be comparatively centralized.

Yet centralized, capital intensive options may not always be the most viable alternatives. Such options often require complicated distribution networks that are subject to systemic disruption. Externalities include those not only associated with water and air pollution but also economic availability and equity issues. Energy use in the built environment has been increasing due to a number of causes. Transportation systems, rapid growth in population and increases in the number and scale of buildings augment demand for usable energy. Energy is required for primary services such as the need for conditioned space that complies with upgraded standards for human comfort, especially in the workplace.

IMPLEMENTING SUSTAINABLE DEVELOPMENT POLICIES

Approaches to implementing sustainable development are varied. The process involves selecting the problems to be resolved, developing a strategy for resolution, implementing interventions and resolving the problems. Problems that have global implications, social or environmental characteristics are easily politicized. For example, in September 2002, the World Summit on Sustainable Development was held in Johannesburg, South Africa. At the conference, U.S. Undersecretary of State Paula Dobriansky indicated that the U.S. was the world's leader in sustainable development, seemingly unaware that the U.S. had no adopted national policy concerning sustainable development.

There exists in the U.S. a policy tradition that frames sustainability as a need to trade economic development for environmental solutions as if both were always mutually exclusive. This sort of illogical "either-or" thinking has stymied development of creative efforts to achieve sustainability since the mid-1970s. Consider industrial air pollution. Crenson studied 18 variables prominent to air pollution problems and found that "the correlation coefficient… shows that where business and industrial development is a topic of concern, the dirty air problem tends to be ignored" [24]. In fact, when objective measures such as energy resource consumption and the emission output of critical pollutants are considered, it seems the U.S. fails to score highly as a sustainable economy, especially when compared to other industrialized nations.

Using a common system input measure such as size of energy footprint (often a more conservative measure than gross energy used or per capita energy consumption), the U.S. is the third least-efficient country in the world, behind the United Arab Emirates and Kuwait, both major producers and net exporters of oil. Based on this measure, Canada and most every other industrialized country in the world is more energy efficient than the U.S. Consider Japan, a country that imports 100% of its oil and taxes it to a far greater extent, and yet consumes only half as much energy per unit of economic output as the U.S. [25:125]. In fact, when emission output measures are considered the U.S. also ranks poorly. Economic policies in the U.S. divert revenues from social issues such as housing, education, health and welfare to supporting fossil-fuel producing companies that (not coincidentally) wield far greater political influence.

It has long been suggested that even a small level of taxation can significantly reduce demand for imported oil if the proceeds are used to promote energy efficiency and energy conservation [26]. However, the opposite often occurs in the U.S. For example, federal gasoline taxes and funding policies have long subsidized highway construction to a far greater extent than mass transit, enabling less efficient transportation systems to be supported. As a result, countries such as the U.K., The Netherlands, Germany, France, Spain, Italy and Japan are among the international leaders in the design, production and export of high-speed regional rail services, magnetic levitation transport, submersible tunneling systems, merchant shipping and cruise ship construction. These advantages provide their economies with expanded manufacturing employment and technological capabilities while the U.S. has excluded itself from these lucrative, high technology markets.

Contrary to assertions by U.S administrations, the U.S. Department of Energy freely admits on its Energy Efficiency and Renewable Energy Network (which it co-sponsors) that, "the complex problems shared by cities throughout the U.S. are evidence of the impacts of urban sprawl—increasing traffic congestion and commute times, air pollution, inefficient energy consumption and greater reliance on foreign oil, loss of open space and habitat, inequitable distribution of economic resources, and the loss of a sense of community" [27]. All of these issues can be framed in the language of sustainable development.

The President's Council on Sustainable Development provided a report in 1996 entitled *Sustainable America—A New Consensus* which included ten sustainable development policies and recommendations for changes at all governmental levels [25:83]. It appeared that the council's recommendations were simply importing to the U.S., albeit belatedly, policies widely understood and being seriously implemented in other developed countries [28]. Despite this effort, achieving sustainable development is not yet an adopted goal of the U.S. federal government, and there are no specific targets for implementation [25:93]. To date, none of the ten recommendations has been effectively implemented by the central government, either internally or externally. The efforts of local and state governments aside, this evidence is compelling—assertions that the U.S. leads the world in sustainable development are unsupportable.

Interestingly, concrete examples of planning for sustainable development are more easily found elsewhere in the world. The development plans of cities in the European Union serve as one example. In China, Kisho Kurokawa, a Japanese Architect, has been a proponent of the planning concept of an "ecological city." His "eco-media city" plans are a more "developed version of the eco-city" [29]. He proposed a plan for Futian in China based on a central eco-media city park concept to symbolize urban sustainability. In a suburb of Linz, Austria, called Pichling, a new solar powered district housing development for 25,000 residents, has been constructed [30]. In the Netherlands, Ecolonia—a demonstration town for ecological development—is yet another example.

It is clear that many policies are being implemented both locally and internationally with a goal of improving local sustainability. While results are mixed, there is an obvious trend towards implementation of the agenda internationally.

CLIMATE CHANGE

Perhaps no single issue demonstrates the linkages between sustainable development, energy use, and policy implementation on an international scale than climate change. An overwhelming scientific consensus has emerged that human beings are substantially contributing to changes in global climate systems. The challenge presented by climate change is daunting and merits immediate attention. For the first time in recorded history, and likely for the first time in a million or so years, we have recorded atmospheric concentrations of CO_2 approaching 410 parts per million (ppm). It is the existing stock of greenhouse gases in the atmosphere that is driving climate change, while the annual flow of emissions exacerbates the associated problems. From 1850 to 2005, the U.S. released more energy-related carbon dioxide into the atmosphere than any other country, an estimated total of 324.9 billion metric tonnes. Germany is second with 117.8 billion metric tonnes. Once released, CO_2 remains in the atmosphere for a number of years.

The Problem

According to many world-renowned scientists, economists and political leaders, it is clearly evident that the surface temperatures on Earth are warming at a pace that signals a decisive shift in the global climate. Scientists warn that current CO_2 emissions must be reduced by half over the next 50 years to avert changes to our climate that will be difficult to mitigate. To make this happen we must reduce carbon emissions.

Certainly, mankind's activities are not the only sources of emissions. There are substantial naturally occurring emissions as well. Yet before the industrial revolution, the Earth's atmosphere contained only about 280 ppm of CO_2. Earth's atmospheric global temperature averaged about 14°C (57°F)—a reasonably acceptable and habitable environment. The industrial revolution and the use of coal for steam power was in part responsible for the rise in atmospheric CO_2 concentrations at a time when those increases were not monitored. When humans began combusting more coal, gas and oil for heat and power, the amount of atmospheric CO_2 concentrations increased. By the time atmospheric CO_2 measurements were first established in the late 1950s, emissions had reached the 315-ppm level. CO_2 emissions are now increasing by approximately 2-ppm annually. Though the rate may seem trivial, the added CO_2 traps enough extra heat to warm the planet significantly. Earth's atmospheric temperature has increased by more than 1°F in the past 30 years.

While it is impossible to precisely predict the consequences of further CO_2 emissions, there are indications in geological evidence and science-based models. The warming as observed to date has triggered glacial retreat and Arctic ice melt. Climate change has distorted seasons, altered rainfall patterns, caused sea levels to rise and triggered species migration. Based on the current scientific data demonstrating global climate change, 450-ppm of CO_2 has been repeatedly and strongly recommended as a maximum threshold. Beyond this, Earth's environment becomes increasing vulnerable to irreversible and detrimental impacts. Averting the trend towards higher global temperatures requires an urgent, worldwide shift to low-carbon economies; this initiative must encompass innovative carbon reduction technologies. At our present trajectory, the Earth's atmosphere will likely reach this threshold in 2040, about ten years earlier than most scientific estimates—if the increases in CO_2 continue unabated. Furthermore, this scenario excludes the impacts of increases in other greenhouse gases such as methane and nitrous oxide, which also contribute to global climate change.

Thus far, the European countries and Japan have begun to trim carbon emissions, though they are not meeting their modest targets. Meanwhile, U.S. carbon emissions—about a quarter of the world's total—have begun to decline. China is limiting its coal use. India is pursuing renewable energy. Yet both countries are generating increasingly larger quantities of CO_2 and continuing to construct more coal-fired power plants. On a per capita basis, these countries continue to emit far less CO_2 than does the U.S. However, their populations are so large and their growth so rapid that lowering worldwide CO_2 emissions will be challenging. Preventing a global catastrophe requires rapid, dramatic cuts in CO_2 emissions by technologically advanced countries. Countries such as the U.S., the European countries, Japan and others must take a leading role in reforms. This can enable developing countries to power their emerging economies without a heavy dependency on CO_2-emitting coal and other fossil fuels.

If left unaddressed, climate change will result in a host of negative outcomes, including severe, long-term, destructive economic impacts. The anticipated effects of climate change pose a serious threat to human societies. Hence, solutions and preventative measures are desperately needed. Relying solely on the hope of successful future mitigation efforts will prove costly.

Creating Solutions

The conceptual framework involves finding real solutions that address climate change and reduce greenhouse gas emissions, particularly those that are causing increases in atmospheric CO_2 concentrations. The issue is not whether CO_2 emissions need to be casually trimmed and monitored. The real questions are how to determine the best pathways to reducing the levels of CO_2 emissions and which strategies will ultimately be most effective. In our assessments, we must consider both the short-term and long-term impacts.

Like many other environmental issues, people are accepting the reality that the long-term costs of response, adaptation, remediation and mitigation associated with climate change will ultimately far exceed the costs of inaction. The Paris Agreement, adopted by representatives of 193 countries in December 2015, is one step toward policy improvement at an international scale. Laurent Fabius, France's foreign minister, said this agreement is an "ambitious and balanced" approach and a "historic turning point" in our goal of reducing global warming. Perhaps its most important goal is to quickly achieve the global peaking of greenhouse gas emissions. Many of the countries participating in this agreement have made substantial gains that illustrate the achievements that have accompanied their progressive measures. Ultimately, we need to create a pathway for Earth's climate resilience. For the Paris Agreement to be effective, all countries must cooperatively participate.

Finding ways to slow the onslaught of global climate change is an important task for the world's architects, planners, engineers, scientists, corporations, communities, governments and decision makers. Ways to mitigate climate change require the prompt implementation of innovative policies, strategies and technologies—those associated with energy efficiency, renewables, fuel substitution, sustainable buildings and alternatively fueled vehicles. Avoiding further impacts from climate change mandates a worldwide shift to a low carbon economic platform. Planning and creating this sustainable platform of initiatives and solutions involves implementing a radical reformation of policies and programs and deploying new technologies.

It is important that the U.S. take a leading role in climate change regulation both nationally and internationally. Presently, the U.S. is second (17.9%) in total world emissions to China (20.1%). As of June 2017, the U.S. has the dubious distinction of remaining the largest emitter of greenhouse gases on a per capita basis. Incremental improvements will occur when more fully engaged. At the Marrakech Conference of Parties 22 held in November 2016, countries agreed to "move forward purposefully to reduce greenhouse gas emissions and to foster adaptation efforts, thereby benefiting and supporting the 2030 Agenda for Sustainable Development and its Sustainable Development Goals" [31].

CONCLUSIONS

Sustainability can be so broadly defined that potential solutions for its core goals remain unfulfilled, leaving a disconnect between needs and the means of fulfilling them. The relationships between sustainability and energy are so interlaced that it is unrealistic to consider sustainability without dealing with the issues and impacts of energy. Urban literature is rich in the discussion of inequities, central city decline, urban housing, social needs, de-industrialization, suburbanization, globalization, urban regimes and economic impacts of development in cities [32-35]. The idea that energy is central to sustainable development tends to be frequently devalued and minimized. Regardless, energy is directly linked to sustainability. In order to gain a broader understanding of sustainable development, energy use, energy conservation, and energy efficiency must be key components of any worthwhile solution.

Likely the greatest strengths of sustainable development include its persistence and its ability to be variously interpreted. It is a flexible concept that seems to redesign itself as needed. The applications of sustainability tend to broaden over time and become more encompassing with each application.

Sustainability has become a core concept in major policy-making processes, for energy management and applications, and in the redevelopment of existing facilities and infrastructure. Sustainable development can be considered to be a higher form of environmental policy. Due to the broad implications of resource management, sustainability is functionally dependent on energy use, on the ways energy is transformed, and the types of energy selected for a given task.

As the world's population expands from today's 7.5 billion people, and as energy use increases, policies diverge and sustainable development becomes increasingly difficult to achieve. Indicators of sustainability can be theorized and measured, and sustainable solutions can be implemented—but sustainable development can also unleash economic and political interests that threaten it. Energy-saving consumer products can be readily marketed to the public, but the benefits from alternative energy sources are sometimes difficult to measure. External costs are often overlooked.

It is not the concept of energy *per se* that is critical to sustainable development. Rather it is the selection of the forms of energy involved and how they are converted to use that impacts sustainable development policies. Energy policies will ultimately prove to be the most important component of sustainable development. For example, understanding of how energy affects cities is critical to comprehending urban sustainability.

Sustainable development is an evolving agenda that has entered the world debate since the 1980s. Achieving sustainability requires consensus and concerted action. Both remain elusive. While sustainability has its critics, it is maturing into mainstream acceptance in energy-intensive countries such as the U.S. It is both alluring and gaining in popularity. Sustainable development certainly provides a fundamental vision and framework for future planning and development.

In the U.S., past policies have been directed toward minimizing energy costs. The policies have been based on the assumption that the cost of goods and services that are dependent on energy will also be minimized. Regrettably, this type of thinking has hindered our resourcefulness in allocating income by diverting attention away from improving efficiencies. It has also reduced international competitiveness. This misdirection has failed at achieving its goal of maintaining low energy cost, contributing to a system of incentives and subsidies that have effectively maintained high-cost infrastructure.

The time has come for an alternative agenda. We can see that sustainable development provides a set of overarching principles that can be used to broadly assault certain pervasive problems, such as imprudent development policies, wasteful energy practices and environmental degradation. Implementing sustainable solutions that involve energy conservation, energy efficiency, and alternative energy production can reduce energy use, improve sustainability, and begin to address pervasive environmental problems. Lowering energy consumption, especially of carbon-based fossil fuels, has beneficial results that include reducing greenhouse gas emissions, improving health and the quality of life. Policy efforts toward such ends are beginning to occur at the local and regional levels in the U.S., Europe and elsewhere.

The Rio Agenda 21 provided a set of principles regarding sustainable development. It is concerned with sustainable economic growth, environmental protection, intergenerational equity, and resource conservation. These are keys to sustainability. The concept, while representing a new paradigm that evolved from a convergence of ideals, can be ambiguous and is subject to misinterpretation. The concept of sustainability can be polarizing, divisive, and politically confrontational.

A long-term view is necessary. Sustainable development "can be seen as based on a new recognition of the complex web of interconnections between different issues, fields, disciplines and actors," requiring a "holistic and interdisciplinary perspective" [15]. Achieving sustainability by modifying energy generation and consumption patterns may be among the most viable alternatives.

With the increased use of carbon-based energy, earlier policies that failed to support sustainable development are now proving unworkable. The idea of doing nothing and maintaining the status quo is increasingly unacceptable. Their supporting arguments are weak and categorically unsupportable. More importantly, advocates of the "do nothing" alternative fail to offer substantive evidence of effectiveness, and are unable to suggest adjustments to existing policies that could resolve their shortcomings.

Actual solutions include improving governance and modifying policy to reduce hydrocarbon-based energy use. Supporting continued development of renewable energy systems is a key component of the solution. Evidence shows that renewable energy technologies are proven job creators, improving economic sustainability. By 2015, employment in renewable energy, including small and large hydropower systems, accounted for 9.4 million jobs worldwide, an 18% increase from 2014 [36]. Renewable energy solutions enable us to improve the economy, improve the environment, and encourage leading technological research.

Sustainable energy implementation creates opportunities for an alternative future, one that is far less dependent on fossil fuels. It is a future that envisions less greenhouse gas emissions, cleaner air and water resources and healthier environments. As costs for renewable energy infrastructure decline, it is increasingly difficult to support counter arguments. Energy policies are a means of achieving local sustainability goals. Solutions must include implementing new technologies and embracing energy conservation and efficiency improvements. It is clear that sustainability, energy, and environmental impacts are inherently linked.

References

[1] Hawken, P. (1993). *The ecology of commerce.* Harper Business: New York. Page 211.
[2] Berke, P. (2002). Does sustainable development offer a new direction for planning? Challenges for the Twenty-first century. *Journal of Planning Literature*, 17(1), page 23.
[3] Low, N., Gleeson, B., Elander, I. and R. Lidskog (2002) *Consuming cities—The urban environment in the global economy after the Rio Declaration.* Rutledge: London, England. Page 37.
[4] Ehrlich, P. (1971). *The population bomb.* Ballantine: London, England.
[5] Commoner, B. (1971). *The closing circle: nature, man and technology.* Knoph: New York.
[6] Meadows, D. et al. (1972). *Limits to growth.* The Club of Rome.
[7] Holland, L., Holland, A., and D. McNeil (2000). *Global sustainable development in the 21st century.* Edinburgh: Edinburgh University Press. Page 10.
[8] Andrews, R. (1999). *Managing the environment, managing ourselves.* Yale University Press: New Haven, Connecticut. Page 307.
[9] Barnett, L. (1957). *The universe and Dr. Einstein.* Bantam Books: New York.
[10] Bertalanffy, L. (1968). *General system theory, foundations, development, applications.* Braziller: New York.

[11] Laszlo, E. (1972). *The relevance of general systems theory*. Brazille: New York.
[12] Pattee, H. (1973). *Hierarchy theory—the challenge of complex systems*. New York: Braziller.
[13] Miller, J. (1978). *Living systems*. McGraw Hill: New York.
[14] Clayton, M. and Radcliffe, N. (1996). *Sustainability—A systems approach*. Westview Press: Edinburgh, U.K. Page 12.
[15] LeGates, R. and F. Stout (ed.) (2000). *The city reader*. St. Edmundsbury Press: Bury St. Edmunds, Suffolk. Pages 436-438.
[16] Panjabi, R. (1997). *The Earth Summit at Rio*. Boston: Northeastern University Press. Page 282.
[17] Perloff, H. (1980). *Planning the post-industrial city*. Planners Press: Chicago. Page 182.
[18] Al-Homound, M. (2000). Total productive energy management. *Energy Engineering Journal*, 98(1), pages 21-28.
[19] Saha, P. (2003). Sustainable energy development: a challenge for Asia and the Pacific region in the 21st century. *Energy Policy*, 31, page 1.
[20] Verrengia, J. (2002, 14 August) U.S. not embracing expanded renewable energy development, *Courier Journal*, page A4.
[21] Guralnik, D. (1972). *Webster's new world dictionary of the American language*. World Publishing: New York.
[22] Morris, W. (ed.) (1969). *The American heritage dictionary of the English language*. The American Heritage Publishing Company: New York.
[23] Cineros, H. (1995, March). *Regionalism: the new geography of opportunity*. U.S. Department of Housing and Urban Development:Washington, D.C. Page 449.
[24] Crenson, M. (1971). *The un-politics of air pollution: a study of non-decision making in the cities*. John Hopkins University Press: Baltimore, Maryland. Page 165.
[25] Holland, L. Holland, A. and D. McNeil (2000). *Global sustainable development in the 21st century*. Edinburgh University Press: Edinburgh, U.K.
[26] Roosa, S. (1988). The economic effects of oil imports: A case for energy management. *Strategic Planning for Energy and the Environment*, 8(1), pages 41-42.
[27] U.S. Department of Energy (2002, May 22). http://www.sustainable.doe.gov/landuse/luintro.shtml.
[28] DeSimone, L. and F. Popoff (1997). *Eco-efficiency—The business of sustainable development*. The Massachusetts Institute of Technology Press: Cambridge, Massachusetts. Page 162.
[29] Kurokawa, K. Architects (1998). Public space conceptual design along central axis in Shenzhenu City Center, Stage II. Page 2.
[30] Beatley, T. (2000). *Green urbanism: Learning from European cities*. Island Press: Washington, D.C. Page 275.
[31] United Nations (2016, November 18). Countries at COP 22 pledge to press ahead with implementation of Paris Agreement. http://www.un.org/sustainabledevelopment/blog/2016/11/countries-at-cop22-pledge-to-press-ahead-with-implementation-of-paris-agreement.
[32] Benfield, F., Terris, J. and N. Vorsanger (2001). *Solving sprawl—Models of smart growth in communities across America*. New York: National Resources Defense Fund.
[33] Bluestone, B. and B. Harrison (1982). *The deindustrialization of America*. Basic Books: New York.
[34] Savitch, H. and P. Kantor (2002). *Cities in the international marketplace*. Princeton University Press: Princeton, New Jersey.
[35] Duany, A., Plater-Zyberk, E. and J. Speck (2000). *Suburban nation: The rise of sprawl and the decline of the American dream*. North Point Press: New York.
[36] International Renewable Energy Agency (2015). Renewable energy and jobs, annual review 2015. http://www.irena.org/menu/index.aspx?mnu=Subcat&PriMenuID=36&CatID=141&SubcatID=585, accessed 29 June 2017.

Chapter 2

Energy Policies and Ethical Notions in Developing Countries

Ali Asghar Pourezzat, Ali Asghar Sadabadi, Mohammad Koohi Khor, Narges Salehi, Shahrabi, Alireza Aslani, and Mohammad Mahdi Zolfagharzadeh

Energy has an important role in the development of societies, an attribute common to all the world's countries. Energy policy makers enact policies intended to increase the security of energy supply and consumption. In most cases the ethical issues of the policies are not considered during the process of policy making and their implementations. This chapter provides one of the first studies which considers energy policies and their ethical notions in developing countries. The focus is to determine and prioritize energy consumption policy criteria in order to achieve intergenerational justice in a case study for a selected country. The interesting results show that the sustainability relative values gained the most priority among the other criteria. Indeed, despite different policy encouragement packages for other energy sources, renewable energy is the best alternative from the perspective of ethical notions.

INTRODUCTION

Today one of the major factors of development concerns the energy sector. Moral issues are often neglected when considering future generations. In contrast, policy makers have an economic perspective regarding the energy sector.

Iran, our proxy for developing countries, is facing challenges in the area of energy policy. The increase in energy usage in Iran is distinctly out of proportion with the development of economic productivity. In the past 15 years, Iran's energy policy has focused on satisfying the growing demand for energy by using oil and by expanding natural gas production [1:233]. As a strategic policy, Iran tries to use all kinds of accessible primary energy resources including natural gas, coal, nuclear, hydro, solar, wind and geothermal energies [2:1]. Iran is rich in fossil fuel resources [1:234] and has progressed in nuclear energy. The country is potentially strong in renewable energy, such as solar and wind energy.

Thinking about public policy development requires an understanding of its framework. It is essential to improve environmental, ethical and social basis, and then assess various policy alternatives for congruency with public interests. Policy making for the future of the energy sector faces many challenges including reduced use of fossil fuels, events in nuclear power plants like Fukushima, and finding capital for investment in renewable energy. Social issues might include considering the impacts on future generations regarding energy. This chapter responds to the question: In Iran, how is the energy consumption policy in accordance with ethical considerations and intergenerational justice? This study's criteria are retrieved from the book; *Climate Justice: Ethics, Energy and Public Policy* by James B. Martin-Schramm [3], a professor of Luther College, and "An Energy Policy for New Brunswick" [4]. The obtained conceptual model is illustrated in Figure 2-1.

Sustainability—long-term supply of resources and conservation of intact natural resources—is one of the ethical notions considered when assessing energy policies. Sustainability considers assessing:

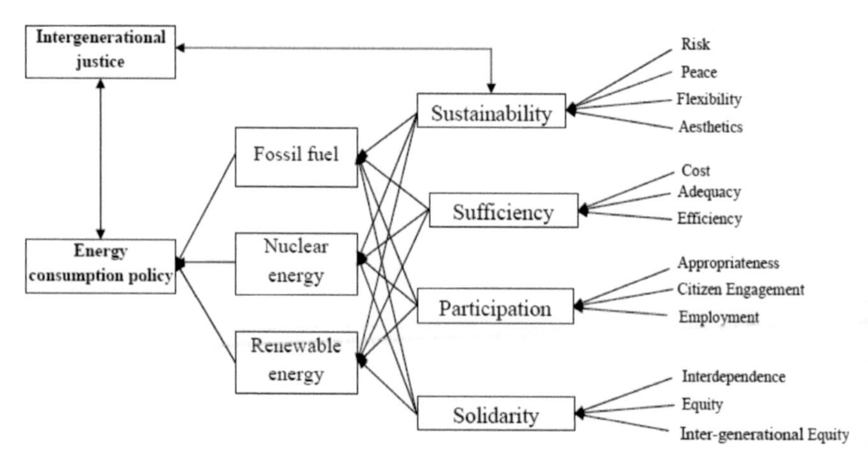

Figure 2-1. Research framework.

- Risk—measures must include the least vulnerability to human health and environmental systems.
- Renewability—showing energy option's capacity to restore its resources.
- Peace—energy policies must prevent resource dependency, which increases the potential armed clashes.
- Flexibility—it is implied policies have potential and options for alteration and reversal. It is notable that greater flexibility is preferable and systems should avoid sudden disruption.
- Aesthetics, one of the major aspects of quality of life, and policies which cause gaps in vision should be avoided.

Sufficiency—policies and replaced energies should be adequate to meet the needs—is the second ethical principle which comprises:

- Cost—all of the financial, social and environmental externalities should be included in energy prices for industry and consumers, instead of imposing these burdens on people's health, quality of life and environment.
- Adequacy—policies should guarantee providing for everyone's needs.
- Efficiency—energy policies should bring power along with lower resource us, pollution and waste with improved consumption patterns.

Participation of all who are able to express their opinion on decisions is the third base which contains:

- Appropriateness—energy systems should be in accordance with contentment of basic needs, human potentials, final usages, local demand and employment levels.
- Citizen engagement—rely on citizen's needs and engage them in developing policies.
- Employment—policies impact recruitment, skills and jobs in demand. It could be cited that systems and policies should stimulate new jobs and skills.

Solidarity—nondiscrimination, considering other species and ecosystems, paying attention to the nature of social life versus individual and not sacrificing weak creatures—is the final principle that consists of:

- Equity—How can the policy impacts on various social segments accommodate the special concerns for poor and vulnerable classes? Interests and the burden of responsibilities should be distributed and assessed in a way that none of the pertinent groups gain disproportionate economic profits or losses.
- Intergenerational equity—How do today's energy policies assure us that we are preventing the transfer to future generations of negative externalities associated with environmental impacts?
- Interdependence—Have energy policies recognized our interdependency and nature?

Energy policies are a means (often governmental) of institutionalizing decisions about issues such as production, distributions and energy consumption. Energy policies are characterized by laws, rules, international commitment, investment drivers, instructions for energy preservation, tax and other public policy techniques [4:1].

Environmental ethics is a sub-discipline of philosophy that deals with the ethical problems surrounding environmental protection. It aims to provide ethical justification and moral motivation for the cause of global environmental protection [5:23].

The list of research efforts similar to our study are demonstrated in Table 2-1.

RESEARCH METHODS

This study is applied descriptive research based on a practical survey. To determine energy criteria's morality, an analytical hierarchy process has been used. Multi index decision making methods are developed to evaluate the indexes and to choose premier alternatives, which the entropy method, LINMAP method, weighted least squares method and Analytical Hierarchy Process (AHP) provide examples.

The study's statistical society are all the expertise and specialists in energy policy making and planning who are involved in Iran's Power Ministry, the International Energy Studies Institute and Sharif University. In respect to the none probable targeted sampling method, 30 people

Table 2-1. Conducted research of energy policies and intergenerational justice.

Results	Researcher(s)	No.
Confine the proposition of fossil fuel Providing energies compatible with environment	Massarrat (2004) [1]	1
Morally desirable options should follow future generation's interests Reduction of fossil fuels reliance of countries	Taebi (2010) [19]	2
- Concern for environmental ethics - The need for environmental justice among the present generation (especially to eliminate absolute poverty) - The need to care for future generations - The need to live harmoniously with nature	Yang (2006) [5]	3
- The effect of present decisions on future generations	Schwarze (2003) [18]	4
- End of nonrenewable energy resources - Decrease of countries reliance on fossil fuel - Consider later generation's interests in choosing a desirable option - Focus on nuclear energy, choose a segmentation method and return to the best desirable alternative	Taebi (2010) [19]	5
- Abundant supply of petroleum in Iran and its environmental harms - Iran's low tendency to discourage the pursuit of alternative renewable energy sources - Lack of definite knowledge and rules in the field of renewable energy and requiring modification in private sectors	Atabi (2004) [12]	6
- Excessive consumption of fossil fuel and environmental risks - Peaceful use of nuclear technology	Dabiri et al. (2009) [8]	7
- Considering the importance of nuclear waste - Concern about later generations - Protect various resources for future - Fair distribution of environmental risks and responsibilities within generations - Offering suitable replacement for nonrenewable resources	Johnson (2002)	8
- Appropriate consumption of nonrenewable resources over time - Specifying cycle values of nuclear energy	Taebi & Kadak (2011) [19]	9
- In regards to the criteria, solar energy is the best choice - Petroleum, neutral gas and nuclear energy are the last resort - Renewability, cost and availability of energy	Zahedi & Karimi (2008) [10]	10
- Emphasis on enhancement of the proportion of renewable energy	Rahimi & Eshghi (2008)	11
- Choosing controlling criteria like interests, opportunities, costs and risks - Biomass as the best alternative for energy - Petroleum, neutral gas and nuclear energy are the last resort	Ulutas (2005)	12
- Utilizing nuclear energy as future energy resource in Sweden - Using cost- effective energies	Viklund (2002) [21]	13

were considered in the sample. When there are limitations on the number of qualified people or requirements in the field of study, this method is applied [6:180]. To deploy this method for data analysis, initially all pairwise comparisons incompatibility rates for each responder are controlled, then after ensuring an acceptable incompatibility rate (less than 0.1), by utilizing Expert Choice software*, respondents' opinions were compounded and the paired group comparison matrix released. Due to high incompatibility rate (more than 0.1), portions of the questionnaire were omitted, and 24 were used in the statistical analysis.

The analytical hierarchy process is a multi-criteria decision-making approach which can be used to solve complex decision problems. This methodology examines complex problems based on their interactions, and converts them to a plain form to be solved. AHP consists of these major steps:

- Modeling—in this step decision, problem and goal are arranged in a hierarchy of pertinent elements of decision.
- Preferred judgment—Pairwise comparisons of each option are implemented in terms of each criterion. These comparisons are performed for decision criteria.
- Calculation of relative weight—the importance and weight of elements of decision are computed relative to each other by means of numerical calculation.
- Combination of relative weights—this step is performed to rank the options for decision making, so for each option, criteria weight matrix should be multiplied in criteria's weight vector.
- The judgment's compatibility are examined. If the relative compatibility is less than 0.1, it is acceptable [7:165].

DISCUSSION

The energy sector is one of the most vital and effective sectors for economic and social development of each country. Therefore, appropriate planning and policy making is essential and requires a systematic approach

*There are various supportive software for the Analytical Hierarchy Process (AHP). Expert Choice is the most popular, and it is produced by Dr. Saati and his colleague. This software is a multi-criteria decision support based on AHP method.

to various relations of the energy sector with other sectors (e.g., social, environment and economic). It is notable that future generations will have energy requirements, and it is the present generation's duty to illustrate its commitment to them by means of precise planning and policy.

Sustainability, sufficiency, participation and solidarity are our criteria in this study. Each has its own sub-criteria which are noted in Figure 2-1. In respect to data that were obtained from the experts' response, the criteria's relative value and sub-criteria are drawn and demonstrated in Table 2-2.

According to Table 2-2, the priority of each criterion is as follows: sustainability with 0.488 obtained the highest value, then sufficiency comes by 0.240, participation by 0.139 and solidarity by 0.132. We believe that in addition to energy consumption policy and implementation of intergenerational justice, sustainability is the most vital criterion in ethics which requires policy makers' future attention. The close relevancy between sustainability and intergenerational justice is due to their use as the primary criteria. After prioritizing the criteria and synthesizing alternatives, final priority energy options resulted as follows (see Table 2-3).

1-Renewable energy by 0.491 relative values
2-Fossil fuel by 0.28 relative values
3-Nuclear energy by 0.229 relative values

Table 2-2. Research's relative values of criteria and sub-criteria.

Relative value of main criteria	Sub-criteria		Criteria
0.488	0.515	Renewability	Sustainability
	0.152	Peace	
	0.123	Flexibility	
	0.122	Risk	
	0.089	Aesthetic	
0.24	0.441	Adequacy	Sufficiency
	0.306	Cost	
	0.252	Efficiency	
0.139	0.407	Citizen Engagement	Participation
	0.315	Appropriateness	
	0.278	Employment	
0.132	0.439	Equity	Solidarity
	0.370	Inter-generational Equity	
	0.191	Interdependence	

For prioritizing, relative values are considered in a distributive state due to our purpose for not achieving the best option. Instead, we focus on the relative value of each option regarding the appropriate future policy and express sub-criteria about them. When considering Table 2-3, it is obvious that the relative value of renewable energies is more than nuclear or fossil fuels. Special attention to renewable forms of energy is needed by policy makers in the future. Additionally, this issue is seen in other countries as being gradually enhanced. Similar to past use of fossil fuel resources as primary energy source, bringing renewable energy resources on-line requires better policies, development time and assessments of resource potential in Iran.

The significance of the criterion for each option is represented in Figure 2-2 and Table 2-3. In Table 2-3 the criteria's relative value on options is determined by the sum of its point value to the total value of each option. To compare options based on criteria, Figure 2-2 could be used. For example, the relative value of participation in fossil fuels is the only criterion which is greater than the others in renewable energy.

In Iranian governmental general policies, all three kinds of energy

Table 2-3. Final ranking of energy options.

Sum	Criteria	Energy
0.051	Participation	Fossil
0.041	Solidarity	
0.066	Sufficiency	
0.124	Sustainability	
0.282	**Fossil total weight**	
0.041	Participation	Nuclear
0.036	Solidarity	
0.06	Sufficiency	
0.093	Sustainability	
0.23	**Nuclear total weight**	
0.047	Participation	Renewable
0.057	Solidarity	
0.115	Sufficiency	
0.272	Sustainability	
0.491	**Renewable total weight**	
1.003	**Total**	

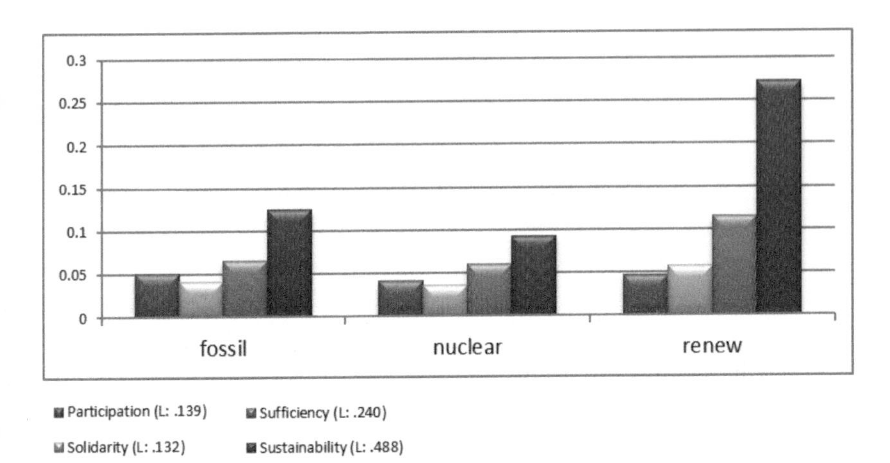

Figure 2-2. Combination of alternatives and criteria.

(fossil, nuclear and renewable) are included. In spite of Iran's location in the Middle East which is one of the richest areas in fossil fuels, the country relies primarily on this type of energy, causing irrecoverable losses to its economy.

After the Fukushima incident, many countries revised their energy policy in the nuclear sector. Also, nuclear waste is one of the major difficulties in the field of nuclear energy. Sanctions have been imposed against Iran for nuclear energy and this sector is influenced by political issues among different countries. It is clear that Iran endures great costs due to unsolved international problems in nuclear energy.

As expressed before, many countries are developing their renewable energy resources and this indicates that in spite of high initial development costs, renewables are a priority for policy making in developed countries. The outcomes of this study represent that energy experts consider renewable energy as the best choice regarding the mentioned criteria.

One of the most challenging problems in developing renewable energy is acceptance of responsibility for investing in this field. In 2001, Iran's Congress set a momentous law to protect renewable energy investment in private economic sectors. These protections were not able to persuade the private sector to invest. However, in recent years this trend is improving. Another reason for the inattention of the authorities toward renewable energy is the low cost of conventional energy in Iran.

RESULTS

Our assessment of Iran's environmental and energy processes indicates that ethical motions have not been integrated in energy policies due to reasons that primarily include: lack of a long-term focus on the problem while implementing energy strategies and policies without considering social, cultural and environmental aspects. Broad measures taken by active international energy organizations (such as World Energy Council, Asia Peace Energy Committee, International Energy Agency) demonstrate that the world has awakened from the dream of unlimited energy resources. Iran's energy policies lack practicality, despite the authority's attention. The most significant results are as follow:

- Appropriate energy policy making to achieve sustainable development requires, political, economic, social and environmental considerations. Moreover, energy resources and available technologies are limited.
- Sustainability is a major factor in choosing renewable energy.

As seen in analytical hierarchy processes, the priority of implementing renewable energies is declining.

- There is an inverse relationship between sustainability and fossil fuel energy and nuclear energy usage.
- The relative value of renewable energy declines as sufficiency, solidarity and participation increase.
- By growth in the relative value of participation, the priority of renewable energy is reducing and is being added to the priority of fossil fuel.

According to the data analysis, to guide Iran's energy policy making, the following suggestions are recommended:

1. Endeavor to protect fossil fuels as fiduciary which are transmitted to the current generation by previous generations.

2. Encourage public participation in protecting accurate consumption of fossil fuel.

3. Observe ethical principles in future energy policy making.
4. Develop intergenerational justice by accurate policy making in the field of energy.
5. Protect various sectors to invest in the renewable energy sector.
6. Persuade and protect applicable ideas in districts such as Sistan and Baluchestan, Yazd and potential rural regions.
7. Invest in nuclear energy with a long-term perspective and in safe areas.
8. Note human and environmental risks regarding all three energy options.
9. Consider sustainable criterion in all energy policy making dimensions (financial, ecological and social).

Among the fundamental issues that we encounter are regulations and environmental relationships. With the deterioration of ecological systems which are caused by human activities and environmental crisis intensification, environmental pollution problems and ecological imbalance remain unsolved.

References

[1] Massarrat, M. (2004). Iran's energy policy: Current dilemmas and perspective for a sustainable energy policy. *International Journal of Environmental Science and Technology*, 3(1), pages 233-245.

[2] Manzoor, D. (2004). "Renewable energy development in Iran." Middle East and North Africa Renewable Energy Conference. Pages 1-7.

[3] Martin-Scramm, J. (2010). Climate justice—ethics, energy and public policy. Fortress Press.

[4] New Brunswick Energy Commission (2011). An energy policy for New Brunswick 2010-2020, pages 1-11.

[5] Yang, T. (2006). Towards an egalitarian global environmental Ethics. *Environmental Ethics and International Policy*, 8, pages 23-45.

[6] Khalil, M. (2009). Research, researching, journal writing. Sociologists publisher: Tehran, Iran.

[7] Reza, M. (2004). Advanced operation's research. First Edition. Academic book publisher: Tehran, Iran.

[8] Dabiri, F., Abbas, P. and M. Hosein (2009). The right of peaceful use of nuclear energy in terms of environmental rights. *Environmental Science and Technology*, 43, pages 522-515.

[9] Somaye, R. (2008). Determining an optimal combination of Iran's energy resources, by applying network analysis process. *Journal of Energy Economy's Reviews*, 18, pages 123-160.

[10] Gholamreza, Z. and K. Alireza (2008). "Identifying energy resource's priorities of Iran by utilizing analytical hierarchy process," Twelfth national congress of Iran's chemical engineering.

[11] Mohse, M. (2005). "Iran's energy policy: challenges and supersedes." *Irannameh*, 84, pages 469-488.

[12] Atabi, F. (2004). Renewable energy in Iran: challenges and opportunities for sustainable development. *International Journal of Environmental Science and Technology*, 1(1), pages 69-80.

[13] Comby. B. (1996). Environmentalists for nuclear energy: the benefits of nuclear energy. TNR Editions.

[14] Ghorashi, A. and A. Rahimi (2011). Renewable and non-renewable energy status in Iran: art of know-how and technology-gaps. *Renewable and Sustainable Energy Reviews*, 15, pages 729–736.

[15] Gosseries, A. (2008). Theories of intergenerational justice: a synopsis. Integr.

[16] Karegar, K. (2005). Wind and solar energy developments in Iran. Iran energy reports by IIES, Iran energy report by the Ministry of Petroleum.

[17] Sabetghadam, M. (2005). Energy and sustainable development in Iran. Sustainable Energy Watch, pages 1-17.

[18] Schwarze (2003). Intergenerational justice and sustainability—economic theory and measurement. *Sustainability and Economics Research Programme*, 5, pages 1-56.

[19] Taebi, B. and A. Kadak (2010). Intergenerational considerations affecting the future of nuclear power: equity as a framework for assessing fuel cycles. *Risk Analysis*, 30(9), pages 1,341-1,362.

[20] Ulutas, B. (2005). Determination of the appropriate energy policy for Turkey. *Energy*, 30(11), pages 46-61.

[21] Viklund, M. (2002). Energy policy options—from the perspective of public attitudes and risk perceptions. *Energy Policy*, 32(10), pages 1,159-1,171.

Chapter 3

Facility-scale Energy Storage: Applications, Technologies and Barriers

Jesse S. Remillard

Many large facilities in the U.S. are considering the deployment of energy storage technologies for electric demand response programs. Technologies developed for facility- and campus-scale energy storage show promise for managing short-term electrical demand peaks as well as longer-period demand response events.

This chapter's author has investigated facility and campus-scale energy storage for efficiency program administrators in the U.S. and recently completed a storage technology research report for an international consortium of utilities. This work has identified promising avenues for distributed storage. Facility-scale storage has three primary uses: 1) power quality—the monitoring and regulation of voltage fluctuations, frequency disruptions, and harmonic distortions; 2) bridging power—short-term power supply for critical demands, often used to cover time periods in which emergency generators are powering up; and 3) energy management—energy storage on a scale to support a facility or campus of buildings for extended periods of time. These systems can be responsive to utility demand programs and time-of-use rates to reduce electrical peak demand costs.

All three of these facility-scale applications incorporate the development of strategies to use distributed storage for electric power continuity and demand management strategies.

This chapter considers the technical properties of current storage systems, including flywheel, compressed air, and various battery technologies. The technical and market barriers associated with distributed storage, along with proposed paths for resolving these barriers, will also be discussed.

INTRODUCTION

Modern energy storage originated at the electrical grid scale in the mid-1920s. Pumped hydro storage provided an early means of shifting electricity from periods of low demand to periods of high demand [1]. Little research was devoted to energy storage applications from that time until the 1990s, when the Sandia National Laboratory (SNL) identified and documented thirteen ways that utilities could use energy storage. During the summer of 2013, SNL released an updated electricity storage handbook detailing additional uses for energy storage—specifically, behind-the-meter, customer applications [1]. The article focuses on the practical considerations of customer energy storage applications, the energy storage technologies currently available, and the primary barriers to widespread behind-the-meter energy storage implementation.

Energy storage has become a proven solution for demand response events, peak demand reduction, power quality regulation, and emergency response power supplies. While there are several technologies that support facility-scale energy storage, batteries and flywheels are the most mature and readily available technologies. The advantages that batteries and flywheels offer over competing technologies, such as generators, include millisecond response times and highly accurate load-following capability. Their primary disadvantage is that these are emerging technologies. As such, costs are high and most city building codes accept only the most basic technologies for indoor installations.

There is increasing recognition of the important role of energy storage for electric utility grids that rely on large percentages of renewable generation whether they are distributed or utility scale. The recognition of this imminent need, combined with the growing acceptance of hybrid and electric vehicles, is driving a research boom in battery technologies.

Despite the need for energy storage solutions, there remain barriers to the widespread implementation of energy storage. These include the high costs of energy storage systems, limited lifetimes of storage equipment, public perception of safety, material hazards and building code acceptance. It is estimated that once the costs of installed energy storage systems drop below $300 per kWh, they will begin to displace generation used for peak power requirements [2]. If energy storage providers, utilities and facilities work together to overcome these barriers, energy storage could provide solutions for grid stability, power quality, demand management and renewables integration.

Key Terminology

In order to understand the functions that energy storage serves for utilities and their customers, certain key parameters need to be defined. The first of these is discharge time, which is defined as the amount of time that a battery can maintain its rated power output. Energy storage applications can generally be categorized into one of three groups—power quality, bridging power, or energy management—each requiring increasing response times. Table 3-1 contains a list of key energy storage terms, their definitions and considerations.

Applications for energy storage can be categorized as either power (demand) applications or energy (consumption) applications. A power application refers to a system that is designed to provide power to the system over a short time period in order to reduce peak power demands and/or to improve facility power quality. Energy applications are designed to shift energy usage from one time period to another.

FACILITY APPLICATIONS FOR ENERGY STORAGE

Energy storage equipment is expensive and commercial facility owners will likely install the equipment if it is necessary or provides tangible value. An example is the installation of uninterruptible power supplies (UPS) that keep critical systems operating during brief power disruptions. Incentives through rate structures, or other mechanisms, create added value that outweighs the capital cost and risks associated with installing storage systems. The facility uses for energy storage that are considered in this chapter include:

1. Power quality and dependability
2. Demand charge reduction
3. Demand response
4. Retail energy time shift
5. Renewables integration

These facility uses are discussed in the most recent version of the Department of Energy (DOE) Energy Storage Handbook and are supported by multiple interviews with energy storage system providers. Their comparative parameters are summarized in Table 3-2.

Table 3-1. Energy storage terms, definitions and considerations.

Term	Definition	Considerations
Depth of discharge (DOD)	The percentage of a battery's technical energy capacity that has been discharged.	Batteries are rated for DOD, and cycling beyond this will significantly reduce cycle life for certain battery types.
Power capacity	The equipment's safely rated operating output in kilowatts (kW).	Operating at higher power outputs, relative to the battery's rated power, can cause excessive wear and tear.
Energy capacity	The total amount of energy (kWh) that the storage can hold.	The technical energy capacity of the battery will be greater than the rated energy capacity because of efficiency losses and DOD limitations.
Discharge time	The maximum duration over which a battery can discharge at its rated power.	Discharge time is derived from the ratio of the battery's energy capacity to its power capacity.
Cycle life	The number of charge and discharge cycles that a battery can sustain within its expected useful lifetime (EUL).	Cycle life varies widely by technology, as well as within each technology, by manufacturing quality and operating conditions.
Degradation	The rate of reduction in a battery's technical energy capacity over time or use.	Most batteries are considered at the end of their EUL when they reach 80% of their original energy capacity. Degradation rates are impacted by battery design and operational factors including the DOD, operating temperature, and rate of discharge.
Self-discharge rate	The rate at which batteries lose energy while idle.	Typically 2% to 5% of the total system capacity per month for lithium-ion (Li-ion) and lead acid batteries; in part this rate defines the shelf life of the battery.
Round-trip efficiency	The ratio of usable energy to the energy required to charge the battery.	Round-trip efficiency is a measure of the charging and inverter losses.
Power density	The battery's power output (kW) per unit of the device's physical volume.	Power density defines how much space a battery will need for a given power rating.
Energy density	The battery's energy capacity per unit volume.	Energy density defines how much space a battery will require for a given energy capacity.

Table 3-2. Facility use characteristics [1,3,4].

Facility Use	Power Capacity	Discharge Time	Frequency of Use
Power quality and dependability	100 kW to 1 MW	15 minutes or less	Variable; as needed
Demand charge reduction	50 kW to 1 MW	1 to 4 hours	Daily
Demand response	50 kW to 1 MW	4 to 6 hours	Infrequent
Energy cost savings	100 kW to 1 MW	> 1 hour	Daily
Renewables integration (power quality and dependability)	100 kW to 500 MW	Up to several hours	Daily
Renewables integration (energy shifting)	100 kW to 500 MW	> 1 hour	Daily

Power Quality and Dependability

Energy storage systems that correct poor-quality power also protect equipment such as compressors and servers. Those that provide dependable power prevent business losses caused by equipment downtime. Examples of poor-quality power that can be corrected with an energy storage system include variations in voltage and harmonic distortions. Dependable power solutions are intended to prevent interruptions of service that can be unacceptable to equipment or business operations. Examples of dependable power solutions are systems that provide 15 minutes or less of power during service interruptions while generators are being started.

These types of systems are commonly known as uninterruptable power systems (UPS). Since they are necessary for certain types of equipment, they are the second-most installed category of energy storage systems, measured in total installed kW of capacity, after utility-scale bulk storage [5]. Although businesses that own data centers are the primary purchasers of UPS systems, they are used for other applications, such as:

ergy storage will become a necessity in order to mitigate the intermittent nature of electrical generation from certain types of renewables. What is not clear is the point it will be reached. This will vary depending on the nature of the loads and other variables associated with the electric grid to which they are connected.

Energy storage plays a larger role in distributed renewable generation in Germany, where retail electricity prices are relatively high and wholesale energy prices are relatively low. This gap in pricing makes it particularly advantageous for renewable systems that can store extra energy until it is required [6].

DUAL PURPOSE SYSYEMS

It is common for facilities to install multiple use systems because of the added value they offer. Examples of potential combinations and case studies are discussed in the following subsections.

Peak Demand Reduction and Emergency Backup

Systems designed to reduce peak demand or allow participation in demand response programs typically have the capability to provide ancillary power during power outages that occur during natural disasters. The Barclay Tower in New York City (NYC) used its demand charge reduction system to power service elevators and emergency lighting during Hurricane Sandy in 2012 [7].

Peak Demand Reduction and Demand Response

While systems designed for peak demand reduction may not be configured to simultaneously deploy during demand response events, owners can choose strategies that provide the most value. Glenwood, the company that owns the Barclay Tower, announced that it plans to install 1 MW of energy storage across its portfolio. The company plans to use its energy storage capacity to participate in the NYC Indian Point Demand Management program during the summer and for daily demand reduction during the winter [8].

Power Quality and Grid Support

While UPS systems are necessary for certain facilities they are not used the vast majority of the time. The Pennsylvania, Jersey, Maryland (PJM) Interconnection took advantage of these standby resources by offering an incentive of $40/MWh for energy supplied to the grid from energy storage for frequency regulation [9]. This resulted from the Federal Energy Regulatory Commission (FERC) Order 745, which allowed utilities to pay for performance for frequency regulation services from fast-responding energy storage technologies such as batteries [10]. Regrettably, Order 745 was vacated due to a case brought by the Electric Power Supply Association, which determined that the FERC lacked jurisdiction over demand response. FERC has appealed this decision and oral arguments were made for both sides to the Supreme Court on October 14, 2015 [11]. The implications of this decision have far reaching impacts regarding the availability of demand response incentives offered by utilities.

Peak Demand Reduction and Retail Energy Time Shift

Taking advantage of differences in peak and off-peak energy prices is a benefit of peak demand reduction or demand response systems since battery systems are typically charging during the off-peak hours and discharging during peak hours.

FACILITY SCALE ENERGY STORAGE TECHNOLOGIES

A variety of technologies are being developed for energy storage applications at all scales. Mature technologies are those that are widely available and generally accepted by building codes for installation. Currently, the mature energy storage technologies that suit the commercial needs for businesses include the following:

- Lead acid (Pb)
- Lithium ion (Li-ion)
- Sodium sulfur (Na-S)
- Flywheels

While other technologies may be available, facility owners are less likely to install such systems because of the risks associated with emerging technologies that are less available and not as widely accepted by building authorities. Technologies that are poised to enter the energy storage market in the near future include:

Table 3-3. Comparison of battery technical parameters.

Market	Battery Type	Installed Energy Cost ($/kWh)		Roundtrip Efficiency (%)	Useful Life	
		Suburban (Outdoors)	Urban (Indoors)		Cycle Life	Expected Lifetime (Years)
Commercial Technologies	Lead acid	$400 – $700	$600 – $1,000	70% – 80%	500 – 1,500	3 – 5
	Lithium ion	$1,000 – $2,000	$1,500 – $2,500	85% – 98%	2,000 – 5,000	10 – 15
	Sodium sulfur (salt)	$750 – $900	$1,000 – $2,000	70% – 80%	2,500 – 4,500	10 – 15
	Flywheels	N/A	N/A	90% – 98%	10,000+	15 – 20
Near Commercial	Advanced lead acid	$900 – $1,500	$1,200 – $1,800	80% – 90%	1,000 – 2,000	5 – 7
	Vanadium redox (flow)	$1,000 – $1,500	$1,500 – $2,000	60% – 70%	10,000+	10 – 20
	Zinc bromine (flow)	$750 – $1,250	$1,250 – $1,750	60% – 70%	10,000+	5 – 10
	Sodium nickel chloride	$1,000 – $1,500	$1,300 – $1,800	80% – 90%	2,500 – 4,500	10 – 15

- Sodium nickel chloride (ZEBRA)
- Flow (vanadium redox or zinc bromine)

There are multiple companies working on emerging storage technologies that are poised to enter the commercial market. Companies that have received funding for demonstration projects include Ambri (liquid metal), EOS (zinc air) and Aquion (magnesium salt). Each has its own proprietary technology and seeks to provide systems at groundbreaking costs and lifetimes competitive with customary peak generation sources [12-14]. Table 3-3 presents a summary of technical parameters for each of the technologies reviewed by the author using published literature and battery manufacturer interviews.

Lead Acid

Lead acid batteries are the most mature battery technology available and they are used worldwide in motorized vehicles [15]. They are typically the standard by which other batteries are compared due to their reliability and low cost, but they provide only mediocre energy/power density and short expected lifetimes. Importantly, they are capable of only a limited depth of discharge (DOD); full discharges will damage the battery and shorten its life.

Due to their prevalence and cost advantages, lead acid batteries will continue to be a staple of energy storage projects worldwide until the costs of other technologies are reduced.

Advantages

- They are the least expensive option per installed kW and kWh.
- They are highly modular.
- They are easily recyclable.
- They are accepted by building codes.

Disadvantages

- Their cycle life is short even under optimal conditions (<2,000 cycles). Under high temperatures or especially deep DODs (>50%) their cycle life is reduced to as few as 500 cycles.
- Lead acid batteries are comparatively heavy, restricting their usage due to practical and building code considerations.

Lithium Ion

Lithium-Ion (LI) batteries are typically constructed of carbon and metallic electrodes with a lithium-based electrolyte. There are a variety of subtly different cell chemistries that can be used to construct these batteries that are often proprietary to specific manufacturers. The market for LI batteries continues to grow due to their excellent energy and power densities. They weigh less and are more compact than other commercial battery technologies. Unlike lead acid batteries, they can be fully discharged and recharged without reducing the battery's cycle life. LI batteries typically last about twice as long as lead acid batteries when operated under optimal conditions. However, LI batteries currently cost about twice as much as lead acid batteries on a power capacity basis. These characteristics give them an edge in situations where space or weight is more highly valued than cost, such as small businesses with space limitations.

The costs of LI batteries are expected to decrease more than any other commercial technology in upcoming years, which makes them a likely candidate to become the dominant battery technology in the next 5-10 years.

Advantages

- They have a long cycle life that is not affected by DOD.
- They can be arranged to provide the same voltages as lead acid batteries.
- They have a high energy/power density.

Disadvantages

- They are costly.
- They have the potential to cause runaway fires if not properly maintained and operated.

Sodium Sulfur

Sodium sulfur (NaS) batteries were developed in the 1980s by NGK Insulators, LTD., the primary manufacturer of the technology and Tokyo Electric Power Co. NaS batteries have favorable characteristics for larger-scale energy storage, such as low cost and high energy capacity. They are often referred to as molten salt batteries because during operation they are composed of molten sulfur and liquid sodium separated by a ceramic electrolyte [15].

Sodium sulfur batteries are primarily installed in controlled outdoor locations because of their high operating temperatures. They are often used in energy arbitrage or other uses that require long discharge times. They have been looked upon unfavorably by building- and fire-code enforcement agencies, limiting their deployment potential in urban areas. For certain niche applications requiring lengthy discharge duration and with ample outdoor space, this technology offers competitive solutions.

Advantages

- They have a long shelf life.
- They are well-suited to energy applications.

Disadvantages

- They require high operating temperatures (>300°F).
- There are very few manufacturers of this technology.
- They are heavy and require a lot of space.

Flywheels

Advancements in flywheel technologies during the last 15 years have enabled them to compete with batteries in the power quality and dependability markets. There are many grid scale flywheel demonstration projects. Companies that offer flywheels include Active Power, Beacon Power, Vycon Energy and PowerThru [16-20].

Flywheels are marketed for high-cycle, low-energy applications such as frequency regulation, and offer distinct advantages over battery systems. These include a very high cycle life (greater than 10,000 cycles), low maintenance and high energy density. Unfortunately, flywheels have struggled in the commercial market because they lack flexibility for energy applications and their future capital costs are unlikely to be competitive with battery systems.

Advantages

- They have long lifetimes.
- They are well suited to applications requiring frequent cycling.

Disadvantages

- Comparatively higher capital costs.

- They are not suited to energy applications.
- There are very few manufacturers of this technology.

EMERGING AND COMPETITIVE

The potential value of energy storage is increasing with the widespread emergence of hybrid vehicles and renewable electricity generation. These applications are causing an explosion of interest in supporting these markets with the development of new storage technologies. While many of these emerging technologies are promising in regard to cost-effectiveness and performance, they often lack established manufacturing practices and safety protocols. It is likely that these barriers will fall in response to the need for lower costs and improved performance. Some of the emerging technologies that may be prominent in future storage applications are discussed next.

Sodium Nickel Chloride

Sodium nickel chloride—also called ZEBRA—batteries are high temperature (>300°C) batteries that are similar to sodium sulfur technologies but with improved safety characteristics. Only two manufacturers are currently making these batteries. They have long lifetimes and generally better performance characteristics than traditional lead acid batteries without some of the safety concerns associated with sodium sulfur batteries [15].

Flow Batteries—Vanadium Redox (VRB) and Zinc Bromine (ZnBr)

Flow batteries rely on a liquid electrolyte that flows through the battery. This means that the energy storage capacity of the battery can be increased or decreased by adding or removing electrolyte. This allows the energy storage capacity to be decoupled from the number of cells. Sumitomo Electric Industries is the main investor in vanadium flow batteries and ZBB Energy Corporation is the primary manufacturer of zinc bromine batteries. Both have package options available for purchase although the number of deployments is limited [21]. Another variation of flow batteries, using iron-chromium chemistry, is being demonstrated in California with support from the DOE, which could prove to be quite inexpensive due to the use of abundant, low-cost materials [22].

Recently, a startup company named Imergy Power Systems has made progress on a cost-effective energy storage solution with their vanadium flow batteries utilizing a proprietary electrolyte developed in collaboration with the Pacific Northwest National Laboratory. They offer several packaged low-power (<250 kW) 4-hour discharge solutions and claim to have more than 200 residential, commercial, and utility grid-scale systems installed [23,24].

The pumps, storage, and piping required by flow batteries reduce their overall energy density (thus requiring expanded footprints for the equipment) and entails operations and maintenance responsibilities that exceed those of other technologies. Although these systems are not yet readily obtainable, they may become more cost-effective in the future than conventional batteries. Vanadium batteries have the potential for very long shelf life (>10 years) and cycle life (>10,000 cycles) [25,26].

TECHNICAL AND MARKET BARRIERS

There are a number of market barriers preventing energy storage systems from reaching their full distribution potential. There is a consensus among energy storage system designers that one of the biggest issues is a negative public perception toward energy storage technologies. Building owners often lack an understanding of how energy storage systems operate and the value they can offer. Improved marketing and better education of building owners can help overcome these market barriers.

Performance

The two primary technical barriers for batteries at this time are:

1. Limited cycle life and shelf life.
2. High costs of systems.

It is essential for battery systems to have longer lifetimes and better warranties in order to gain greater acceptance. A lead acid battery's lifetime of three to five years is very short compared to most commercial equipment's estimated useful life (EUL). Progress is being made, primarily with flow batteries and LI chemistries, which boast lifetimes of seven to ten years but suffer from high cost and lack of technological maturity.

Longer cycle and shelf live increases the value of batteries but the primary barrier to their widespread use remains capital costs. Increasing demand charges and demand management programs (e.g., those available in NYC that provide incentives of $2,100/kW for battery storage) improve the economics of projects. Facilities require high electrical demand charges to provide a revenue stream that supports the installation costs for energy storage systems.

Material Hazards and Siting Barriers

All battery technologies have inherent risks to human and environmental health and safety. They often contain toxic chemicals in their electrolytes and have the potential to overheat, catch fire and explode.

For commercially available storage technologies the risks are generally well understood. Risks can largely be mitigated through appropriate installation and fire protection, rendering batteries safe in urban environments. Many energy storage systems are packaged in containers. These are durable, weatherproof and adequately secure to allow for outdoor installations.

There are two primary construction complexities when installing facility-scale battery storage systems:

1. Size—The facility needs to find suitable unoccupied, dry space designated to permanently site these systems.
2. Weight—Due to the nature of the materials used in their manufacture, most batteries are quite heavy. For instance, sodium sulfur batteries weigh upwards of 500 lbs./ft^2 (21kg/m^2) or about five times the design standard of a normal commercial floor.

Often the best location for these systems is outside on a poured concrete slab in a sheltered enclosure. Because of this, many companies sell their equipment packaged in shipping containers.

Permitting and Codes

Local building fire codes and construction permitting hinder the adoption of specific technologies. Permitting is the largest barrier to adoption of storage technologies excluding costs. Code requirements for battery storage are designed to ensure safe installation and operation of battery systems. Their focus is on safety precautions to mitigate impacts of spills, fires, natural disasters and unauthorized access. Barriers are falling. NYC recently incorporated lead acid battery storage systems and LI batteries in its fire code [27].

SUMMARY

No single battery technology has proven superior to all others. Specific battery technologies are designed to be suitable for varying requirements and applications. Today's competitive markets support a broad range of developmental storage solutions and it is unclear if the marketplace will remain highly competitive or yield to a dominant brand or chemistry.

Energy storage technologies address specific needs for both facility owners and electrical generation and distribution managers. Facility owners can utilize storage to provide resiliency for critical operations, thereby protecting profitability. Utility companies and electric system operators increasingly need storage capability to effectively use renewable and other variable generation resources. By offering electric customers educational opportunities in power management and financial incentives, the value of electrical energy storage systems for both campus and facility scale applications will increase in the future.

References

[1] P. Denholm, E. Ela, B. Kirby and M. Milligan (2010, January). The role of energy storage with renewable electricity generation. NREL—DOE.

[2] Register, C. (2015, January 13). The battery revolution: a technology disruption, economics and grid level application discussion with Eos Energy Storage. Forbes. http://www.forbes.com/sites/chipregister1/2015/01/13/the-battery-revolution-a-technology-disruption-economics-and-grid-level-application-discussion-with-eos-energy-storage/3, accessed 2015.

[3] APS Physics (2007). *Challenges of energy storage technologies*. APS. Ridge, New York.

[4] Erey, G. (2010). Energy storage for the electricity grid: benefits and market potential assessment guide. U.S. DOE, Albuquerque, New Mexico.

[5] PV Magazine (2014). Trends in energy storage markets. http://www.pv-magazine.com/archive/articles/beitrag/trends-in-energy-storage-markets-_100010780/86/?tx_ttnews%5BbackCat%5D=217&cHash=dc05a369712231e32e05bd86fb4a8c86, accessed 2014.

[6] Deign, J. (2015, March 13). German energy storage: not for the fainthearted. Green Tech Media. http://www.greentechmedia.com/articles/read/german-energy-storage-not-for-the-faint-hearted, accessed 2015.

[7] Demand Energy (2014, July 8). Energy storage saves money, helps Con Edison at Manhattan's Barclay Tower. Demand Energy. http://buyersguide.renewableenergyworld.com/demand-energy/blog/energy-storage-saves-money-helps-con-ed-at-manhattans-barclay-tower.html,accessed 2015.

[8] Hardesty, L. (2015, March 17). Glenwood deploys 1 MW of energy storage in NYC Buildings. Energy Manager Today. http://www.energymanagertoday.com/glenwood-deploys-energy-storage-systems-nyc-buildings-0110234.

[9] Kanellos, K. (2013, September 9). Why data centers could be good for the grid.

Forbes. http://www.forbes.com/sites/michaelkanellos/2013/09/13/why-data-centers-could-be-good-for-the-grid/?ss=business:energy,accessed 2015.
[10] Federal Energy Regulatory Commission (2011, March 15). Demand response compensation in organized wholesale energy markets. Available: https://www.ferc.gov/EventCalendar/Files/20110315105757-RM10-17-000.pdf, accessed 2015.
[11] Supreme Court (2015, October 14). Federal Energy Regulatory Commission v. Electric Power Supply Association. http://www.supremecourt.gov/oral_arguments/argument_transcripts/14-840_5ok6.pdf, accessed 2015.
[12] Fehrenbacher, K. (2014, October 21). Startup Aquion Energy shows off the next generation of its battery for solar and the grid. Gigaom. https://gigaom.com/2014/10/21/startup-aquion-energy-shows-off-the-next-generation-of-its-battery-for-solar-the-grid, accessed 2015.
[13] John, J. (2013, May 2). Eos puts its zinc-air grid batteries to the test with Con Ed. Green Tech Grid. http://www.greentechmedia.com/articles/read/eos-puts-its-zinc-air-grid-batteries-to-test-with-coned, accessed 2015.
[14] New York State (2014, March 5). Governor Cuomo awards funding to NY-BEST Companies to develop advanced energy storage technologies. New York State. Available: http://www.governor.ny.gov/news/governor-cuomo-awards-funding-ny-best-companies-develop-advanced-energy-storage-technologies, accessed 2015.
[15] Akhil, A., Huff, G., Currier, A., Kaun, B., Rastler, D., Chen, S. et al. (2013). DOE/EPRI 2013 Electricity storage handbook in collaboration with NRECA. Sandia National Laboratories, Livermore, California.
[16] Active Power (2012). Corporate overview. Available: http://www.activepower.com, accessed 2015.
[17] Beacon Power (2014). Proven Success. Bean Power. http://beaconpower.com/proven-success, accessed 2015.
[18] Vycon Energy (2011). Flywheel technology. http://www.vyconenergy.com/pages/flywheeltech.htm, accessed 2015.
[19] PowerThru (2014). Company history. Phillips Service Industries. http://www.power-thru.com/company_history.html, accessed 2015.
[20] Nelder, C. (2013, April 10). Turn up the juice: new flywheel raises hopes for energy storage breakthrough. *Scientific American*. Available: http://www.scientificamerican.com/article/new-flywheel-design, accessed 2015.
[21] Klein, A. and T. Maslin (2011). U.S. utility scale battery storage market surges forward. IHS Emerging Energy Research.
[22] LaMonica, M. (2014, May 21). Iron-chromium flow battery aims to replace gas plants. IEEE Spectrum. http://spectrum.ieee.org/energywise/energy/the-smarter-grid/new-flow-battery-aims-to-replace-gas-plants, accessed 2015.
[23] Meza, E. (2014, November 13). Imergy signs flow battery deal with Hawaii's Energy Research Systems. PV Magazine. http://www.pv-magazine.com/news/details/beitrag/imergy-signs-flow-battery-deal-with-hawaiis-energy-research-systems_100017157, accessed 2015.
[24] Imergy Power Systems (2015). Products. http://www.imergy.com/products, accessed 2015.
[25] Rastler (2010). Electricity Energy Storage Technology Options. Electric Power Research Institute. Palo Alto, California.
[26] Chen, H., Cong, T., Yang, W. et al. (2009). Progress in electrical energy storage system: a critical review. *Progress in Natural Science*, 19, pages 291-312.
[27] Cerveny, J. (2014, March 21). NY Best. Available: http://www.ny-best.org/blog-entry/update-nyc-%E2%80%93-commissioner%E2%80%99s-forum-energy-storage-new-york-city-department-buildings, accessed May 2014.

Chapter 4

Production of Clean Energy from Cyanobacterial Biochemical Products

Md. Faruque Hossain, New York University

In this chapter, an improved technology to produce hydrogen biologically will be discussed as a source of clean energy. The photochemical reaction among photons, ultraviolet (UV) light, and cyanobacterial biomaterials in photobioreactors offer a unique methodology for producing hydrogen energy. A photobioreactor is a type of bioreactor that utilizes a light source to cultivate phototrophic microorganisms. Using this technology, hydrogen production is significantly higher than for any other technology that has ever been used. This hydrogen evolution is a product of the ultimate reaction of agitated photon electrons into the cyanobacterial biomolecules, where hydrogenase enzymes function as an active catalyst. The evolved hydrogen is then clarified using an electronic semiconductor-based sensor gas chromatograph with the efficiency recorded using a computerized data acquisition system. The results confirmed that this larger amount of hydrogen formation could be an interesting source of clean energy production. It is suggested that producing hydrogen using cyanobacteria could be a method of meeting future global energy demand. The purpose of this chapter is to describe this process and discuss its benefits.

INTRODUCTION

During the last 150 years, the combustion of fossil fuels has resulted in more than a 25% increase in the levels of atmospheric carbon dioxide. Climate scientists predict that if atmospheric carbon dioxide levels continue to increase, Earth's atmosphere will become much warmer during the next century. This will likely result in severe human health issues, melting glaciers, sea-level rise, floods, air pollution and ecosystem destruction [27].

Unlike fossil fuels, renewable energy comes from natural resources (i.e., wind, sunlight, gravity, heat from the earth and biomass materials) that are environmentally benign. Using these resources to supply our energy needs will not only meet global energy demand, it will further support sustainability by lowering greenhouse gas emissions. The development and use of renewable energy provides benefits to the world's nations including incremental energy production, environmental protection, and reduced pollution. Renewable energy sources such as solar (thermal or photovoltaic), wind, hydroelectric, biomass, and geothermal energy constitute the most common sustainable sources of energy. The characteristics of specific energy resources can be evaluated in terms of sustainability indicators [2]. In 2006, sustainable energies represented about 18% of the global total energy consumption [33]. They substitute for traditional fuels (e.g., coal, natural gas, petroleum, etc.) providing power generation, heating and transportation fuels. Due to its common use in developing countries for local energy supplies, biomass represents a major source of renewable energy constituting as much as 75% of the renewable energy used today [18,19].

Cyanobacteria could produce third generation eco-energy since its photo biological hydrogen (H_2) production is considered to be a candidate for renewable energy production [15,18]. Cyanobacteria possess certain properties which have entitled them to be one of the most promising feedstocks for energy generation. Cyanobacteria, being photosynthetic organisms, use the sun's energy, water (H_2O) and carbon dioxide (CO_2) to synthesize their energy storage components. These energy storage components form a potential feedstock which can be converted into bioenergy [35]. Cyanobacteria possess unique properties which make them a promising model to transform carbohydrate sources into valuable fuels.

Hydrogen is produced by many strains of cyanobacteria by the reversible activity of hydrogenase. When cyanobacteria are grown under nitrogen (N_2)-limiting conditions, H_2 is formed as a byproduct of N_2 fixation by nitrogenase. Several reports have considered cyanobacterial species capable of producing H_2 [1,14,15]. Cyanobacteria should not only be used as a biocatalyst of sunlight. They possess other properties which make them ideal candidates for the development of bio-friendly systems for the generation of clean fuels for 21st century.

METHODS AND MATERIALS

Central Metabolism of Photosynthetic Cyanobacterial for Hydrogen Production

The main constraint for H_2 production in cyanobacteria is that hydrogenases are highly intolerant to the oxygen (O_2) produced during photosynthesis. I have used the reducing agents ferredoxin and nicotinamide adenine dinucleotide phosphate (NADPH) as these are involved in respiration to reduce the O_2 by cyanobacteria. In order to enhance H_2 production, it is important to redirect part of the electron flow towards the H_2-producing enzymes and to oxygen-tolerant hydrogenases [5,43]. An attempt to eliminate pathways that consume reducing agents has also been performed.

Figure 4-1 shows the oxygenic "light reactions" of photosynthesis driven by the solar energy captured by the light-harvesting complexes of PSI and PSII. Electrons extracted from H_2O by the oxygen-evolving complex of PSII are passed along to the photosynthetic electron transport chain via plastoquinone (PQ), the cytochrome b6f complex (Cyt b6f), plastocyanin (PC), photosystem I (PSI), and ferredoxin (Fd), then by ferredox-

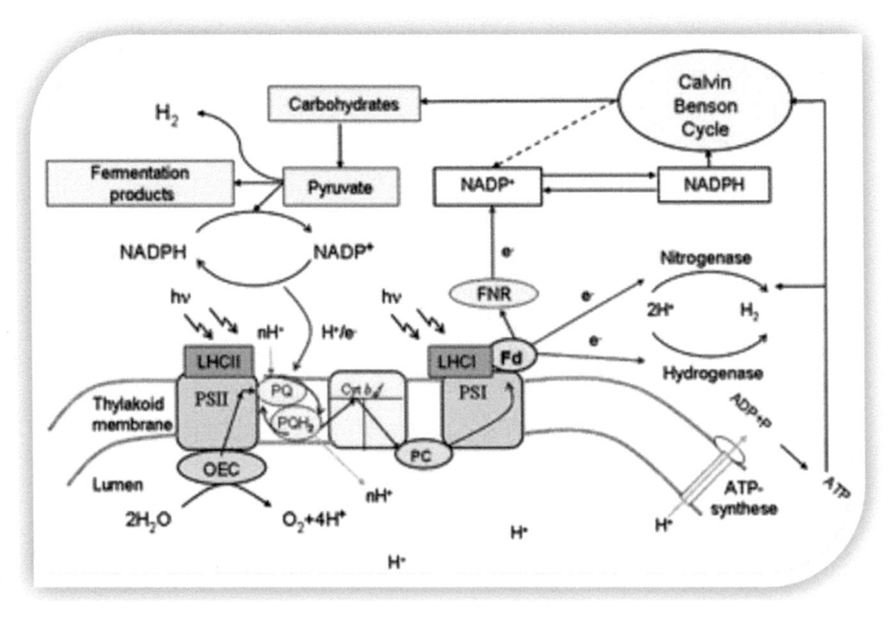

Figure 4-1. The scheme of H_2 production during oxygenic photosynthesis and subsequent formation of carbohydrates in microalgae [from 5,6,9,13,26].

in-NADP+ oxidoreductase to NADP+ ultimately producing NADPH. H+ are released into the thylakoid lumen by the PSII and PQ/PQH_2 cycles and used for adenosine triphosphate (ATP) production via ATP synthase.

The ATP and NADPH generated during primary photosynthetic processes are consumed for CO_2 fixation in the Calvin-Benson cycle which produces sugars and starch. Under anaerobic conditions, hydrogenase accepts electrons from reduced Fd molecules and uses them to reduce protons to H_2. Certain algae under anaerobic conditions can use starch as a source of H+ and e– for H_2 production (via NADPH, PQ, cytb6f, and PSI) using the hydrogenase. In cyanobacteria, the H+ and e– derived from H_2O can be converted to H_2 via a nitrogenase. Nitrogenase is responsible for nitrogen-fixation and hydrogenase is responsible for hydrogen production. Molecular nitrogen is reduced to ammonium with consumption of reducing power (e' mediated by ferredoxin) and ATP. The reaction is substantially irreversible as follows:

$$N_2 + 6H1^+ + 6e^- = 2HN_3$$
$$12ATP \rightleftharpoons 12(ADP+Pi)$$

However, nitrogenase catalyzes proton reduction in the absence of nitrogen gas (i.e., in an argon atmosphere).

$$2H^+ + 2e^- = H_2$$
$$4ATP \rightleftharpoons 4(ADP+Pi)$$

Hydrogen production catalyzed by nitrogenase occurs as a side reaction at a rate of one-third to one-fourth that of nitrogen-fixation, even in a 100% nitrogen gas atmosphere. It is important to analyze the nitrogenase function on cyanobacterial metabolism for the proper clarification of hydrogen production. As Figure 4-2 shows, the nitrogenase itself is extremely oxygen-labile which ultimately consumes O_2 and accelerates H_2 production. Unlike hydrogenase, cyanobacteria have developed mechanisms for protecting nitrogenase from oxygen gas and supplying it with energy (ATP) and reducing power.

Vegetative cells (ordinary cells) in filamentous cyanobacteria perform oxygenic photosynthesis. Organic compounds are produced by carbon monoxide (CO); reductions are transferred into cyanobacterial heterocyst and are decomposed to provide nitrogenase with reducing power. ATP

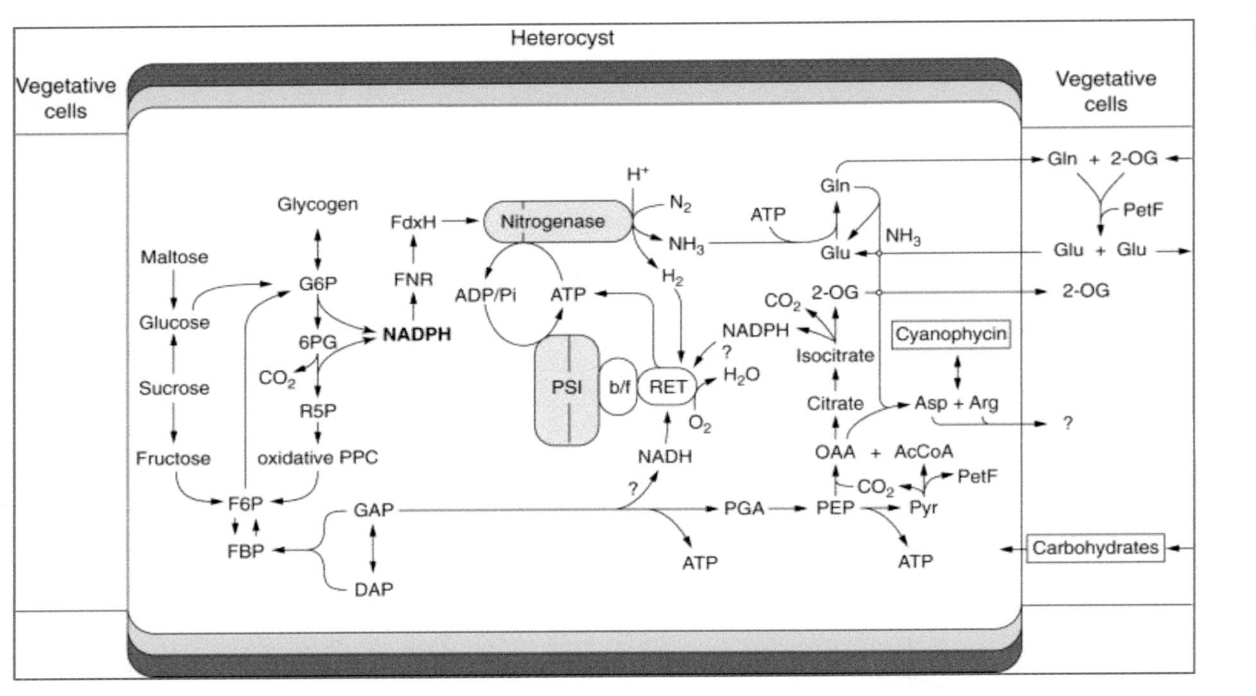

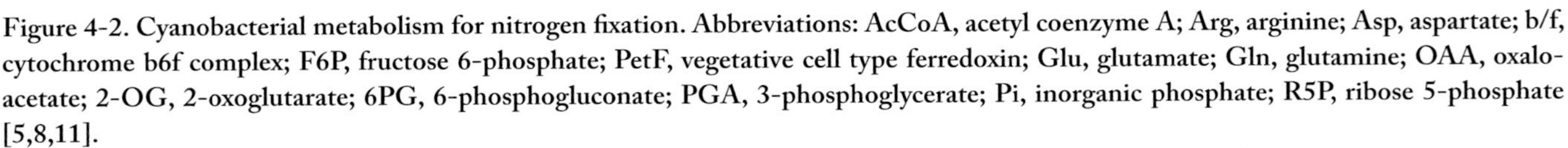
Figure 4-2. Cyanobacterial metabolism for nitrogen fixation. Abbreviations: AcCoA, acetyl coenzyme A; Arg, arginine; Asp, aspartate; b/f, cytochrome b6f complex; F6P, fructose 6-phosphate; PetF, vegetative cell type ferredoxin; Glu, glutamate; Gln, glutamine; OAA, oxalo-acetate; 2-OG, 2-oxoglutarate; 6PG, 6-phosphogluconate; PGA, 3-phosphoglycerate; Pi, inorganic phosphate; R5P, ribose 5-phosphate [5,8,11].

can be provided by PSI-dependent and anoxygenic photosynthesis within heterocyst. Investigations into prolongation and optimization of hydrogen production revealed that the hydrogen-producing activity of cyanobacteria was stimulated by nitrogen starvation. Therefore, we continuously sparge argon-based gas while the hydrogen content of the effluent gas was measured. The average conversion efficiency over a period of one month (combustion energy of hydrogen gas produced by cyanobacteria/incident solar energy into the photo-bioreactor area) significantly produced the highest amount of H_2 (68 μmol mg^{-1} chl a h^{-1}).

Cyanobacteria are photoautotrophic microorganisms [23] that use two sets of enzymes to generate hydrogen gas. The first is ***Nitrogenase***. It is found in the cyanobacterial heterocyst of filamentous cyanobacteria when grown under nitrogen limiting conditions. Hydrogen is produced as a by-product of fixation of nitrogen into ammonia. The reaction consumes ATP and has the general form:

$$16ATP + 16H_2O + N_2 + 10H_2 + 8e^- \xrightarrow{\text{Nitrogenase}} 16ADP + 16Pi + 2NH_4^+ + H_2 \quad [24]$$

A Nitrogenase enzyme consists of two parts: one is dinitrogenase (MoFe Protein, encoded by the genes *nifD* and *nifK*, α and β respectively) and the other is dinitrogenase reductase (Fe Protein, encoded by nifH). Dinitrogenase is a $\alpha_2\beta_2$ heterotetramer, having molecular weight of 220 to 240 kDa respectively, that breaks apart the atoms of nitrogen. Dinitrogenase reductase is a homodimer of about 60 to 70 kDa and mediates the transfer of electrons from the external electron donor (a ferredoxin or a flavodoxin) to the dinitrogenase [17,25,30]. There are three types of dinitrogenase found in Nitrogenase, which vary depending on the metal content. Type I contains molybdenum [39], Type II contains vanadium (V) instead of molybdenum [22,39], and Type III has neither molybdenum nor V but contains iron [10,22].

The other hydrogen-metabolizing/producing enzymes in cyanobacteria are *Hydrogenases*; they occur as two distinct types in different cyanobacterial species. One type uptakes hydrogenase encoded by *hupSL* [38] and has the ability to oxidize hydrogen. The other type of hydrogenase is reversible or bidirectional hydrogenase (encoded by *hoxFUYH*) and it either absorbs or produces hydrogen. *Uptake hydrogenase* enzymes are found in the thylakoid membrane of cyanobacterial heterocyst, where they transfer electrons from hydrogen for the reduction of oxygen via the respiratory chain in a reaction known as oxyhydrogenation or the Knallgas reaction. The enzyme consists of two subunits. The *hupL*-coded protein uptakes hydrogen and the smaller *hupS*-coded subunit is responsible for reduction. Under ambient conditions, the hydrogen formed is reoxidized by an uptake hydrogenase via a Knallgas reaction. This is counterproductive when the goal is to produce hydrogen on a commercial scale. The reaction catalyzed by the uptake hydrogenase takes the following form:

$$H_2 \xrightarrow[\text{Nitrogenase}]{} 2H + 2e^-$$

The biological role of *bidirectional* or *reversible hydrogenase* is thought to control ion levels in the organism. Reversible hydrogenase is associated with the cytoplasmic membrane and likely functions as an electron acceptor from both NADH and H_2 [12]. The reversible hydrogenase is a multimeric enzyme consisting of either four or five different subunits depending on the species [12,34]. On a molecular scale, it is a NiFe-hydrogenase of the $NAD(P)^+$ reducing type which consists of a hydrogenase dimmer coded by *hoxYH* gene. Maturation of reversible hydrogenases requires the action of several auxiliary proteins collectively termed as hyp (products of genes: *hypF, hypC, hypD, hypE, hypA*, and *hypB*) [44]. Unlike uptake hydrogenase,

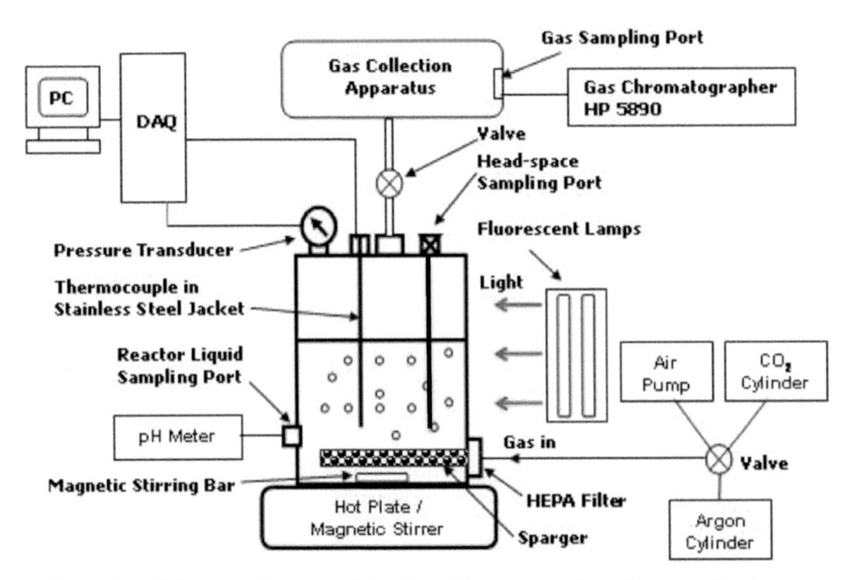

Figure 4-3. Schematic diagram of the photobioreactor and production of hydrogen.

reversible hydrogenases are helpful in hydrogen production. Most cyanobacterial species preferentially absorb red light near 680 nm [31], the light required for hydrogen production.

Photobioreactor Design

The effects of light on nitrogenase mediated hydrogen production by most types of cyanobacteria are well studied [36]. Since light is essential for cyanobacterial growth, a unique photo bioreactor was designed for large-scale hydrogen production [3,7]. Our photobioreactors require sunlight along with some controlled illumination (i.e., florescent light).

Inside photobioreactors there are photic zones, close to the illuminated surface and dark zones, those further away from this surface. The hydrogen productivity of a photobioreactor is light limited and tends to decrease at higher light intensities. Photosynthesis diverts the hydrogen production pathway. The light regime is determined by the light gradient which must be diluted and distributed as much as possible since the highest production levels occur in the darkest conditions. Rates of aeration (hydrogen producing enzymes are oxygen susceptible; anaerobic conditions or inert gas environments are preferred) impact hydrogen productivity. The red light needed by cyanobacteria is generated by panels constructed in specialized bioreactors [37].

The photobioreactor (PBR) was mainly divided into two parts: a vertical column reactor (VCR), and a tubular type and flat panel photobioreactor. The reactor for photobiological hydrogen production that was used met two conditions:

1. The photobioreactor is enclosed so that the produced hydrogen may be collected without any loss.

2. To maximize the area of incident light (thus allowing high growth and hydrogen production) the photobioreactor design provides a high surface to volume ratio.

The vertical column reactor (air-lift loop reactor and bubble column) consists of a transparent column usually composed of high quality glass and surrounded by a water jacket. This configuration allows the temperature to be maintained with circulating water and provides adequate light entry. The reactor has medium inlets and outlets for the gases such as argon and hydrogen. Fresh medium is added from a reservoir from above the VCR [6,28]. Cyanobacteria are inoculated through a septum that helps maintain sterility and prevents contamination. The bottom section of the VCR column retains outlets for the culture and an inlet/outlet for argon gas. In bubble columns using sunlight as light source, the presence of gas bubbles enhances internal irradiance at sunset and sunrise. As the position of sun changes from the horizon in the morning to overhead at noon, the bubbles diminish the internal column irradiance relative to the un-gassed state.

Biomass productivity varies substantially during the year. The peak productivity in the summer may be several times greater than the lower productivity in the winter. An example of this type of VCR is hydrogen production using cyanobacteria [6,7,28]. This reactor column consists of a glass cylinder with an inner volume of 400 ml surrounded by a water jacket. The optimal dimensions of the vertical column are about 0.2 m in diameter and 4 m in column height. The optimal column height depends on factors including the wind speed and the strength of optically transparent materials (e.g., glass or thermoplastics).

A typical flat-panel photobioreactor consists of a stainless-steel frame with polycarbonate panels. These sections of reactor are placed side by side. Water is circulated via a temperature controlled water bath through the bioreactor compartment to maintain the desired temperature of the culture. This design for a PBR often uses direct photon particles from sunlight. The average light intensity provided at the reactor surface is 175 W/m^2. A red light emitting diode (LED) that peaks at 665 nm is used as the light source on one side. In addition, the extreme ultra-violet (wave length 10 to 121 nm) and photon energy (see calculation described in Figure 4-4) is introduced.

A membrane gas pump circulates the gas through the spargers (hypodermic needles) at the bottom of the reactor. The gas produced is collected in a gasbag. In this reactor system, pressure vessels prevent pressure fluctuations in the gas recirculation system and a pressure valve maintains a constant input pressure to the mass flow controller. A condenser prevents water vapor from entering the gas recirculation system. The reactor is autoclaved prior to cyanobacterial cultivation and hydrogen production. The culture medium is separately autoclaved and fed to the reactor. Sampling is performed via the sample port which is attached to the outflow tube. Bacterial growth is monitored by a computer.

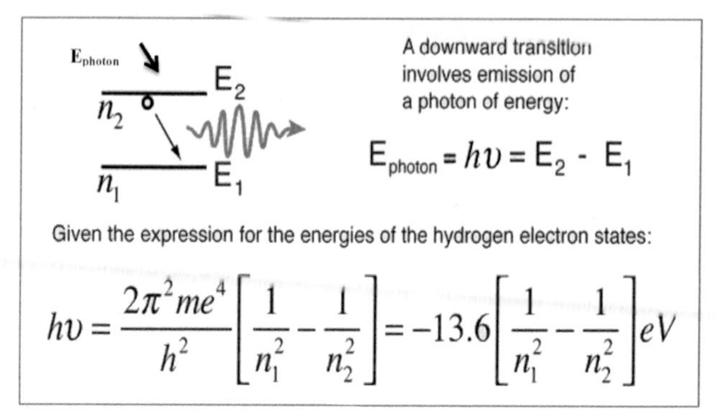

Figure 4-4. Photo reaction over the hydrogen electron states confirms that photobacteria hydrogen is a way to produce energy. The equation shows that the high rate of electron deliberation occurs once the photon energy passes through the hydrogen electron state.

Results and Discussion

Substantial progress has been made over the last decade in understanding the fundamental reaction of photosynthesis that evolved in cyanobacteria 3.7 billion years ago. This process uses water molecules as a source of electrons to transport energy derived from sunlight. Light energy (E_{photon}) introduced to cyanobacteria give bacteria their blue (cyano) color, enabling plants to evolve by "kidnapping" bacteria for their photosynthetic engines to produce H_2. The emission of photon energy into the hydrogen electron state is higher than any other technology previously used for renewable energy production.

It is known that cyanobacteria possess thylakoid, granum and other pigments, capturing the energy from sunlight using photosynthetic systems (PSII and PSI) to perform photosynthesis [20,21,39,41]. The pigments in PSII (P680) absorb the photons, generating a strong oxidant capable of splitting water into protons (H^+), electrons (e^-) and O_2 as shown in Figure 4-5. The electrons or reducing equivalents are transferred through a series of electron carriers and cytochrome complex to PSI. The pigments in PSI (P700) absorb the photons, which further raises the energy level of the electrons to reduce the oxidized ferredoxin (Fd) and/or nicotinamide adenine dinucleotide phosphate ($NADP^+$) into their reduced forms. The proton gradient formed across the cellular (or thylakoid) membrane drives adenosine triphosphate (ATP) production via ATP synthase.

Consequently, biophotolysis produced hydrogen within 24 hours of light irradiation. This occurred when light energy was absorbed by the

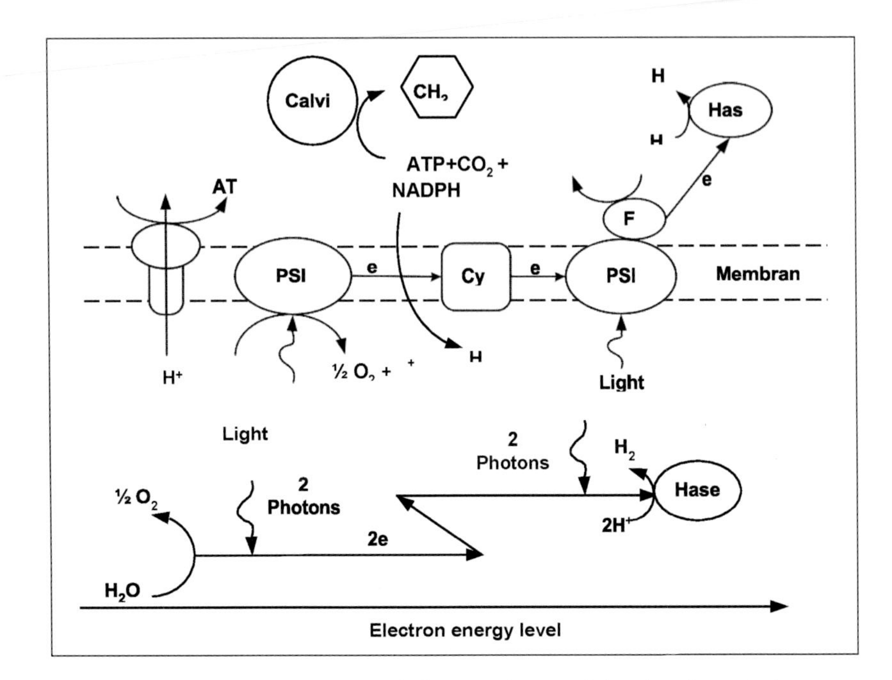

Figure 4-5. Schematic mechanisms of photosynthesis and biophotolysis of photoautotrophic cyanobacteria. The energy level of electrons or reducing equivalents from water oxidation is raised by the adsorbed photons at PSII and PSI. The reducing equivalent (NADPH) is used for CO_2 reduction in photosynthesis and carbohydrates (CH_2O) are accumulated inside the cells. The reducing power (Fd) could also be directed to hydrogenase (Hase) for hydrogen evolution.

pigments at PSII, or PSI or both, raising the energy level of electrons from water oxidation when transferred from PSII via PSI to ferredoxin. This biochemical process in cyanobacteria provides an oxygen-free environment to the oxygen-sensitive nitrogenase reducing molecular nitrogen into NH_2 as well as protons into H_2 [23,39,40] and results in much higher hydrogen formation in the absence of molecular nitrogen (Figure 4-6).

The hydrogen productivity was then calculated based on the reactor surface area from the volumetric productivity and a critical optical length. The latter has a range from 2 to 6 cm [41], depending on factors that include cell density, cell size, light intensity and light saturation [16,29,34,44]. The hydrogen collected is indicated by gas chromatography. The H_2 energy productivity is calculated by multiplying the volumetric productivity (mmol H_2/L/hour) by the heat of combustion of hydrogen at 25°C.

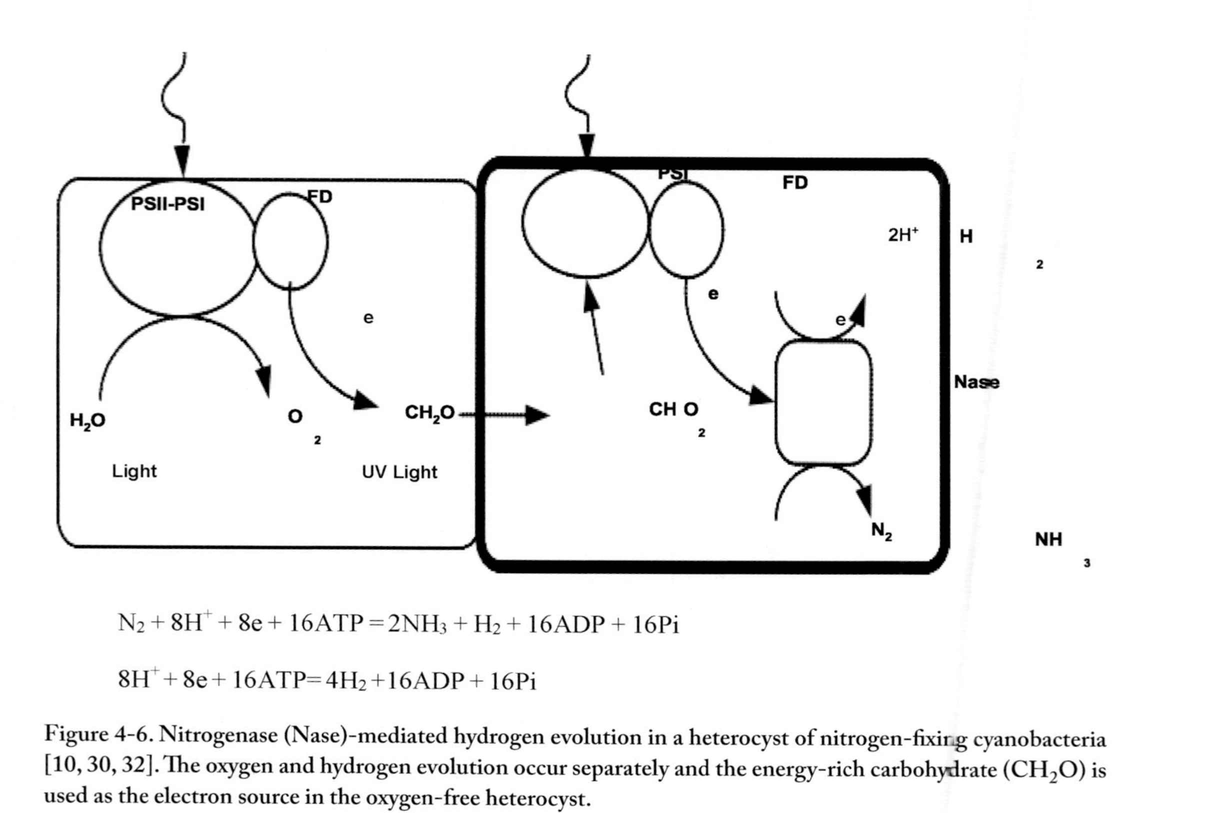

$$N_2 + 8H^+ + 8e + 16ATP = 2NH_3 + H_2 + 16ADP + 16Pi$$

$$8H^+ + 8e + 16ATP = 4H_2 + 16ADP + 16Pi$$

Figure 4-6. Nitrogenase (Nase)-mediated hydrogen evolution in a heterocyst of nitrogen-fixing cyanobacteria [10, 30, 32]. The oxygen and hydrogen evolution occur separately and the energy-rich carbohydrate (CH_2O) is used as the electron source in the oxygen-free heterocyst.

Table 4-1.
Design and construction cost of the photobioreactor for hydrogen production.

List of Component	Materials Cost	Labor Cost	Equipment Cost	GC & OH Cost	Total Cost
Site Preparation	\$5,000	\$3,000	\$2,000	\$2,000	\$12,000
Photo Bioreactor Equipment	\$20,000	\$5,000	\$2,500	\$5,400	\$32,900
Instrumentation	\$2,000	\$1,000	\$2,000	\$1,000	\$6,000
Electrical, Mechanical, Plumbing and Control	\$2,500	\$1,000	\$1,000	\$ 900	\$5,400
Supply for 30 Years cost at \$0.05/kWh for monthly 4000 kWh for 100 people					\$72,000
				Total Cost	\$128,300

This estimate was prepared using January 2016 material costs from leading manufacturers and labor installation costs using union labor wages. The equipment rental was calculated using market rental costs in conjunction with production rate standard construction practice.

Conversion of H_2 into Electricity

I have used a fuel cell to produce hydrogen as a fuel along with electrons, protons, heat and water. Fuel cell technology is based upon the following combustion reaction:

$$2H_2 + O_2 \leftrightarrow 2H_2O$$

The proton exchange membrane (PEM) fuel cell produces electricity by chemical reaction. Hydrogen and oxygen pass over the electrodes producing electricity, heat and water. Hydrogen fuel is supplied to the anode (negative terminal) of the fuel cell while oxygen is supplied to the cathode (positive terminal) of the fuel cell. Through a chemical reaction, the hydrogen is split into an electron and a proton. Each takes a different path to the cathode. The electrons take a path bypassing the electrolyte to produce electricity. The proton passes through the electrolyte and both are reunited at the cathode. The electron, proton, and oxygen combine to form water as a byproduct. Significantly cleaner emissions result from using a PEM fuel cell rather than from a fossil fuel combustion process. PEM fuel cells are considered one of the best technologies to produce electricity from H_2 due to their high efficiency (80%) of electricity production.

Energy Cost Savings

The total cost for 30 years of electrical energy consumption from a conventional source for a standard industry (100 people capacity) at 0.12/kWh using 4,000 kWh per month is equal to $172,800 (30 years x 12 months x 4,000 kWh x 0.12/kWh). This comparison between conventional energy use and cyanobacterial energy production clearly indicates a cost savings of $44,500 when cyanobacterial energy is substituted as the energy source.

CONCLUSIONS

Cyanobacteria provide unique opportunities for researching the biological production of hydrogen. Cyanobacterial hydrogen production is poised to be a very useful and effective method of producing hydrogen. Hydrogen produced by cyanobacteria as described in this chapter is economically preferable when compared to traditional hydrogen production technologies (fermentative, photocatalytic water splitting). Hydrogen gas can be a future energy provider since it does not emit greenhouse gases when combusted. It frees large amounts of energy per unit weight in combustion, is easily converted to electricity, and offers an inexhaustible energy resource.

Cyanobacterial hydrogen production has several advantages over traditional hydrogen generation processes. The photobioreactor is a closed transparent box with low energy requirements. The process is very cost effective. The most appealing aspect of biological hydrogen production is the simplicity of this technology, using nothing but photons, water and cyanobacterial biochemical reactions. The cyanobacterial biochemistry and photosystems use light-driven electrochemical devices to produce high efficiency hydrogen. As sunlight is abundant, our insight into light conversion and hydrogen production by cyanobacteria is essential for creating renewable alternatives to fossil fuels. Biomechanisms and photochemical reactions achieve high rates of hydrogen production, thus providing environmentally benign energy [24].

The best methodologies for photobacterial hydrogen production yield zero emissions of greenhouse gases and compare favorably to other renewable technologies such as hydroelectric plants and solar photovoltaics over their production life-cycle. Hydroelectricity plants and solar photovoltaics emit as much as 0.7% greenhouse gases during their reaction process while phtotobioreators emit 0.0% greenhouse gases. Considering the costs of biological hydrogen energy production, this is another reason to deploy this technology globally. The U.S. Energy Information Administration (EIA) in its 2014 forecast depicts photobiogically produced hydrogen energy power as the only generation technology that has a levelized avoided cost of electricity (LACE) greater than levelized cost of electricity (LCOE). It is the only technology competitive in electrical supply systems that have excess capacity or stable demand for new power. The EIA's analysis includes the transmission and integration costs imposed by intermittent technologies deployed by the companies and agencies in Table 4-2.

Table 4-2. Comparative costs of electrical energy generation (cents/kWh).

PG&E	SCE	SDG&E	Average
Biologically H_2	**5.94**	**6.82**	**6.98**
Geothermal	7.19	6.75	7.03
Wind	8.40	9.77	8.68
Small Hydro	8.72	8.91	8.66
Solar Thermal	14.23	13.48	13.52
Solar PV	15.18	11.90	13.96
UOG Solar PV	16.21	47.00	21.65

If the economic potential of biologically produced hydrogen energy resources can be realized, it would represent an enormous source of energy production with the capacity to produce 14,830.8 billion MWh of energy annually, or three times more than present global energy needs. This energy technology not only reduces greenhouse gas emissions, but also increases profitability by lowering the cost of energy production. With rapid increases in global population and continuing environmental problems, developing alternative energy sources offers solutions. The capability of individual nations to produce hydrogen would eliminate oligopolies in the fossil fuel industries and energy price volatility. Hydrogen produced by cyanobacteria offers the promise of a cleaner renewable energy resource.

References

[1] Abed, R., Dobretsov, S. and K. Sudesh (2009). Applications of cyanobacteria in biotechnology. *Applied Microbiology,* 106, pages 1-12.

[2] Afgan, N. and M. Carvalho (2002). Multi-criteria assessment of new and renewable energy power plants. *Energy,* 27, pages 739-755.

[3] Akkerman, I., Janssen, M., Rocha J. and R. Wijffels. (2002). Photobiological hydrogen production: photochemical efficiency and bioreactor design. *International Journal of Hydrogen Energy,* 27, pages 1,195-1,208.

[4] Ananyev, G., Carrieri, D. and G. Dismukes (2008). Optimization of metabolic capacity and

flux through environmental cues to maximize hydrogen production by the cyanobacterium "Arthro- spira (Spirulina) maxima". *Applied Environmental Microbiology,* 74, pages 6,102-6,113.

[5] Angermayr, S., Hellingwerf, K., Lindblad, P. and M. de Mattos (2009). Energy biotechnology with cyanobacteria. *Current Opinion Biotechnology,* 20, pages 257-263.

[6] Arik, T., Gunduz, U., Yucel, M., Turker, L., Sediroglu, V. and I. Eroglu (1996). Photoproduction of hydrogen by Rhodobacter sphaeroides O.U.001. Proceedings of the 11th World Hydrogen Energy Conference. Stuttgart, Germany. Pages 2,417-2,424.

[7] Asada, Y. and J. Miyake. (1999). Photobiological hydrogen production. *Journal of Bioscience Bioengineering,* 88, pages 1-6

[8] Atsumi, S., and Hanai, T. and J. Liao. (2008). Non-fermentative pathways for synthesis of branched-chain higher alcohols as biofuels. *Nature,* 451, pages 86-89.

[9] Behera, B., Balasundaram, R., Gadgil, K. and D. Sharma (2007). Photobiological production of hydrogen from Spirulina for fueling fuel cells. *Energy Source,* 29. pages 761-767.

[10] Bishop, P. and R. Premakuma (1992) Alternative nitrogen fixation systems. In: Stacey, G., Burris, R., Evans, H., editor. *Biological nitrogen fixation.* New York: Chapman and Hal. Pages 736-762.

[11] Böhme, H. (1998). Regulation of nitrogen fixation in heterocyst-forming cyanobacteria. *Trends in Plant Science,* 3, pages 346-351.

[12] Boison, G., Bothe, H., Hansel, A. and P. Lindblad (1999). Evidence against a common use of the diaphorase subunits by the bidirectional hydrogenase and by the respiratory complex I in cyanobacteria. *FEMS Microbiology Letter,* 174, pages 159-165.

[13] Dahlqvist, A., Stahl, U., Lenman, M., Banas, A., Lee, M., Sandager, L., Ronne, H. and S. Stymne (2000). Phospholipid: diacylglycerol acyltransferase: an enzyme that catalyzes the acyl-CoA-independent formation of triacylglycerol in yeast and plants. Proceedings of National Academy of Science, U.S. 97, pages 6,487-6,492.

[14] Das, D. and T. Veziroglu. (2001). Hydrogen production by biological processes: a survey of literature. *International Journal of Hydrogen Energy,* 26, pages 13-28.

[15] Dutta, D., De, D., Chaudhuri, S. and S. Bhattacharya (2005). Hydrogen production by Cyanobacteria. *Microbial Cell Fact,* 4, pages 1-11.

[16] Fay, P. (1992). Oxygen relations of nitrogen fixation in cyanobacteria. *Microbiology Review,* 56, pages 340-373.

[17] Flores, E. and A. Herrero (1994). Assimilatory nitrogen metabolism and its regulation. In: Bryant D., editor. *The molecular biology of cyanobacteria.* Kluwer Academic Publishers: Dordrecht. Pages 487-517.

[18] Hall, D. and P. Moss (1983). Biomass for energy in developing countries.

[19] Hall, D., Markov, S., Watanabe, Y. and K. Rao (1995). The potential applications of cyanobacterial photosynthesis for clean technologies. *Photosyntetic Research,* 46, pages 159-167.

[20] Heyer, H., Stal, L. and W. Krumbein (1989). Simultaneous heterolatic and acetate fermentation in the marine cyanobacterium Oscillatoria limosa incubated anaerobically in the dark. *Arch Microbiology,* 151, pages 558-564.

[21] Hoekema, S., Bijmans, M., Janssen, M., Tramper, J. and R. Wijffels (2002). A pneumatically agitated flat-panel photobioreactor with gas re-circulation: anaerobic photoheterotrophic cultivation of a purple non-sulfur bacterium. *International Journal of Hydrogen Energy,* 27, pages 1,331-1,338.

[22] Kentemich, T., Haverkamp, G. and H. Bothe (1991). The expression of a third nitrogenase in the cyanobacterium Anabaena variabilis. *Z Naturforsch,* 46, pages 217-222.

[23] Kentemich, T., Danneberg, G., Hundeshagen, B. and H. Bothe (1988). Evidence for the occurrence of the alternative, vanadium-containing nitrogenase in the cyanobacterium Anabaena variabilis. *FEMS Microbiology Letter,* 51, pages 19-24.

[24] Madamwar, D., Garg, N. and V. Shah (2000). Cyanobacterial hydrogen production. *World*

Microbial Biotechnology, 16, pages 757-767.
[25] Masepohl, B., Schoelisch, K., Goerlitz, K., Kutzki, C. and H. Böhme (1997). The heterocyst-specific fdxH gene product of the cyanobacterium Anabaena sp. PCC 7120 is important but not essential for nitrogen fixation. *Molecular Gen Genetics,* 253, pages 770-776.
[26] McNeely, K., Xu, Y., Bennette, N., Bryant, D. and G. Dismukes (2010). Redirecting reductant flux into hydrogen production via meta- bolic engineering of fermentative carbon metabolism in a cyanobacterium. *Applied Environmental Microbiology,* 76, pages 5,032-5,038.
[27] Hossain, Md. (2016). In situ geothermal energy technology: an approach for building cleaner and greener environment. *The Journal of Ecological Engineering* (17), pages 49-55.
[28] Miro'n, A., Go'mez, A., Camacho, F., Grima, E. and Y. Chisti (1999). Comparative evaluation of compact photobioreactors for large-scale monoculture of microalgae. *Journal of Biotechnology,* 70, pages 249-270.
[29] Masukawa, H., Nakamura, K., Mochimaru, M. and H. Sakurai (2001). Photobiological hydrogen production and nitrogenase activity in some heterocystous cyanobacteria. In: Miyake, J., Matsunaga, T., San Pietro, A., editor. *BioHydrogen II Elsevier,* pages 63-66.
[30] Orme-Johnson, W. (1992). Nitrogenase structure: where to now? *Science,* 257, pages 1,639-1,640
[31] Pinzon-Gamez, N., Sundaram, S. and L. Ju (2005). Heterocyst differentiation and H_2 production in N_2-fixing cyanobacteria. Technical program.
[32] Phillips, E. and A. Mitsui (1983). Role of light intensity and temperature in the regulation of hydrogen photoproduction by the marine cyanobacterium Oscillatoria sp. Strain Miami BG7. *Applied Environmental Microbiology,* 45, pages 1,212-1,220.
[33] REN21 (2007). Global status report. *Ren21,* pages 1-52.
[34] Schmitz, O., Boison, G., Hilscher, R., Hundeshagen, B., Zimmer, W., Lottspeich, F. and H. Bothe (1995). Molecular biological analysis of a bidirectional hydrogenase from cyanobacteria. *European Journal of Biochemistry,* 233, pages 266-276.
[35] SERI (1984). Fuel option from microalgae with representative chemical composition.
[36] Stal, L. and W. Krumbein (1985). Oxygen protection of nitrogenase the aerobically nitrogen fixing, non-heterocystous cyanobacterium Oscillatoria sp. *Archives of Microbiology,* 143, pages 72-76.
[37] Suh, S. and S. Lee (2003). A light distribution model for an internally radiating photobioreactor. *Biotechnology and Bioengineering,* 82, pages 180-189.
[38] Tamagnini, P., Axelsson, R., Lindberg, P., Oxelfelt, F., Wunschiers, R. and P. Lindblad (2002). Hydrogenases and hydrogen metabolism of cyanobacteria. *Microbiol Molecular Biology Review,* 66, pages 1-20.
[39] Thiel, T. (1993). Characterization of genes for an alternative nitrogenase in the cyanobacterium Anabaena variabilis. *Journal of Bacteriology,* 175, pages 6,276-62,86.
[40] Thiel, T. and B. Pratte (2001). Effect on heterocyst differentiation of nitrogen fixation in vegetative cells of the cyanobacterium anabaena variabilis ATCC 29413. *Bacteriol,* 183, pages 280-286.
[41] Tsygankov, A., Serebryakova, L., Rao, K. and D. Hall (1998). Acetylene reduction and hydrogen photoproduction by wild type and mutant strains of anabaena at different CO_2 and O_2 concentrations. *FEMS Microbiology Letter,* 167, pages 13-1.7.
[42] Oost, J., Bulthuis, B., Feitz, S., Krab, K. and R. Kraayenhof (1989). Fermentation metabolism of the unicellular cyanobacterium cyanothece PCC 7822. *Arch Microbiology,* 152, pages 415-419.
[43] Weyman, P. (2010). Expression of oxygen-tolerant hydrogenases in synechococcus elongatus. 10th Cyanobacterial Molecular Biology Workshop. June 11-15, Lake Arrowhead, USA.
[44] Wünschiers, R., Batur, M. and P. Lindblad (2003). Presence and expression of hydrogenase specific C-terminal endopeptidases in cyanobacteria. *BMC Microbiology,* 3, page 8.

Chapter 5

Bridging Gaps in Energy Planning for First Nation Communities

Roopa Rakshit, Chander Shahi, M.A. (Peggy) Smith, and Adam Cornwell

There is a link between energy security, economic prosperity, sustainability and sovereignty for indigenous communities in Canada. Geographically remote locations, absence of all-season roads, off-grid status, diesel dependency and lack of alternative energy access causes energy insecurities along with economic, social, and local environmental problems for the Keewaytinook Okimakanak (KO) First Nation communities in northwestern Ontario. Being free of diesel dependency and scoping sustainable energy solutions are immediate priorities. Both are key motivational factors for effective community energy planning (CEP). However, most CEP is based on top-down decision-making approaches which lack effective community engagement to design culturally appropriate, community-centric energy plans. Such approaches fail to acknowledge local socio-cultural drivers as indicators of energy planning.

This chapter details the disconnects between theory and practice in energy planning for First Nation communities. The overarching purpose of this chapter is to bridge knowledge gaps regarding socio-cultural requirements, discuss the social costs in energy planning, and advance academic literature about indigenous perspectives on energy planning.

A literature review, key informant interviews and in-field observations in KO First Nation communities form the basis of our study. This chapter examines community insights on local energy planning to elicit drivers and determinants for a conceptual, bottom-up energy planning framework. It offers recommendations to integrate socio-cultural factors as part of a sustainable and functional energy planning approach for the KO communities. It provides justification that this process ensures multiple benefits such as buy-in by the communities, acceptance, and readiness for CEP implementation which fosters community ownership, self-determination, pride and empowerment. The research findings are timely. There is growing interest in ensuring local energy security amidst longstanding colonial marginalization of indigenous communities in Canada.

INTRODUCTION

Keewaytinook Okimakanak (KO) First Nation* communities are nestled in the boreal landscapes of northwestern Ontario. Among the first things that one notices when landing at one of their airports are the rows of fuel storage tanks or large tank farms with the adjacent diesel-powered generating stations. These tanks store fuel that is airlifted or trucked at an annual cost of approximately $1 million† per community of 500 people. It is estimated that 115 liters of diesel fuel are consumed every minute in these remote First Nation communities, adversely affecting the environment, individual health and socio-economic opportunities [1]. Geographically remote locations, the absence of all-season roads, off-grid utility status, diesel dependency and non-accessibility to alternative energy sources have contributed to acute energy insecurities, and serious local environmental, social and economic concerns [2,3].

The historical context of First Nations in Canada is one of repeated assimilation, marginalization, deprivation and isolation. However, overcoming past hardships and engaging with local energy planning for self-sufficiency is high on development agendas of First Nation communities [4]. The active "voices" of the KO communities and their aspirations for community-driven energy planning are important to achieving sustainable energy solutions. We recognize and interpret energy planning in the KO communities as having multiple purposes—a planning tool defined by their interactions with the environment, their values and need for economic development plus a means of creating pathways for self-determination, pride and empowerment. Since energy planning for these communities is

---

*There are several terms used to describe indigenous peoples in Canada. "First Nations" are those Status Indians governed under the Indian Act and residing on federally owned reserve lands set aside for First Nations. "Aboriginal" is the term used in section 35 of the Constitution Act, 1982 inclusive of "Indians, Inuit and Métis." "Indigenous" is commonly used in the international arenas (e.g., United Nations declarations). It is becoming more popular, is often preferred by indigenous peoples and now acknowledged by the Canadian federal government.

†All dollar values are provided in Canadian dollars.

formative, this chapter draws from community motivations to demonstrate the need to include sociocultural factors as key drivers in local energy development.

The academic approach to community energy planning integrates policy, urban planning, and energy management components into a single model called community energy management [5]. In contrast, Ontario Power Authority's (OPA) generic indigenous CEP concept includes an understanding of community energy usage, identifying conservation opportunities, scoping renewable energy sources, understanding the risks and rewards, and establishing energy goals for the community [6]. Rizi lists various organizations that have contributed to the concept and practice of CEP from 1997-2010 [7]. Present-day CEP practices are more evident in a non-academic landscape and through various lenses—economic, technology, policy, renewable energy, and greenhouse gas emission reductions. St. Denis and Parker examined ten local action plans in remote, rural and urban Canada affirming that CEPs have limitations when applied in local contexts [8]. Each CEP needs to be individualized to the attributes of each local community and must therefore use unique approaches, applications, assessments and contexts. Necefer et al. note that local indigenous community contexts include socio-cultural factors—historical, cultural, artistic, and religious or sacred beliefs both in tangible attributes (e.g., land, sites, lakes, rivers, waterfalls and mountains) and in intangible forms (e.g., practices, cultural norms, representations, expressions, knowledge and skills) [9]. These deep-rooted values, identity, and the stewardship of the land need recognition, acceptance and integration into modern energy systems and development [10].

This chapter elicits drivers and determinants for integrated CEP using a literature review, in-field observations, community member interactions and key informant interviews. The analysis is also informed by our participation in the annual Northern Ontario First Nations Environment Conference (NOFNEC) in 2015 and 2016 that provided open dialogues. We then offer a conceptual, bottom-up framework and make recommendations to integrate socio-cultural drivers for efficient energy planning for the KO communities.

The research findings are timely as there is growing interest in ensuring local energy security amidst longstanding colonial treatment and marginalization of indigenous communities. In a broader context, it is appealing to the Canadian government's greenhouse gas reduction commitments.

LITERATURE REVIEW

A review of both grey and academic literature was undertaken on the scope and motivations for energy planning by off-grid First Nation communities. For background and historical contexts, we approached the Keewaytinook Okimakanak Research Institute (KORI) and the Nishanawbe Aski Nation (NAN) for institutional reports and visited community websites. Grey literature included international, government and organizational reports and other studies.

The Web of Science database was queried for multi-disciplinary publications including areas of natural sciences, social sciences, archaeology, economics and sustainable development. The database search from 2000 to 2016 found 45 published papers relating to CEP in Canada with very few specifically addressing indigenous communities and none focused on integrating socio-cultural factors in energy planning. Diesel dependency is acknowledged by all the key informants and KO institutional and community reports. All reference to the KO communities is through non-academic documentation except for academic articles about information and communication technology [11]. The review identified substantial academic references on local energy planning for renewable energy development by First Nation communities in Canada. Sources on energy planning in urban settings were excluded since the context of this chapter focuses on remote communities.

A study about CEP in remote, off-grid situations provides theoretical knowledge of concepts, definitions, programs, tools and approaches in Canada. At the national level, aboriginal community energy plans in 2015 focused on improving energy efficiency, reducing electricity consumption, and assessing clean energy solutions. A total of 55 indigenous communities from remote northwestern and southern Ontario with $3.9 million in funding benefitted from aboriginal community energy plan programs [12]. The initiative, though well-received, was a top-down program offered without sufficiently focusing on strengthening community capacities to undertake energy planning. This led to consultant dependency, hefty costs and non-functional reports. Rizi echoes that better understanding of "on the ground" needs are necessary to increase adoption and implementation of energy plans in First Nation communities [7]. There is little academic research on the effectiveness of the aboriginal community energy plan programs for the First Nation communities. This poses a knowledge gap in assessing the success of energy planning for these communities.

An analysis of ten of the first CEPs in Canadian communities included two First Nation communities, emphasizing a participatory bottom-up approach with outcomes that addressed local needs, values and resources [8]. The Hupacasath First Nation in British Columbia undertook energy planning and attributed its success to their chief. The community's efforts led to the development of the 5.2 MW China Creek hydroelectricity project followed by a community-led energy planning process that resulted in a greener approach to energy [13,14].

An example of bottom-up energy planning is provided by the Tlicho (Dogrib) people in Wha Ti, Northwest Territories, who developed their energy plan by assessing their energy use. This exercise involved the entire community including tribal elders, youth committees and the local government, subsequently leading to a successful project [15]. The desire for community self-sufficiency was identified as a primary motivation for developing energy projects in the First Nation communities in British Columbia [16].

The NRCan's guide and the Artic Energy Alliance's toolkit both recognize the role of community members, not local governments, as key initiators of CEP [17,18]. Local or community level energy planning is both desirable and useful. Lerch noted that when local people are engaged, they invest in the outcomes, thus making community energy initiatives more than just plans [19]. This engagement is important because indigenous communities aspire to become more resilient and free of top-down, institution-driven systems. They take pride in embracing innovation and integrating development in their ways of life. Energy security is critical to their aspirations for self-determination and sovereignty. It is also necessary to enhance the capacities and capabilities of indigenous peoples in Canada [20].

Field observations and reviews suggest that interrelated and interdependent factors in CEP have favorable impacts when socio-cultural factors are considered; for which the central element must be based on the fundamental and underlying philosophy of indigenous people that all things—animals, the elements, people and nature—are connected, instructive and illustrative [21]. Elias emphasized that economic development needs must consider cultural consequences [22]. Aboriginal Affairs and Northern Development Canada (AANDC) in its Comprehensive Community Planning Handbook for First Nations in British Columbia indicate that celebrating traditions and cultures are important factors in planning processes [23].

Academic literature, references and documentation concerning off-grid energy planning integration in the indigenous context is limited. This leads to a gap in understanding and acknowledging socio-cultural drivers as measurable and potential motivational factors for indigenous energy planning. In a broader context, indicators and targets to integrate cultural factors are omitted in well-intentioned development programs and policies [24]. The effects of culture were not included in the elaboration of the millennium development goals [24,25]. Such oversights may be due to the subjective nature of culture [26]. Evidence suggests that First Nation communities that are firmly grounded in their culture and secure in the legitimacy of their traditions and social institutions are happier and more economically successful [27]. Tangible and intangible cultural forms can drive sustainable development and serve as powerful socio-economic resources [25]. The Mackenzie Valley Pipeline project guidelines strongly suggests making cultural impact assessments an integral, consistently applied and transparent part of community development planning [28].

The KO communities are presently in the formative stages of CEP, carving their energy development paths, and transforming energy plans into practice [2]. Their endeavors for energy security reflect self-determination and empowerment. Using community insights, we draw upon drivers and determinants for integrated CEP arguing for a conceptual, bottom-up framework. Our recommendations include integrating socio-cultural drivers as part of a sustainable and functional energy planning approach for the KO communities.

BACKGROUND & CONTEXT

Keewaytinook Okimakanak First Nation Communities

We focused on the energy situation of the Keewaytinook Okimakanak (Northern Chiefs Council in Oji-Cree language) First Nation communities in northwestern Ontario as representative of remote indigenous communities in northern Ontario. The six communities under the KO tribal council are Deer Lake, Fort Severn, Keewaywin, McDowell Lake, North Spirit Lake and Poplar Hill. These First Nations comprise a population ranging from 60 in McDowell Lake to almost 1,000 in Deer Lake located over an area of 300,000 square kilometers [2]. The geographic locations of the KO communities are shown in Figure 5-1.

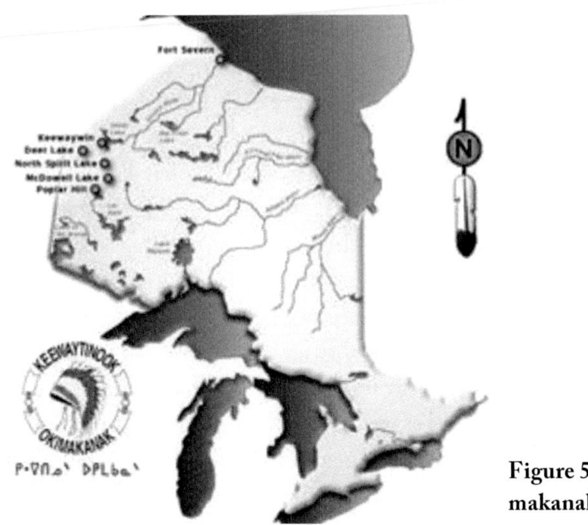

Figure 5-1. Keewaytinook Okimakanak First Nations [2].

Electrification in Remote First Nation Communities

The present dire electrification situation and the motivations for sustainable energy planning are next considered in the historical context of northern First Nations communities.

As stated in the Constitution Act, 1982, section 35, electrification in northern First Nations was long considered the responsibility of the federal government. Ontario's First Nation communities were electrified, mostly in the 1960s and 1970s, through "electrification agreements" between the Province of Ontario and the federal government [29]. The responsibility for providing electricity was shared, with capital costs for generation and distribution equipment provided by the federal government with ongoing operation, maintenance, and equipment replacement provided by Ontario Hydro [30]. Some First Nations chose to independently operate and maintain their electricity systems [31]. Both federal and provincial governments have been involved in the electrification of First Nation communities, creating ambiguity and complexities for these communities. In 1992, Ontario Hydro changed its policy to allow for unrestricted service to First Nation households. While lifting a 20-amp limit increased energy consumption, the cost of generating electricity remained high. Before the deregulation of Ontario Hydro in 1998, the cost of diesel for remote communities was included in the cost of fuel for the whole corporation. After deregulation, and the creation of Hydro One Remote Communities, Inc. (HORCI), diesel fuel became the single largest cost in community budgets [31].

Meeting Current Energy Needs and Service Providers

The two energy service providers for the KO communities are HORCI and independent power authorities (IPA). The HORCI communities from the KO tribal council are Fort Severn and Deer Lake [32]. The IPA communities are Keewaywin, Poplar Hill and North Spirit Lake. Energy planning outcomes vary between IPA and HORCI communities due to the distinct characteristics and operations of the energy providers.

HORCI, a subsidiary and not-for-profit company of Hydro One Network Inc., is owned by the Province of Ontario. HORCI operates and maintains the generation and distribution assets used to supply electricity across northwestern Ontario to communities not connected to the province's electricity grid [33]. HORCI's operations are unlike other generators or distributors in Ontario [34]. They require a subsidy so that electricity can be provided to its customers at a comparable cost to the rest of Ontario. This subsidy is provided by the Rural and Remote Rate Protection (RRRP) fund that is collected from consumers across the province. This subsidy is key to the success of the communities of Fort Severn and Deer Lake since it helps to maintain low electricity prices and ensures cost control and maintenance. Capital agreements with indigenous and Northern Affairs Canada (INAC) are HORCI system subsidies through which INAC recovers the costs of new electrical generation. The final subsidy for HORCI communities is through Standard A interest rates that are charged to accounts that are receiving funding or subsidies from INAC or some other government agency.

The IPAs are non-regulated power authorities. As unlicensed operators, they are not bound by the regulations applied to HORCI. IPAs serve the KO First Nations of Keewaywin, Poplar Hill and North Spirit Lake. Each IPA is unique and provides distinct benefits to its owners such as control, employment and community awareness [35]. As independent entities, IPAs are unable to access provincial subsidies to maintain power prices at artificially low levels as is done with HORCI communities. This key difference negatively affects the profitability and viability of IPAs leading to compromises on renovations and new infrastructure development. Since IPAs offer a higher potential for employment of local community members, they can be a source of community pride. Arrearages on residential utility accounts are lower in IPA communities, suggesting that a community approach to accommodation of payments results in stronger community support for the IPA than for a HORCI utility [35].

The single largest cost for both IPAs and HORCI communities is diesel fuel, due to the high cost of airfreight and the decreasing winter road seasons caused by climate change. In HORCI communities, all power is generated, distributed and sold by HORCI. In the IPA communities, the situation is the same in that the IPA is the sole supplier, distributor and retailer of electricity. This monopoly situation is not entirely without benefit as HORCI operates as a break-even business and does not seek to profit from services provided to remote First Nations [35].

The nature, structure and functionality of IPAs and HORCI as energy service providers affect energy planning in the KO communities. Cost recovery is more difficult for IPAs without RRRP subsidies. Community programs subsidize customer utilities with many IPAs charging a flat rate or an affordable amount. IPAs maintain lower safety standards for domestic hookups and diesel plants. First Nations have limited funding for technicians under IPAs. Finally, IPAs depend on INAC to cover operational losses after auditing has been completed.

For both HORCI and IPA-served First Nations, INAC funding is required to construct, expand and maintain infrastructure. INAC does not fund short-term upgrades. First Nations are responsible for purchasing, shipping and storing fuel for the generators. Liability for injuries, fuel spills, and contaminated sites remains with both the First Nations and HORCI, depending on who owns the fuel tanks. Both HORCI and IPAs purchase fuel from First Nations. In the case of HORCI, the diesel generation systems are built and maintained to higher standards. Many HORCI First Nation communities face restrictions on their energy use and customers pay based on their energy consumption. The RRRP offers some incentives for conservation. Additionally, there are HORCI conservation programs that are underutilized [35].

METHODOLOGIES

A multitude of qualitative approaches were used to understand and analyze the energy challenges of KO First Nations. Participatory research methods were used in planning and conducting the research with the authors and participants equally generating knowledge. Interviewees' perspectives were essential for the processes of discovery, accumulating knowledge and fostering empowerment for energy planning. There was mutual curiosity between the participants and the authors to understand motivational

Drivers and Determinants for an Integrated CEP

Socio-cultural: The people's histories, literature, language, religions, traditions, ceremonies, ancient beliefs, and present lifestyles are integrated within KO communities. Interpretation of the intimate relationships between the people and their lands helped us discern perspectives on energy planning, energy options, ability to control energy consumption and lifestyle choices.

Anishiniini Gayenaabuhstooauch Akheenih (indigenous or First Nation use of the land) includes activities that are recognized as aboriginal and treaty rights under Canada's Constitution. Besides hunting, fishing, trapping, rights-based activities include travel on waterways, occupation and maintenance of portages, access to trails and campsites, planting and harvesting, gathering traditional foods, cutting wood for community use, building shelters, recreational access, and visitation and maintenance of cultural sites. Rights-based features include rock paintings, burial sites, historical campsites, settlement locations, quest sites and ancient villages.

Preservation and protection of history and cultural legacies have deep connections to community well-being over time. From an indigenous people's perspective, no compromise is to be made with any heritage and archeological resources [40]. Careful planning and management of indigenous cultural values, and promoting healthy lands are of high priorities. Thus, integrating land and resource use in energy development is essential whether it is planning new transmission, considering alternative energy sources, deploying technologies or undertaking conservation efforts.

Aboriginal languages are a powerful means of understanding indigenous ways of life. There is community significance and importance in local indigenous languages. Using local language to translate technical energy terminologies enhances energy literacy and community-based knowledge. Collectively, traditional knowledge, historical facts and communication through local language contribute to identifying community-wide energy planning assets. Engaging with the elders and members of the KO communities increases local capacities to conduct energy surveys and create baselines for sustainability assessments.

Marginalization has left devastating and ongoing multi-generational impacts on the health and welfare of individuals, families, and communities that challenge community development efforts. Efforts to engage with the community, especially the youth, in accepting stewardship roles in energy planning poses challenges due to ongoing personal struggles. This was made evident during discussions with key informants. Understanding and sensitivity to the community's historical contexts are social determinants for effective energy planning.

Environment: Waterways (e.g., rapids, rivers, falls and lakes) are regarded by KO communities as important travel routes and sources of subsistence upon which hunting, fishing, trapping and other activities are based. The waters support high quality fish habitats and spawning areas. The lakes and rivers have connected people for trading, marriage and historic events since the earliest times. Waterways would be given offerings to demonstrate respect and safe passage. Protecting culturally significant ecological systems—waterways, aquatic habitat, fisheries, wetlands, wildlife and forests—is important as KO people believe them to be interconnected life forms. Understanding this philosophy is crucial for efficient energy planning. The needs for community energy access, infrastructure, transmission corridors, and renewable energy can be realized while respecting aboriginal rights and minimizing ecosystem disturbances. This maintains a balanced perspective on development processes.

Socio-economic: Any economic development activity in a community, be it renewable energy, tourism, mineral development, commercial forest harvesting, fishing or trapping is directly or indirectly dependent on energy supplies. Energy planning must consider developmental activities, community demographics, gender distribution, social cohesion, energy consumption trends and employment opportunities. Efficient energy planning supports existing and potential economic and resource development initiatives such as community housing and infrastructure needs. CEPs incorporating community land use plans and economic development plans are more inclusive than producing energy plans in isolation.

Technology: A key determinant in energy planning is that the technologies to be adopted should dovetail with KO communities' plans for renewable energy. Feasibility studies and benefit and cost assessments are required to determine the potentials. New technological development needs to be aligned with conservation and natural resource protection efforts. How indigenous people relate to tangible and intangible attributes of nature can help promote the adoptability of technologies such as run-of-the-river hydropower and solar power. Biomass is presently a less viable option due to a narrow view that biomass might require felling trees which is counter to indigenous values.

The lack of energy experts in the community to perform post-installation maintenance and sustainable operations is challenging. There is evidence of projects being stalled or dependent on consultants. Energy planning requires addressing the technological service gaps when considering energy options.

Governance and institutional mechanisms: Progressive leadership enhanced by inclusive, transparent, and robust community engagement is a pivotal determinant in energy planning. KO communities strive to provide opportunities for their members to have multiple roles and benefit from their ideas, talents, skills and resources. These roles include advisors, advocates, problem definers, solution identifiers, evaluators, documenters and surveyors. KO communities recognize the need to identify and mentor local energy coordinators while supporting local individuals to participate. Engaging and training youth for such roles is beneficial at many levels. Institutional mechanisms to transition between energy services (i.e., from IPA to HORCI operations) by appointing local talent promotes effective planning. The active participation of women choosing multiple and influential roles is progressive.

KORI provides a supportive and intermediary role by understanding the community contexts and external stakeholders. They drive community actions and promote community change by mobilizing, addressing opposition and resistance, maintaining efforts, influencing systems, achieving community-level improvements and providing tools for CEP. They encourage building local capacities to conduct surveys and collate baseline data for energy planning. This is a key consideration for sustainable outcomes necessary to wean communities from dependencies on external consultants. A team comprised of the chief, tribal council members, community advisory members, federal and provincial ministries, external stakeholders, and technical experts can form advisory groups to achieve desired results and promote accountability. Elders have an important role in the decision-making process and are very influential in KO communities. Shared challenges including generational gaps and heterogeneity in decision making are affected by work cultures, dynamics and transitional acceptance. The communities are in continuous "election mode," due to Indian Act mandates for elections every two years, creating institutional challenges. This results in a rapid turnover of chiefs and council members, impacting motivation, time and resource commitments to energy planning.

Elements of an Integrated CEP Framework

CEP is in its formative stages in KO communities. An integrated approach engages community participants, partners with external stakeholders, incorporates cultural values and manages external stressors. Each of these elements is addressed below.

Community participants: Socio-cultural knowledge is best learned through community engagement which is pivotal to functional energy planning. Participants include elders and spiritual leaders. Their guidance offers valuable insights, wisdom and life lessons on conservation and sustainability. Engaging a community's tribal chiefs and council members, community members, women, youth councils, teachers, education boards, and utility representatives, aids in defining the objectives of energy planning and assessment tools. CEP must aim to open communication channels and provide information concerning energy development initiatives. This facilitates the objective of empowering community members to contribute to informed energy planning.

Partnering with external stakeholders: Understanding the roles and responsibilities of companies, agencies, and regulators in Ontario's electricity sector is intimidating. Federal and provincial regulators include the Ministry of Energy and the Ontario Energy Board (OEB). Indigenous and Northern Affairs Canada (IESO) is the organization tasked with system operation, planning and procurement. HORCI is the agency for local distribution and generation. INAC, a funding agency, along with the chiefs and the tribal councils have important roles in energy planning in KO communities. Partnerships should be founded on the principles of protecting the lands, waters, and ways of life of indigenous peoples and ensuring sustainable benefits for future generations. Access to grants, funding, conservation incentives, energy efficiency programs, and renewables requires cultural sensitivity and inclusiveness. Awareness and education will help with application and practice, plus generate mutual understanding, respect, and trust among stakeholders and their communities.

Using tools and resources: Traditional ecological knowledge must be considered for environmental assessments and for developing criteria for sustainability. They are powerful tools for managing environmental risks [41]. Traditional knowledge, when well documented, interpreted and applied, benefits and complements western worldviews. Incorporating both western science and traditional knowledge into studies, maps, planning, and assessment tools facilitates knowledge that is technically sound and connected to local values, needs and priorities. KO community resources such as economic, infrastructure, development plans, and traditional land use studies provide information to inform the goals and objectives of energy planning. They also provide baseline data for financial and logistic assessments to establish sustainability criteria.

Managing external stressors: Energy regulatory and partnership development complexities, consensus building challenges, corporate interests, community needs, budget uncertainties, and scheduling variability are external stressors that are linked to energy planning. Stressors can be diminished with better understanding of local contexts and socio-cultural factors. Approaching the KO communities with respect and listening to their concerns has positive impacts and facilitates acceptance.

Based on an understanding obtained from community interviews and the literature review, a conceptual framework is suggested for the KO communities (see Figure 5-2). A reductionist approach is used to design the conceptual framework that was predominantly driven by researcher's empirical observations and informants' insights. The detailed descriptions in the results demonstrate the elements of the framework and their linkages. They reflect the voices of the community with a functional, bottom-up approach to energy planning. Figure 5-2 offers a structure to explain the observations, provides context, and suggests direction for culturally appropriate energy planning in KO communities. It is independent of previous models and provides fresh perspectives that include social-cultural factors.

RECOMMENDATION: LONG-TERM ENERGY PLANNING

The remote off-grid energy crisis of KO communities presents unique challenges and opportunities. Energy systems have tremendous impact on communities, affecting their traditional ways of life and activities. Planners need to be sensitive and aware that the adoptability and acceptance of energy plans lies with the communities as its beneficiaries. Based on interview responses, researcher's observations, and the conceptual framework, the following recommendations are proposed:

- Energy surveys, baselines, and assessments should be undertaken by the KO communities to ensure integration of their needs, priorities and ways of life. This results in enhanced implementation and com-

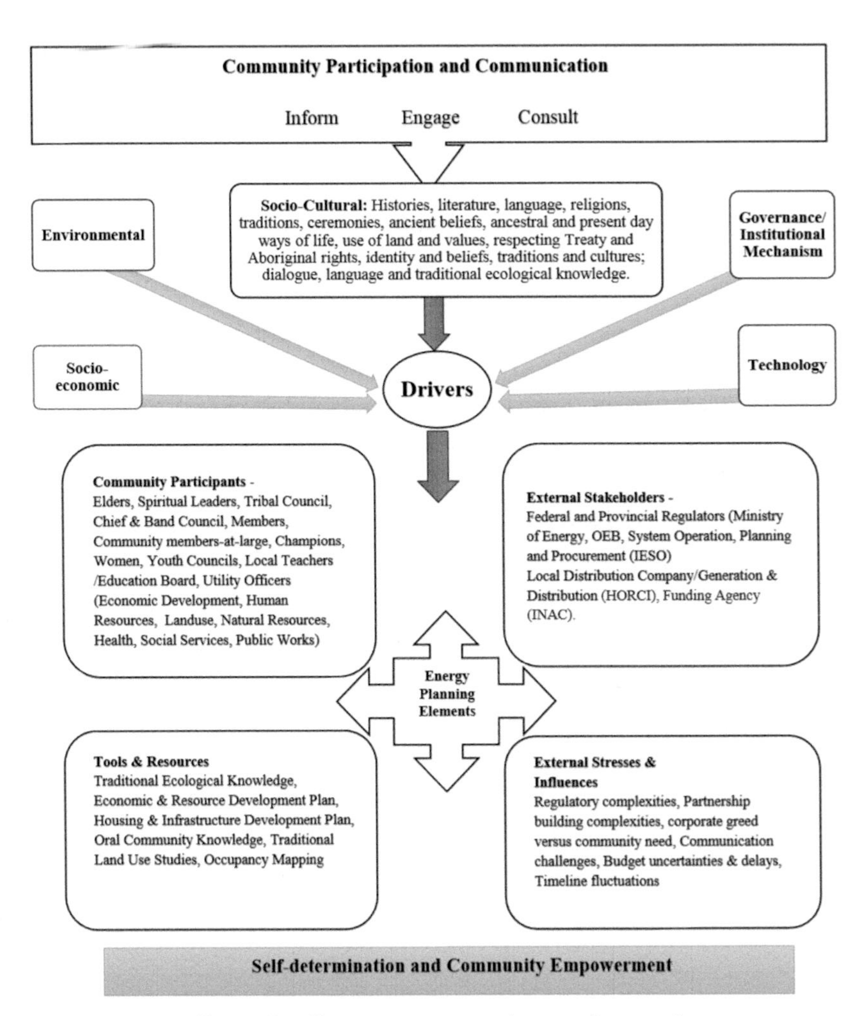

Figure 5-2. Community energy planning framework.

munity-based energy solutions incorporating traditional interpretations of energy systems. This also facilitates building local capabilities.

- The energy agenda can be made more relevant when long-term institutional mechanisms are established at the community level. For example, Watay Power has provided funding support for a KO community to establish an energy coordinator position for two years to coordinate sustainable energy efforts and assist in building local capacities.

- CEP must emphasize engagement of youth and women. Modern economies are changing lifestyles, especially for youth, children and women. KO communities, while striving to provide a promising future for their youth with career pathways, are also keen for them to remain in their communities. Youth perspectives on adopting new, modern lifestyle choices rooted in traditional values helps retain young people in their communities. Women are beneficiaries of energy technologies and are positioned to lead educational, awareness and outreach efforts. They can become energy entrepreneurs, producing or selling improved equipment, marketing renewable energy products, or providing after-sales services. Female entrepreneurs can lower customer acquisition and servicing costs and lead efforts to support decentralized, small-scale power generation projects.
- Energy plans should be prepared with KO communities as primary beneficiaries. This will result in more equitable access and narrow energy distribution imbalances.
- There is an urgent need to integrate energy planning in the broader framework of development planning because energy is essential for sustainable resource development.
- Diversifying energy supplies by deploying renewable energy is linked to achieving sustainable energy solutions. For example, energy production can be augmented by exploring the use of biomass as a renewable resource. More research is needed to assess local forestry and other resources for biomass applications in KO communities.

CONCLUSIONS

The overarching purpose of this chapter was to bridge the knowledge gap about socio-cultural requirements for energy planning and to advance academic literature on indigenous perspectives on energy planning. The voices of First Nations peoples provide us with new perspectives toward implementing energy planning, not simply focusing on energy supply options but also efficient energy use. The emergent, custom-designed, community-driven energy development approach provides a more effective planning methodology. It creates self-determined and empowered indigenous KO communities.

Our research found that electrification is a First Nations right by treaty and is necessary to improve the lives of indigenous people living in remote, economically disadvantaged communities. The off-grid, remote northwestern Ontario KO communities have expressed interest in addressing energy insecurity by enabling and supporting energy development through functional and relevant energy plans based on their singular history, values, culture, spirituality, language and knowledge. Each KO community is unique in its composition, size, energy consumption and needs. Custom, rather than generic, energy planning approaches are needed to meet every community's requirements.

Energy planning processes that enable KO communities to participate fully in their social and economic advancement, manage their lands and resources, and maintain their sacred connections to mother earth, lead to sustainable energy solutions. While indigenous cultures have always highlighted the importance of protecting nature, the rest of the world is now grappling with climate change by proposing cleaner, greener energy systems to reach the same goal.

Their present-day energy crisis, challenges due to historical contexts, diesel-dependency, lack of local capacities, tools, resources, top-down energy planning guidelines, and minimal community engagement require urgent action by the KO communities to ensure their future energy security.

Drivers and determinants—socio-cultural, environment, socio-economic, technology, governance and institutional mechanisms—derived by active and informal consultations and engagement with the community are vital for developing culturally appropriate energy planning. Integrating them is beneficial in building robust community-centric energy plans and to enhance local capacities, capabilities and confidence. The efforts are undoubtedly a self-determined approach in finding sustainable energy solutions. This chapter emphasizes that energy development for the First Nations communities is linked to community sovereignty and sustainability.

The proposed conceptual energy planning framework is drawn from participatory action and indigenous research methods. Development plans including energy solutions must be derived through local participation that can eventually guide energy conservation and renewable energy development. When properly applied, community well-being is enhanced by innovation, technology, local traditions and culture. This is accomplished by considering the unique needs and socio-cultural frameworks of each community.

Further, as the provincial government is strengthening and transforming its electricity and fuels systems, indigenous communities are emerging as partners in the broader energy planning context. They bring singular perspectives, knowledge and leadership to energy projects and systems development. Their involvement supports improved data gathering, outcome and services. This is crucial for advancing responsive energy development for the diverse First Nation communities.

This chapter examined community insights on local energy plan development for the KO communities. It identified drivers and determinants for conceptual, bottom-up energy planning frameworks and suggests recommendations to integrate socio-cultural factors within sustainable and functional energy planning approaches. It provided justification that this process ensures multiple benefits for the communities, including acceptance and readiness for CEP implementation.

Potential future research opportunities include developing more culturally appropriate tools and platforms to support indigenous communities in their engagement and leadership efforts. Research is needed to explore consistent and systemic energy policy approaches to complex socio-cultural economic and socio-technical systems within local indigenous contexts.

References

[1] Wataynikaneyap Power (2016, July 16). http://www.wataypower.ca.

[2] Keewaytinook Okimakanak Research Institute (2013, April 10). About us. http://www.kochiefs.ca/kori.

[3] Nishanawbe Aski Nation (2015, July 10). About us. http://www.nan.on.ca/article/about-us-3.asp.

[4] Krupa, J. (2012). Identifying barriers to aboriginal renewable energy deployment in Canada. Energy Policy, 42, pages 710-714. doi:10.1016/j.enpol.2011.12.051.

[5] Jaccard, M., Failing, L. and T. Berry (1997). From equipment to infrastructure: community energy management and greenhouse gas emission reduction. *Energy Policy*, 25(13), pages 1,065-74.

[6] Ontario Power Authority (2010, July 4). Aboriginal community energy plans. www.aboriginalenergy.ca/pdfs/MNO_AGA_Aug19_2010_ACEP.ppt.

[7] Rizi, B. (2012, July 29). Community energy planning: state of practice in Canada. Major paper. York University: Toronto, Ontario. http://sei.info.yorku.ca/files/2013/03/Bahareh-MP-Final.pdf.

[8] St. Denis, G. and P. Parker (2009). Community energy planning in Canada: The role of renewable energy. *Renewable and Sustainable Energy Reviews*, 13(8), pages 2,088-2,095. doi:10.1016/j.rser.2008.09.030.

[9] Necefer, L., Wong-Parodi, G., Jaramillo, P. and M. Small (2015). Energy development and native Americans: values and beliefs about energy from the Navajo Nation. *Energy Research and Social Science*, 7, pages 1-11. http://dx.doi.org/10.1016/j.erss.2015.02.007.

[10] Tobias, J. and C. Richmond (2014). That land means everything to us as Anishinaabe: environmental dispossession and resilience on the North Shore of Lake Superior. *Health*

and Place, 29, pages 26-33. doi:10.1016/j.healthplace.2014.05.008.
[11] Beaton, B. (2009). On-line resources about Keewaytinook Okimakanak, the Kuhkenah Network (K-Net) and associated broadband applications. *The Journal of Community Informatics*, 5(2).
[12] Independent Electricity System Operator (2009-2015). Aboriginal energy partnerships program. Aboriginal community energy plans. http://www.aboriginalenergy.ca/about-aboriginal-energy-partnerships-program, accessed, April 30, 2016.
[13] Hupacasath First Nation (2015, August 2). http://hupacasath.ca.
[14] Heap, N. and T. Weis (2015). Hupacasath First Nation community energy plan. Pembina Institute, Calgary, Alberta. http://hupacasath.ca/wp-content/uploads/2015/02/Community-Energy-Plan.pdf, accessed July 29, 2016.
[15] Aboriginal Affairs and Northern Development Canada (2011). Archived-community energy plan for Wha Ti, Northwest Territories. https://www.aadnc-aandc.gc.ca/eng/1312212959922/1312213056686, accessed, June 15, 2016.
[16] Rezaei, M. and H. Dowlatabadi (2016). Off-grid: community energy and the pursuit of self-sufficiency in British Columbia's remote and First Nations communities. *Local Environment*, 21(7). http://dx.doi.org/10.1080/13549839.2015.1031730.
[17] Church, K. (2007). Natural resources Canada community energy planning—2007. CANMET Energy Technology Centre. Natural Resources Canada, Government of Canada. Ottawa, Ontario.
[18] Arctic Energy Alliance (2009). Community energy planning toolkit. http://aea.nt.ca/communities, accessed July 30, 2016.
[19] Lerch, D. (2007). Post carbon cities: planning for energy and climate uncertainty. Post Carbon Institute: Sebastopol, California.
[20] Henderson, C. (2013). Aboriginal power—clean energy and the future of Canada's First Peoples. Rainforest Editions: Erin, Ontario.
[21] Chretien, A. (2010). Resource guide to aboriginal well-being in Canada. Nuclear waste management organization. Toronto, Ontario. http://dgr.nwmo.ca/uploads_managed/MediaFiles/2166_a_resource_guide_to_aboriginal_wellbeing.pdf, accessed, August 30, 2016.
[22] Elias, P. (1995). Northern economies in Elias, P.D. (ed.) Northern aboriginal communities: economies and development. Captus Press: Toronto, Canada. Page 15.
[23] Aboriginal Affairs and Northern Development Canada (2013). The CCP handbook: comprehensive community planning for First Nations in British Columbia. 2nd ed. Public works and services Canada, Ottawa. https://www.aadnc-aandc.gc.ca/DAM/DAM-INTER-BC/STAGING/texte-text/ccphb2013_1378922610124_eng.pdf, accessed March 25, 2016.
[24] United Nations Research Institute for Social Development (2014). Social drivers of sustainable development. Beyond 2015, brief 04. UNSRID, Geneva, Switzerland. http://www.unrisd.org/80256B3C005BCCF9/(httpAuxPages)/BC60903DE0BEA0B8C-1257C78004C8415/$file/04%20-%20Social%20Drivers%20of%20Sustainable%20Development.pdf, accessed March 28, 2016.
[25] UNESCO (2012). United Nations System Task Team on the Post-2015 U.N. Development agenda. Culture: driver and enabler of sustainable development. Thematic think piece. UNESCO. Paris, France. https://en.unesco.org/post2015/sites/post2015/files/Think%20Piece%20Culture.pdf, accessed July 30, 2016.
[26] UNESCO (2001). Universal declaration on cultural diversity. http://unesdoc.unesco.org/images/0012/001271/127162e.pdf, accessed July 29, 2016.
[27] Kant, S., Vertinsky, I., Zheng, B. and P. Smith (2014). Multi-domain subjective wellbeing of two Canadian First Nations communities. *World Development*, 64, pages 140-157. doi:10.1016/j.worlddev.2014.05.023.

[28] Mackenzie Valley Review Board (2009). Developing cultural impact assessment guidelines: a Mackenzie Valley Review Board initiative (867), pages 1-18. http://reviewboard.ca/upload/ref_library/may_2009_cultural_impact_assessment_guidelines_status_report_1242859917.pdf, accessed July 25, 2016.

[29] Simeone, N. (2001). Federal-provincial jurisdiction and aboriginal peoples. Library of Parliament, Parliament of Canada. Ottawa, Ontario. http://www.parl.gc.ca/Content/LOP/ResearchPublications/tips/tip88-e.htm, accessed July 30, 2016.

[30] Canadian Off-Grid Utilities Association. History Hydro One remote communities. Document1http://www.cogua.ca/history/hydroone_systems.htm, accessed September 20, 2016.

[31] NAN (2012a). NAN Chiefs energy conference presentations. http://www.nan.on.ca/article/nan-chiefs-energy-conference-presentations-591.asp, accessed August 10, 2016.

[32] Hydro One, Inc. (2009). Remote communities. http://www.hydroone.com/ourcommitment/remotecommunities/Pages/home.aspx, accessed July 29, 2016.

[33] Ontario Energy Board (2008). Hydro One Remote Communities, Inc. for an order or orders approving rates for the distribution of electricity. EB-2008-0232. http://www.ontarioenergyboard.ca/oeb/Industry/Regulatory%20Proceedings/Decisions/2009%20Decisions, accessed July 2, 2016.

[34] Aboriginal Affairs and Northern Development Canada and Natural Resources Canada (2011). Status of remote/off-grid communities in Canada. Government of Canada. Ottawa, Ontario. https://www.nrcan.gc.ca/sites/www.nrcan.gc.ca/files/canmetenergy/files/pubs/2013-118_en.pdf, accessed July 29, 2016.

[35] NAN (2012b). NAN Chiefs energy conference presentations. Energy overview in Nishnawbe Aski Nation Territory-Edward HoshiZaki. http://www.nan.on.ca/upload/documents/energy2012-pr-ed-hoshizaki---community-energy-needs.pdf, accessed July 10, 2016.

[36] First Nations Centre (2005, October). Ownership, control, access, and possession or self-determination applied to research. http://www.naho.ca/documents/fnc/english/FNC_OCAPCriticalAnalysis.pdf, accessed September 2, 2016.

[37] Community research planning guidebook (2012). http://research.knet.ca/?q=system/files/2012_Community%20Consultation%20Guidelines.pdf, accessed September 2, 2016.

[38] Fischhoff, W., Bostrom, A., Lave, L. and C. Atman (1992). Communicating risk to the public. First, learn what people know and believe. *Environmental Science and Technology*, 26(11), 2,048-2,056. DOI: 10.1021/es00035a606.

[39] INAC (2012). Off-grid communities. https://www.aadnc-aandc.gc.ca/eng/1314295992771/1314296121126, accessed July 12, 2016.

[40] Joseph, B. (2016). Indigenous Corporate Training Inc. First Nation relationship to the land. http://www.ictinc.ca/blog/first-nation-relationship-to-the-land, accessed April 20, 2016.

[41] Berkes, F., Colding, J. and C. Folke (2000). Rediscovery of traditional ecological knowledge as adaptive management. *Ecological Applications*, 10(5),1,251-1,262. http://gettingtoimplementation.ca/wp-content/uploads/2014/11/CommunityEnergyPlanningGuide_en.pdf, accessed, July 30, 2016.

Chapter 6

Hybrid Electrical Generation for Grid Independent Oil and Gas Well Fields

Francisco Porles, M.Sc., CEM, PMP

In Perú today, fossil fuels including petroleum, diesel fuel and natural gas are used to produce electrical energy for oil field extraction purposes. Due to their remote geographic locations, oil and gas fields are not linked to the national interconnected electricity system (Sistema Eléctrico Interconectado Nacional or SEIN). Therefore, high investment costs are required to produce electricity in their remote locations.

Wind power generation is a mature technology used worldwide. Perú has a substantial potential to produce electricity using wind turbine generators. Evidence of this includes the county's 164 MW of installed capacity, distributed among four operating wind farms that provide power to the electricity market. In 2018, three additional wind parks are expected to be commissioned. They will add another 162 MW of pollution-free energy, in total representing 3.8% of the power supplied to Perú's national power grid. The purpose of this chapter is to present the research that assessed the implementation of a hybrid wind-thermal natural gas system to provide electrical power for the Block X-Talara oil field and estimate the reductions in fuel consumption and CO_2 gas emissions.

USING WIND POWER FOR OIL AND GAS PRODUCTION IN PERU

The subject of this chapter is an oil field (Block X-Talara) located in the El Alto district, province of Talara, region of Piura in northwestern Perú. The El Alto Thermal Plant (TP) operates independently from SEIN and produces 8.3 MW of electricity using nine single cycle generating sets.

These use internal combustion engines fired by natural gas produced in the oil field at a charging factor of roughly 50%.

According to studies developed by the Ministry of Energy and Mines of Perú (MEM, Spanish acronym for Ministerio de Energía y Minas), Piura has a total wind potential of 17,628 MW with 7,554 MW usable. This research considers the design of a hybrid wind and thermal natural gas system for electricity generation. An optimally sized wind park would be connected to the power grid of the oil field with a possible interconnection to SEIN, enabling excess electricity to be sold. This wind park will operate together with the El Alto TP. The available power resources (both wind and natural gas) in the oil and gas field can be optimized, thus reducing the cost of electricity, fuel gas consumption, and CO_2 emissions. It will effectively provide the electricity needed for the oil field.

Using the oil field electricity demand requirements (5.0 MW peak demand/4.2 MW average demand), its distribution of electric charges, the current architecture of the oil field's power grid and DIgSILENT software, scenarios with different combinations of power supply were developed for the El Alto TP (as base charge) and wind power (as backup source). These scenarios were analyzed by means of electrical transient stability and reliability studies. The results confirmed that a hybrid wind-thermal system with a maximum penetration of wind power of 14.3% and a minimum of 3.75 m/s (12.3 feet/second), offers the stability and reliability required for the power supply of the Block X field. The data used for the hybrid system assessment included: 1) the wind data 4.88-7.01 m/s (16-23 feet/second) for the oil field's zone; 2) the electricity demand requirements; 3) the maximum wind power penetration of 14.3%; 4) the current electricity production costs (0.139 U.S.$/kWh, including fuel gas); 5), the total investment (U.S. $3.96 million); and 6) the projected operation and maintenance costs of the wind systems. The optimal configuration of the hybrid system yielded a reduction of 11.2% in electricity production costs, a 36.4% reduction in CO_2 emissions (equivalent to 6,633 tons of CO_2/year) and a 24.2% reduction in fuel gas usage.

The results indicated that the hybrid wind-thermal system as proposed is an alternative for electricity generation for the onshore oil and gas field on the northwest side of Perú. The economics of developing electricity generation infrastructure in the country using wind technologies are encouraging.

RESEARCH GOALS AND OBJECTIVES

Research Goals

The research goals were to assess the implementation of a hybrid wind-thermal natural gas system to supply electrical power for the Block X-Talara oil field (Figure 6-1), and to reduce natural gas consumption and CO_2 emissions.

Objectives

- To assess the present condition of natural gas consumption for the generation of electricity and the number and duration of the interruptions in the power supply of the Block X oil field.
- To determine the optimal operation of Block X oil field's power system with the implementation of a hybrid wind-thermal system. For

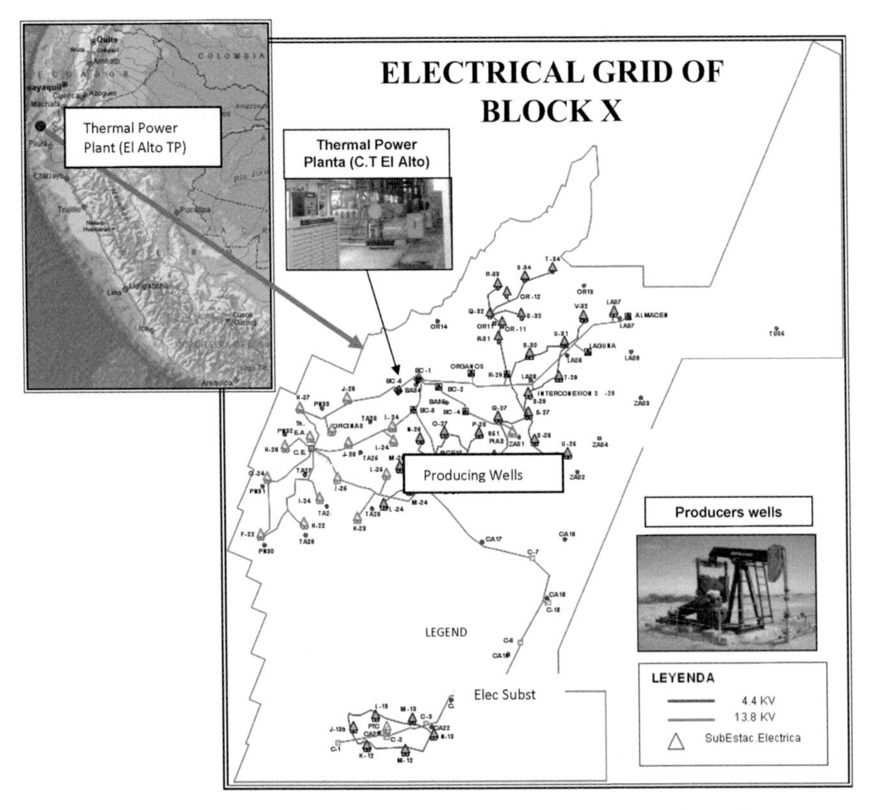

Figure 6-1. Geographical location of Block X oilfield (study zone).

this, the Hybrid Optimization Model for Electric Renewables (HOMER) software was used.

- To assess Block X's power system stability with the implementation of a hybrid wind-thermal system. For this, DIgSILENT Power Factory software was used.
- To assess the reliability of Block X's power system with the implementation of a hybrid wind-thermal system. For this, the DIgSILENT Power Factory software was used.
- To show the reduction of natural gas consumption as fuel gas, of CO_2 gas-emission production from power generation and the interruptions (outages) in the power supply.
- To show that it is possible to achieve an equivalent cost reduction of auto-production of electricity (U.S.$/kWh) using wind power as an alternative source for electricity generation.

APPROACH AND FORMULATION OF THE PROBLEM

The main electric charges of the Block X oil field are asynchronous three-phase electric motors with variable power between 5 hp and 75 hp. An electrical supply of 460 volts is used to activate the beam pumping unit for more than 2,100 producing wells located throughout the oil field in an area of more than 47,000 hectares (470 km^2) that form small- and medium-charge groups of roughly 80 kVA. This oil field continuously produces 11,000 barrels of petroleum and 15.5 million of standard cubic feet of natural gas each day. Nevertheless, the field's beam pumping units have been adversely affected by interruptions in the power supplied by its El Alto TP.

Due to increased drilling activities, the addition of new producing wells (many of them remote from the existing wells), and the volume of fluids being processed, the consumption of natural gas to produce electrical energy has increased. This increased consumption by natural gas-fired generating sets with internal combustion engines has reduced the volume of natural gas available for sale to third parties.

Interruptions in the electricity supply have increased due to the theft of electric copper cables, periodic voltage drops and power losses within the distribution system. The extensive length of the distribution networks (214 km) continues to expand as producing wells are added to the system.

HYPOTHESIS

General Hypothesis

The implementation of a hybrid wind-thermal natural-gas system in the Block X-Talara oil field will enable a reduction in the amount of fuel gas required for the supply of electrical power.

Specific Hypotheses

H1: The present consumption of natural gas for electricity generation is significant and the number and duration of interruptions exceeds the normal conditions.

H2: The optimal operation of the power system with the implementation of the hybrid wind-thermal system promotes reduced natural-gas consumption.

H3: The assessment of the stability of the electrical power system with the implementation of the hybrid wind-thermal system indicates that the system is stable over time and recovers quickly before fault conditions.

H4: The assessment of the reliability of the power system with the implementation of the hybrid wind-thermal system ensures reduced electrical power outages.

THEORETICAL FRAMEWORK OF THE RESEARCH

Hybrid Energy Systems

The increasing demand for energy and its associated environmental concerns have awakened interest in the development of hybrid renewable energy systems for electricity generation. Wind and solar energy generation potentials are dependent on the weather conditions. There is no single renewable energy source available in the region capable of supplying energy economically and reliably. Combining multiple energy sources can be a viable way to achieve reliable and marketable solutions [1]. With the combination of various energy sources, it is possible that power fluctuations will occur. To mitigate or possibly eliminate these fluctuations, energy storage systems such as batteries can be used [2].

Hybrid renewable energy systems are widely used for electricity generation in locations not connected to power grids. This is due to improvements in renewable energy technologies and the higher cost of petroleum fuels delivered to remote locations. An electricity generation system that uses a combination of different sources has the advantage of electrical supply balance and stability [3]. Typical hybrid renewable energy systems might include combinations of solar, wind energy and hydropower. Hybrid systems might also combine energy sources such as wind-diesel, diesel-battery-wind, photovoltaic-diesel-wind, photovoltaic-diesel-battery, photovoltaic-diesel or photovoltaic-diesel-battery [4]

Numerous publications describe the optimization of systems using one type of renewable energy source. Systems with solar energy and thermal energy storage with photovoltaic systems are one example [5]. Complex hybrid energy systems are optimized (simulated) by means of computer programs due to availability and improvements in computer software.

Computer modeling allows the optimization of various economic and engineering parameters that are considered in order to plan, design and construct a hybrid energy system. In particular, computer simulations can be used to perform feasibility studies of new systems. These simulations can also be used to diagnose problems that might occur during system operation. Research on the use of computer modeling has been performed [6,7].

A detailed analysis of Saudi Arabian wind data was performed to assess wind energy potential in five coastal locations. Rehman et al. estimated the cost of energy generation in 20 locations in Saudi Arabia by assessing their net present values [8]. Rehman and Halawani presented statistical characteristics of wind speed and daytime variations [9]. Autocorrelation coefficients that allow matching the actual daytime variation of the hourly average wind speeds were determined. They also calculated the Weibull parameters (a continuous probability distribution) for 10 locations and found that the wind speed was well represented by the distribution function.

Rehman et al. conducted an economic feasibility study of an existing electric grid for a remote location of 750 inhabitants which used a diesel thermal plant—adding wind turbines reduced diesel consumption and environmental pollution [4]. The Hybrid Optimization Model for Electric Renewables (HOMER) model was used as a dimensioning and optimization tool. This software contains a series of energy components and evaluates the appropriate options based on price and availability of energy resources.

Electrical system reliability is important. Reliability is defined as "the probability that a device carries out its purpose properly for a period of time under the foreseen operating conditions" [10]. Several reliability indexes are introduced in the literature [11,12]. Some of the most common indexes used in assessing the reliability of generation systems are the loss of load expected (LOLE), the loss of energy expected (LOEE), the expected energy not supplied (EENS), the loss of power supply probability (LPSP) and the equivalent loss factor (ELF). The ELF is considered as the primary index of reliability. The ELF index is the ratio of effective hours of load interruption for the total number of operating hours. It contains information on the number and magnitude of outages. In rural areas and stand-alone installations, an ELF <0.01 is considered acceptable.

APPLIED RESEARCH METHODOLOGY

The research performed is analytical, explanatory and considers the descriptive, explanatory and co-relational levels. It uses 2011 and 2012 data for the analysis period using the following units of analysis:

- Present electrical system of the Block X oil field located in Talara.
- Natural gas electrical generators located in El Alto TP of Block X.
- Availability of natural gas in the oil field for electricity generation.
- Electrical motors (loads) that activate the beam pumping units (for extraction of petroleum and gas).
- Wind resources available in the study area of Talara.
- Previous electrical studies (stability, reliability, load flow, shortcut, harmonics, among others).
- Wind technology available in the market.
- Historical record of the number and duration of interruptions in the present electrical system.
- The structure of the scientific research model used (see Figure 6-2).

The independent or explanatory variable (X) is the "implementation of a hybrid wind-thermal system." Its indicators are:

X1 Wind potential available in the study area
X2 Dimensioning of the hybrid wind-thermal system
X3 Optimal location of the wind generators
X4 Investment, operation and maintenance cost of the hybrid wind-thermal system

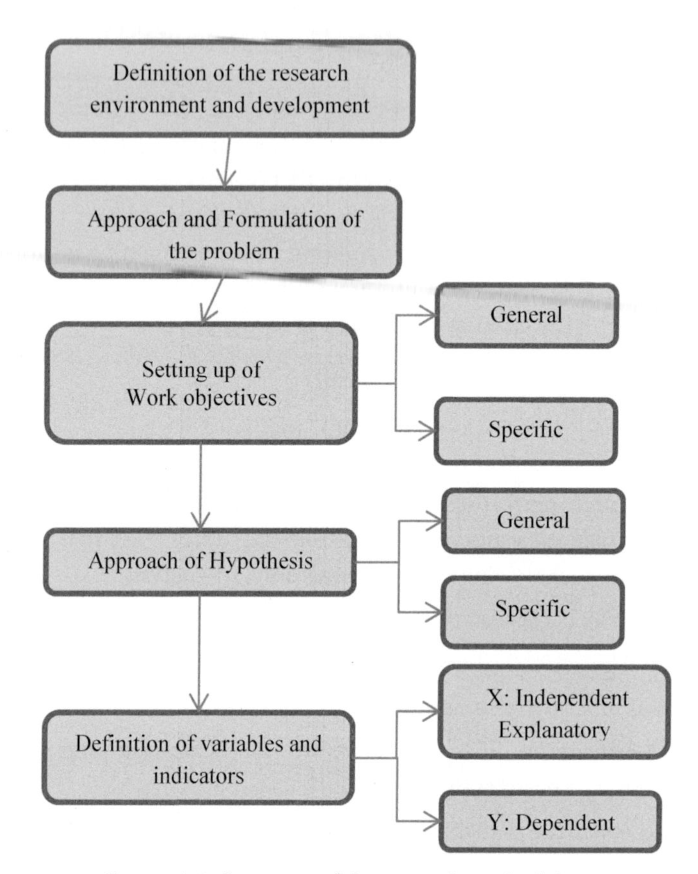

Figure 6-2. Structure of the research methodology.

X5 Stability of the electrical system with the commissioning of the hybrid wind-thermal system

The dependent variable (Y) is the "reduction of the natural gas consumption for the electricity generation and reduction of the interruptions in the power supply of Block X oil field." Its indicators are:

Y1 Reduction in natural-gas consumption for electricity generation.

Y2 Reduction in operating costs (operating hours) of the gas generating sets (thermal generation)

Y3 Increase in the reserve margin available in the electrical system

Y4 Reduction in the number and duration of interruptions in the power grids (operational reliability)

TOOLS USED

Table 6-1. Tools and techniques used in the study.

Name	Type	Description
DIgSILENT PowerFactory	Software	Software for electrical power systems applicable to generation, transmission, distribution and industrial-system studies. Used in this work for the development of the following electrical studies: • Transient stability analysis • Electrical reliability analysis
HOMER (Hybrid Optimization Model for Electric Renewables)	Software	Most used optimization models for the design, modeling, optimization and feasibility analysis of hybrid electrical systems based on renewable energies, developed by the National Renewable Energy Laboratory. Its algorithm is based on three main tasks: • Simulation • Optimization • Sensitivity Analysis
Database of wind measurements in Talara (CORPAC S.A).	Field research	From the weather station of CORPAC, installed on Air Base Talara.
Database and technical documentation of Petrobras Energia Perú, S.A (CNPC Perú, S.A at present)	Documentary investigation	Monthly production reports. Monthly electrical autogeneration reports Records of natural-gas consumptions Records of electrical energy consumption Electrical faults statistics Electrical faults reports Operation daybooks
Delphi technique (Questionnaires and interviews)	Field research	Judgment of author Judgment of technicians and researchers of Universidad Nacional de Ingeniería

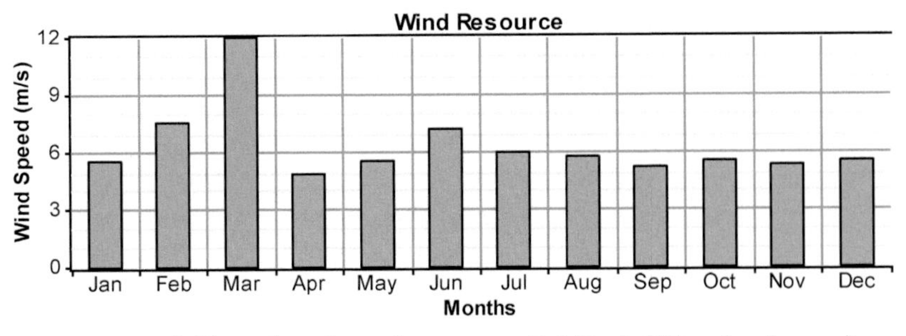

Figure 6-3A. Typical wind speed average availability in Talara (study zone).

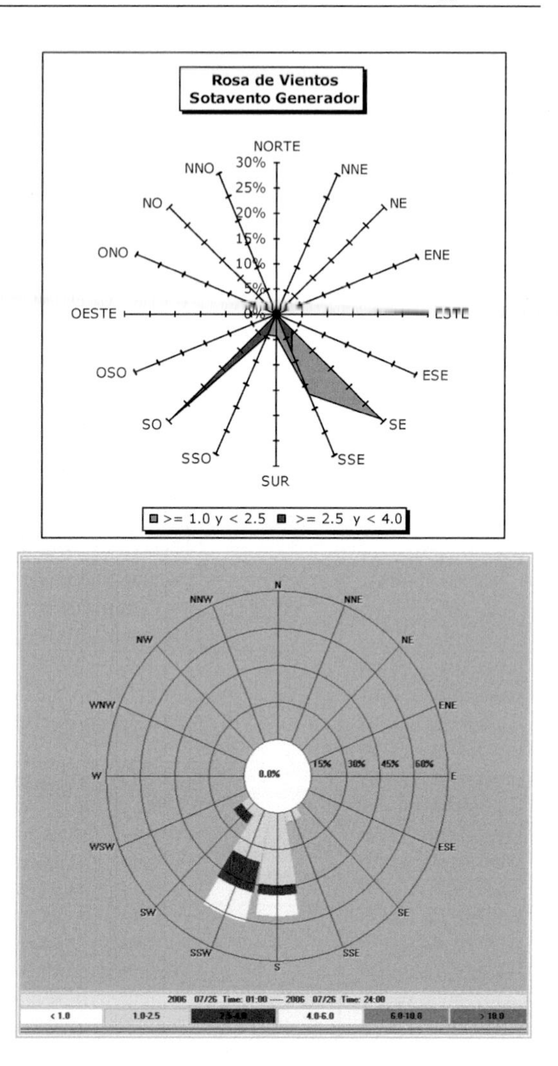

Figure 6-3B. Wind rose for Talara Airport showing wind speed distribution and direction.

ANALYSIS OF OUTCOMES

From the technical and economic perspectives, a hybrid wind-thermal natural-gas system is required for a more reliable supply of electricity (Table 6-2). Since there is no excess electrical energy, battery banks or converters are unnecessary. System components include:

a) Two wind generators Vestas model V100-1.8 MW or similar, which consist of an upwind rotor equipped with three blades plus control and low-voltage connection equipment. Each wind generator provides 500 kW at an average wind speed of 6 m/s (19.7 feet/second). For both, 1,000 kW of electricity is supplied by the wind park.

b) Four 1,028 kW gas generating sets, Cummins model 1250GQNA.

c) Two 428 kW gas generating sets, Caterpillar model 3516LE.

Table 6-2. Generation supply (sub-scenario 3.1).

Effective Power of the hybrid system in kW	Number of effective 450 kW wind generators at 6 m/s.	Wind generation total kW	Supply of thermal natural-gas generators
Wind generation covering 14.3 % of the demand of the power plant in kW.	02	900	5,396 kW (one gas generating set at 1,028 kW plus three at 427.9 kW)
Total Power (kW)		**6,296**	

As a result of the transient stability study, the most feasible scenarios from the point of view of stability are sub-scenarios 3.1 and 4.1. A maximum wind energy penetration of 14.3% is established in the electrical system of the reservoir. These sub-scenarios present frequency and voltage values slightly out of the tolerable range during and after the disturbance. Nevertheless, for the restoring status there are sustained oscillations which are neither damped nor stabilized, thus a power stabilizer or power system stabilizer (PPS) is required in the synchronous generators. The proposed model is the PSS-IEEEST.

The rotor angle of all the synchronous generators increases when the fault on the entry 1 bar (Figure 6-4) is produced and oscillates during the first two seconds after a disturbance and are stabilized subsequently allowing the machines to be synchronized. It can be concluded that the angle of the rotors of the synchronous generators present acceptable values for system stability.

From the technical and economic analysis developed using HOMER, a reduction in the average energy cost (COE, acronym for costo promedio de energia) was 11.2%. Including the cost of the fuel gas, the COE was 0.139 U.S. $/kWh for the base scenario and 0.125 U.S. $/kWh for sub-scenarios 3.1 and 4.1 (Figure 6-5).

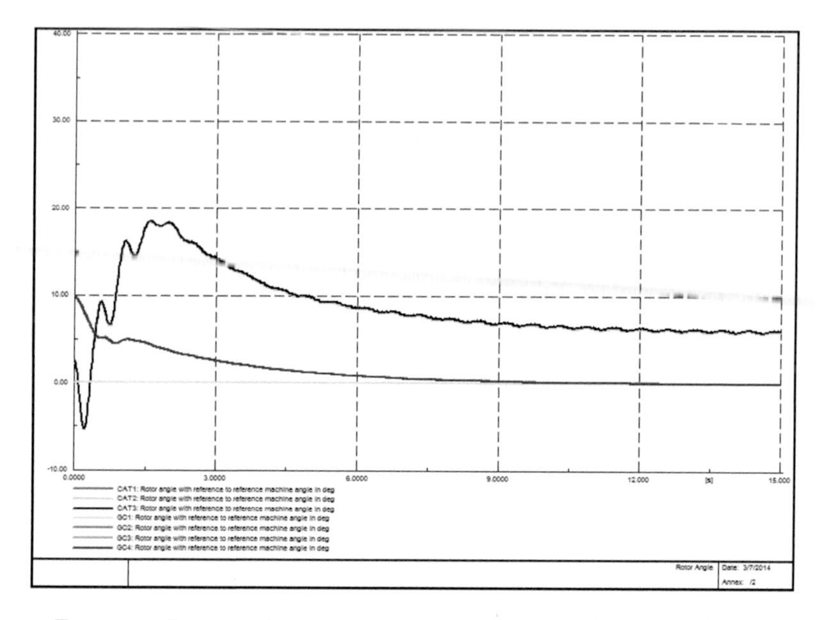

Figure 6-4. Rotor angle of each synchronous generator (sub-scenario 3.1).

From the historical records (2004-2013) of system operation and using the DIgSILENT software, reliability indexes (Table 6-3) of the electrical system were determined. These included the system average interruption frequency index (SAIFI), system average interruption duration index (SAIDI), customer average interruption duration index (CAIDI), average service availability index (ASAI) and the energy not supplied (ENS) to the system.

The following conclusions were derived from a comparative analysis of the simulated scenarios:

- Sub-scenario 4.3 has the lowest values for the SAIFI, SAIDI, CAIDI, ENS indexes and a higher value for the ASAI index. According to the transient stability study, this scenario is not feasible due to the sustained voltage and frequency oscillations after large disturbances. This sub-scenario presents one fault every 263 hours.
- Sub-scenario 4.2 offers the best reliability indexes for the system and represents the wind generation distributed with 35.7% of the electrical demand. According to the transient stability study it is not a feasible option. This sub-scenario presents one fault every 256 hours.

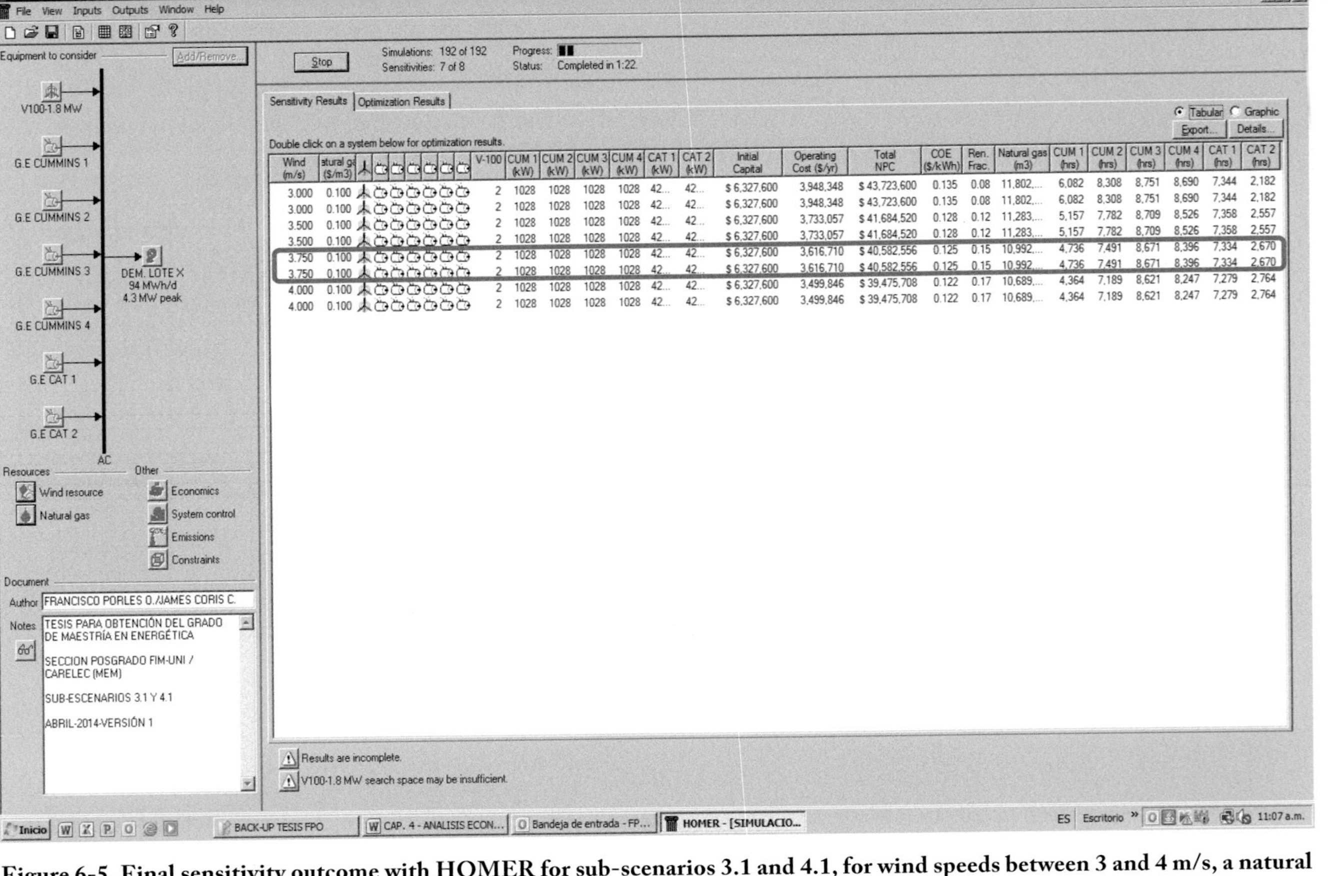

Wind (m/s)	atural g ($/m3)	V-100	CUM 1 (kW)	CUM 2 (kW)	CUM 3 (kW)	CUM 4 (kW)	CAT 1 (kW)	CAT 2 (kW)	Initial Capital	Operating Cost ($/yr)	Total NPC	COE ($/kWh)	Ren. Frac.	Natural gas (m3)	CUM 1 (hrs)	CUM 2 (hrs)	CUM 3 (hrs)	CUM 4 (hrs)	CAT 1 (hrs)	CAT 2 (hrs)
3.000	0.100	2	1028	1028	1028	1028	42...	42...	$ 6,327,600	3,948,348	$ 43,723,600	0.135	0.08	11,802,...	6,082	8,308	8,751	8,690	7,344	2,182
3.000	0.100	2	1028	1028	1028	1028	42...	42...	$ 6,327,600	3,948,348	$ 43,723,600	0.135	0.08	11,802,...	6,082	8,308	8,751	8,690	7,344	2,182
3.500	0.100	2	1028	1028	1028	1028	42...	42...	$ 6,327,600	3,733,057	$ 41,684,520	0.128	0.12	11,283,...	5,157	7,782	8,709	8,526	7,358	2,557
3.500	0.100	2	1028	1028	1028	1028	42...	42...	$ 6,327,600	3,733,057	$ 41,684,520	0.128	0.12	11,283,...	5,157	7,782	8,709	8,526	7,358	2,557
3.750	0.100	2	1028	1028	1028	1028	42...	42...	$ 6,327,600	3,616,710	$ 40,582,556	0.125	0.15	10,992,...	4,736	7,491	8,671	8,396	7,334	2,670
3.750	0.100	2	1028	1028	1028	1028	42...	42...	$ 6,327,600	3,616,710	$ 40,582,556	0.125	0.15	10,992,...	4,736	7,491	8,671	8,396	7,334	2,670
4.000	0.100	2	1028	1028	1028	1028	42...	42...	$ 6,327,600	3,499,846	$ 39,475,708	0.122	0.17	10,689,...	4,364	7,189	8,621	8,247	7,279	2,764
4.000	0.100	2	1028	1028	1028	1028	42...	42...	$ 6,327,600	3,499,846	$ 39,475,708	0.122	0.17	10,689,...	4,364	7,189	8,621	8,247	7,279	2,764

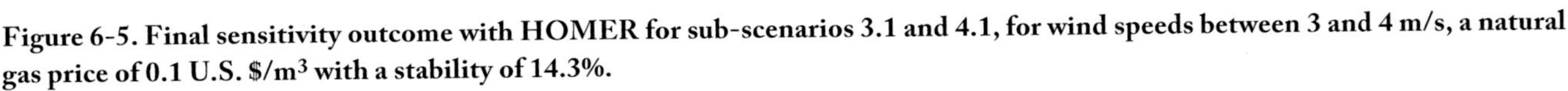

Figure 6-5. Final sensitivity outcome with HOMER for sub-scenarios 3.1 and 4.1, for wind speeds between 3 and 4 m/s, a natural gas price of 0.1 U.S. $\$/m^3$ with a stability of 14.3%.

Table 6-3. Comparative table of reliability indexes of the simulated scenarios

DESCRIPTION	INDEX	UNIT	BASE SCENARIO	SUB-SCENARIO 3.1	SUB-SCENARIO 3.2	SUB-SCENARIO 3.3	SUB-SCENARIO 4.1	SUB-SCENARIO 4.2	SUB-SCENARIO 4.3
System average Interruption Frequency Index	SAIFI	1/a	31.927	38.889	38.889	38.889	36.938	34.246	33.307
System Average Interruption Duration Index	SAIDI	h/a	97.818	114.116	114.116	114.116	104.543	90.628	85.866
Customer Average Interruption Duration Index	CAIDI	H	3.064	2.934	2.934	2.934	2.83	2.646	2.578
Average Service Availability Index	ASAI	-	0.989	0.987	0.987	0.987	0.988	0.990	0.990
Energy Not Supplied	ENS	MWh/a	627.541	746.33	746.328	746.328	676.626	576.085	541.453

- Sub-scenario 4.1 offers the best reliability indexes and represents the wind generation distributed with 14.3% of the electrical demand. It represents the most feasible option for transient stability. This sub-scenario presents one fault every 237 hours.

According to the reliability indexes of the network topology, sub-scenario 4.1 does not significantly affect the present network topology. Using 14.3% of wind generation distributed by zones, operation of the present system will not be affected. It is recommended sub-scenario 4.1 be implemented with wind generation.

High Impact Indicators

High impact indicators were obtained considering a 10-year project useful life, equivalent to the remaining concession period for the Block X oil field. These indicators are shown in Table 6-4.

CONCLUSIONS

The electrical grid for the Block X oil field has always been based on fossil fuels. Renewable energy now provides a viable alternative. A hybrid wind-thermal system is proposed to generate electrical energy for oil field operations in northwestern Perú where oil wells are distant from thermal plants. The progress of renewable energy technologies and the high cost of natural gas generated electricity in Talara provide opportunities to improve the electrical supply system for the oil fields.

Accounting for the outcomes obtained in both the transient stability study and the reliability study, a power supply system with hybrid (wind and thermal) generation, dimensioned to a renewable energy penetration of 14.3%, offers the power supply stability and reliability required by oil fields such as those located in northwestern Perú. With the area's available wind data, it is concluded that a hybrid energy system can be planned, modeled and designed.

The most technically and economically feasible scenarios for system stability are:

- Sub-scenario 3.1—Centralized wind generation providing (from El Alto TP) up to 14.3% of electricity demand.

Table 6-4a. Table of high-impact indicators obtained.

General Objective	Specific Objectives	Impact Indicator	Type of Indicator	Description of the impact obtained
To assess the implementation of a hybrid Wind-Thermal (natural gas) system for the power supply of the oilfield Block X of Talara.	To reduce the own natural-gas consumption used for electricity autogeneration.	Reduced natural-gas volume (in MMm³ or MMpc)	Energy	1.71 million m³ (equivalent to 110.3 million feet³).
				Equivalent to a reduction of 24.6% of the natural-gas consumption.
	To assess wind energy penetration in the energy grid of the oilfield for the electrical energy autoproduction.	Maximum Penetration of wind energy (%), regarding the maximum electricity demand of the oilfield.	Energy	14.3%
	To reduce the electricity auto production cost.	U.S.$/kWh	Economic	From 0.139 U.S.$/kWh the cost was reduced to 0.125 U.S.$/kWh.
				Equivalent to a reduction of 11.2% of the electricity cost (including fuel gas).
				It represents an economic saving in electricity of annual 551.9 M U.S.$
	Level of investment.	Required investment	Economic	3,960,000 U.S.$
		VAN	Economic	4 686,321 U.S.$
		Pay-out	Economic	5.28 years

Table 6-4b. Table of high-impact indicators obtained.

General Objective	Specific Objectives	Impact Indicator	Type of Indicator	Description of the impact obtained
To assess the implementation of a hybrid Wind-Thermal (natural gas) system for the power supply of the oilfield Block X of Talara.	To reduce the number and duration of electrical power outages in the electrical networks of the oilfield.	System Average Interruption Frequency Index (SAIFI) (#faults/year)	Operational	An acceptable value of 36.9 is obtained.
		Customer Average Interruption Duration Index (CAIDI) (in h/year)	Operational	It is reduced from 3,064 to 2.83.
	To determine the most optimal configuration and dimensioning of the hybrid wind-thermal to be implemented.	Number of wind generators to be installed	Operational	2 units VESTA 1.8MW
		Total usable power (kW) of the wind generators	Operational	900 KW
	To reduce the level of CO_2 emissions generated with the present thermal plant.	t CO_2e reduced/year	Environmental	6,633 tons CO_2/year
				Equivalent to a reduction of 36.4% of CO_2 emissions.

- Sub-scenario 4.1—Wind generation providing up to 14.3% of demand, but distributed strategically to better use the region's dominant winds, reduce unbalancing and improve electrical system instability.

It is clear that wind speeds are the thermodynamic determinant variable for the configuration of the hybrid systems studied when considering the natural potential energy in Talara. The average wind speed available in the region is from 5 to 6 m/s (16.4-19.7 feet/second). To meet the restriction of 14.3% of maximum wind energy penetration (MWEP), a minimum of 3.75 m/s (12.3 feet/second) is required. It is fully covered by the average speed available in the zone.

The economic aspects of wind technologies in Perú are sufficiently promising to continue the development of wind powered electrical generation capacity in the county's northwest regions. The outcomes obtained in February 2016 in the fourth auction for electricity generation with renewable energy resources by Osinergmin offer proof. The minimum price offered for wind technology dropped 42.3% compared to the minimum price in the first auction in 2010 and the maximum price offered was reduced by 26%.

References

[1] Jahanbani, F. and G. Riahy (2001). Optimum design of a hybrid renewable energy system. *Intech*, page 233.

[2] Wang, L. and C. Singh (2009). Multicriteria design of hybrid power generation system based on a modified particle swarm optimization algorithm. *IEE*, 24, pages 163-72.

[3] Ding, J. and J. Buckeridge (2000). Design considerations for a sustainable hybrid energy system. UNITEC Institute of Technology, Auckland University, New Zealand, pages 1-2.

[4] Rehman, S., El-Amin I., Ahmad, F., Shaahid, S., Al-Shehria, A., Bakhashwain, J. and A. Shash (2005). Feasibility study of hybrid retrofit to an isolated off-grid diesel power plant. *Elsevier*, page 642.

[5] Vries, H. and J. Francken (1980). Simulation of a solar energy system by means of electrical resistance. *Solar Energy*, 25, pages 275-81.

[6] Barnard, J. and L. Wendell (1997).

[7] Simmons, A. (1996). Grid-connected amorphous silicon photovoltaic array. *Progress in Photovoltaics: Research and Applications*, 4, pages 381-388.

[8] Rehman, S., Halawani, T. and M. Mohandes (2003). Use of radial basis functions for estimating monthly mean daily solar radiation. *Solar Energy*, pages 161-168.

[9] Rehman, S. and T. Halawani (1994). Weibull parameters for wind speed distribution in Saudí Arabia. *Solar Energy*, pages 473-479.

[10] Billinton, R. and R. Allan (1992). Power system reliability and its assessment: part 1—background and generating capacity. *Power Engineering Journal*, pages 191-196.

[11] Billinton, R. (1994). Reliability assessment of electrical power system using Monte Carlo methods. Plenon Publishing, pages 936-941.

[12] Xu, D., Kang, L. and B. Cao (2005). The elitist non-dominated sorting GA for multiobjective optimization of standalone hybrid wind/PV power system. *Journal of Applied Science*, page 6.

Chapter 7

Barriers to Renewable Energy Development in Trinidad and Tobago

Zaffar Khan, Ph.D. and Atiyyah A.Khan, BSc, MIIEM, CEM
Atiyyah A.Khan, BSc, MIIEM, CEM

The Republic of Trinidad and Tobago is a country with two primary islands located in the southernmost part of the Caribbean, just 11 km (6.8 miles) off the coast of Venezuela. The economy is essentially hydrocarbon based with the energy industry contributing approximately 45% of the country's gross domestic product (GDP). However, this tropical country faces a number of challenges. These include suppressed oil and natural gas prices, falling hydrocarbon production, and gas curtailment in its petrochemical sector. The latest Ryder Scott hydrocarbon audit indicated that the country has proven natural gas reserves to last to only 2019. It is imperative that the country diversifies its economy including its power sector and transitions to sustainable energy systems. This chapter discusses the primary barriers to renewable energy (RE) development and implementation in Trinidad and Tobago. It also identifies the hurdles the country must overcome in order to formulate strategies to deal with these problems in the context of renewable energy, energy efficiency and greenhouse gas emissions reductions.

INTRODUCTION

Trinidad and Tobago (T&T) has the second highest gross domestic per capita income in its region [1]. The country's economy is driven by its energy sector which includes activities from crude oil production, natural gas production, liquefied natural gas production, asphalt production, petrochemicals, compressed natural gas and electric power [1]. The energy sector of Trinidad and Tobago is the largest and important contributor to the country's government revenues, foreign exchange and gross domestic product (GDP) [2]. Figure 7-1 shows the contribution of the energy sector to Trinidad and Tobago's economy in 2012. It accounts for approximately 45% of GDP, 50% of total revenues, 80% of exports but just 1% of employment [3].

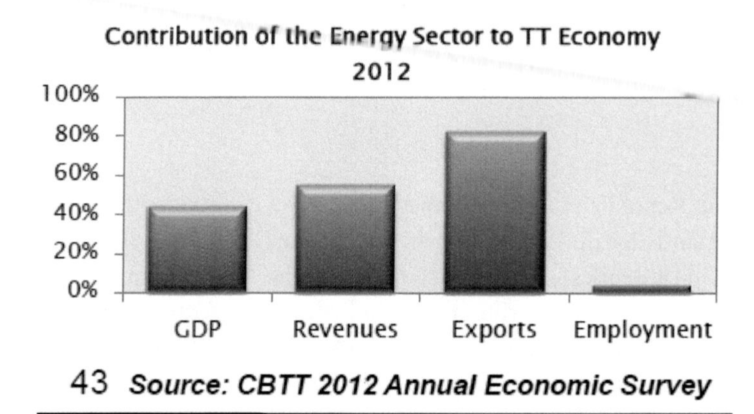

Figure 7-1. Energy sector contribution to Trinidad and Tobago's economy in 2012 [3].

Trinidad and Tobago has a rich history in oil production traversing some 100 years. Its economy has evolved from an oil-based to a natural gas-based economy. This has fueled its development and is responsible for T&T having one of the highest growth rates in Latin America. Other island nations in the region have pursued renewable energy as a means of meeting their energy demands and easing the burden of high petroleum imports. These countries have higher electricity rates, higher fuel prices and higher costs for imported oil and gas.

However, Trinidad and Tobago is a net exporter of petroleum products and enjoys low electricity rates and low fuel subsidies. Thus, Trinidad and Tobago has lacked the economic motivation to introduce and develop RE technologies.

The main barriers to RE development and implementation in Trinidad and Tobago are noted in Table 7-1. Afterwards suitable strategies are recommended to overcome these barriers.

Table 7-1. Primary barriers to RE development in Trinidad and Tobago.

Type of Barrier	Description
Institutional	Limitations (research and development), demonstration and implementation.
Market	Small size of market, limited access to markets, limited involvement of private sector.
Awareness/Information	Lack of awareness, lack of access to information on RE and energy efficiency (EE).
Financial	Inadequate financing available (local, national, international) for RE projects.
Economic	Unfavorable costs, taxes (local and import). Government subsidies offer cheap electricity and fuel.
Technical	Lack of access to technology and inadequate. maintenance facilities.
Capacity	Lack of skilled labor force and training facilities.
Social	Lack of social acceptance and local interest and participation.
Legislative	Lack of formal policy, regulatory and legal frameworks necessary for RE development.

DRIVERS FOR DEVELOPING RENEWABLE ENERGY

There are many reasons for Trinidad and Tobago to adopt RE technologies into its energy mix. Some of the drivers [4] for encouraging RE in T&T are:

- To provide energy security.
- T&T is rich in underutilized natural resources (solar, wave, tidal, geothermal).
- To mitigate impacts of climate change.
- Access to a reliable source of energy.
- To generate employment.
- Economic diversification and improving energy alternatives.
- National and industrial development.
- Reduced dependence on finite fossil fuels for energy.
- Conservation of petroleum resources.

BARRIERS TO RENEWABLE ENERGY INVESTMENTS

Though there are many advantages for including RE in T&T's energy matrix, the introduction and development of RE technologies has been lacking due to numerous barriers. Barriers affecting RE development in T&T include:

- Competition from fossil fuels subsidized by government (low cost fuel and electricity).
- Lack of legal framework, policy and instruments to encourage RE development.
- Limited access to capital, fiscal incentives and enabling financial environment.
- Commercialization barriers.
- High initial cost of technologies and implementation.
- Lack of education and awareness.
- Lack of locally available RE resource data.
- Inadequate institutional capacity.
- Environmental concerns with RE technologies.
- Lack of political will.

These barriers tend to increase the financial risks associated with RE investments. The primary ones that will be discussed in this chapter are economic, legislative, financial, market, technical, institutional, awareness/information, capacity and social barriers.

Economic Barriers

According to the International Energy Agency (IEA), energy subsidies are "any governmental action that primarily concerns the energy sector which lowers the cost of energy production, raises the price received by energy producers or lowers the price paid by energy consumers" [5].

Fuel subsidies were introduced in the 1970s under the Petroleum Production Levy and Subsidy Act of 1974. This Act was amended in 1992 to limit the levies paid by producing companies to a maximum of 3% of gross income. The Act was amended again in 2003 to 4% of the company's gross income. The remainder of the subsidy is paid by the government and these payments have become burdensome in recent years. In the 2014 budget, the government allocated $7 billion for fuel subsidies, accounting for 36% of total expenditures in 2014 [6]. Figure 7-2 shows that the allocation of fuel subsidies since 2010 has substantially increased with 2014 having the greatest allocation to date.

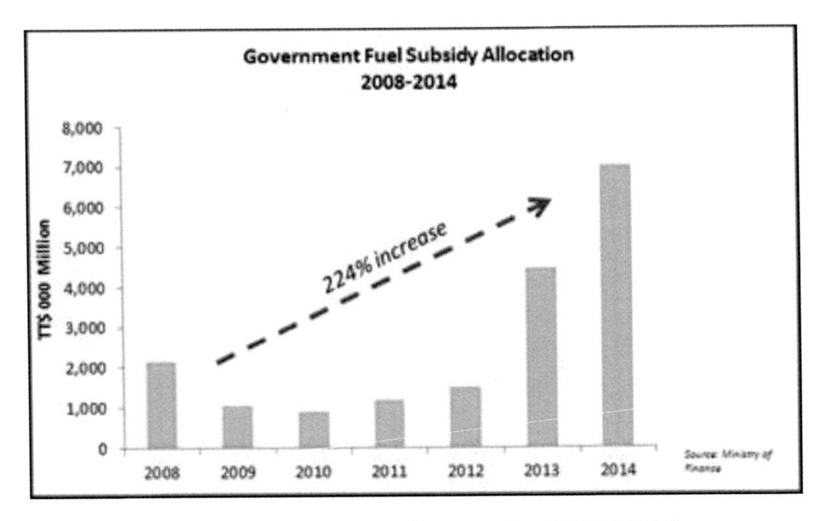

Figure 7-2. Fuel subsidy allocations for 2008-2014 [6].

Heavily subsidized fuel and electricity creates an uncompetitive economic environment for renewable energy technologies. It is a huge barrier against RE market penetration since there is inadequate economic incentive to invest in RE technology development. It limits the market for RE since inexpensive fossil fuels are readily available, thus preventing the demand for cleaner more expensive forms of energy from developing. Fuel subsidies have also led to an influx of vehicles for such a small country [7]. Diesel fuel enjoys the greatest share of the fuel subsidy followed by gasoline. This has encouraged greater vehicle use in the country causing traffic and environmental problems. T&T has approximately 650,000 vehicles, far too many for a country with population of only 1.4 million [7].

Trinidad and Tobago currently use natural gas to generate most of its electricity [2]. Low electricity rates have helped develop the country and its economy. In 2014, T&T was recognized in the World Bank Doing Business Report as the 10th highest ranked of 189 countries for its ease of electricity connectivity [8].

Trinidad and Tobago Electricity Commission (T&TEC), the state-owned utility company, is the country's sole retailer of electrical power which is supplied to customers via a single electrical grid [2]. The company has evolved from power generation, transmission and distribution to include generation by independent power producers (IPPs) [2]. The current IPPs are The Power Generation Company of Trinidad and Tobago (Powergen), Trinity Power Limited (TPL) and Trinidad Generation Unlimited (TGU).

There are concerns associated with power generation in T&T. While natural gas is the cleanest of all the fossil fuels, it emits atmospheric greenhouse gases such as CO_2 and CO when combusted [9]. Natural gas is a non-renewable resource with exhaustible reserves. T&T is wholly dependent on natural gas for electricity generation which creates dependency problems. Services are unreliable at times when there are shortages of natural gas or gas leaks in the pipelines. The country's infrastructure is aged, resulting in low conversion efficiencies. When electricity rates are low, consumers tend to use electricity inefficiently.

Table 7-2. Fossil fuel emission levels 1998 [9].

Pollutant	Natural Gas	Oil	Coal
Carbon Dioxide	117,000	164,000	208,000
Carbon Monoxide	40	33	208
Nitrogen Oxides	92	448	457
Sulfur Dioxide	1	1,122	2,591
Particulates	7	84	2,744
Mercury	0.000	0.007	0.016

Trinidad & Tobago enjoy one of the region's lowest electricity rates. Table 7-3 shows the average tariffs for the period 2012 for selected countries in the region with Trinidad and Tobago (US$0.06/kWh) and Suriname (US$0.05/kWh) having the lowest average tariffs [10].

Low electricity rates have made RE technologies noncompetitive for electricity generation. Technical, institutional and legislative changes are needed for RE technologies to be widely implemented.

Table 7-3. Utility retail tariffs in 2012 [10].

AVERAGE RETAIL TARIFFS PER UTILITY

COUNTRY	AVERAGE TARIFF (2012)	
Antigua and Barbuda	$ 0.43	
Bahamas	$ 0.26	(2010)
Barbados	$ 0.32	
Dominica	$ 0.43	
Dominican Republic (east)	$ 0.20	
Dominican Republic (north)	$ 0.20	
Dominican Republic (south)	$ 0.22	
Grenada	$ 0.40	
Guyana	$ 0.32	(2011)
Jamaica	$ 0.36	
Haiti	$ 0.38	
St. Lucia	$ 0.38	
St. Vincent and the Grenadines	$ 0.36	(2011)
Suriname	$ 0.05	(2011)
Trinidad and Tobago	$ 0.06	(2011)
AVERAGE	$ 0.33	

© Inter-American Development Bank
Infrastructure & Environment / Energy

Renewable energy technologies are made unattractive due to their higher upfront costs and transaction costs when compared to the cost of traditional fossil fuels. For example, the installed cost for a solar water heater ranges from $7,500 to $13,000 [11]. Recently, costs for renewable energy technologies have declined making them more competitive [12]. For example, the International Renewable Energy Agency (IRENA) estimates that "onshore wind costs US$0.06/kWh in Asia and US$0.07/kWh in North America" [13]. In the 'Renewable Power Generation Costs in 2014 Report', renewables such as biomass, hydropower, onshore wind, and geothermal are mentioned as being competitive with or less expensive than fossil fuels without subsidy support [13].

Legislative Barriers

While there are no formal policies for renewable energy and energy efficiency (EE), there has been a framework for the development of a renewable energy policy in T&T since 2011 [2]. However, the Trinidad and Tobago Electricity (T&TEC) Act Chapter 54:70 does not enable the licensing of independent power producers (IPPs) to produce renewable electricity to sell to the national grid [2]. The Regulated Industries Commission (RIC) Act Chapter 54:73 lacks provisions for IPPs to connect to the electricity grid. This lack of connectivity precludes open access, net-metering and feed-in-tariffs [2]. The Ministry of Public Utilities needs to review existing legislation to enable open access, net metering, grid interconnection, feed-in-tariffs, and net billing so that renewables such as solar power can make an impact.

Financial Barriers

One of the major barriers is the lack of financing available for those willing to invest in RE development as entrepreneurs or consumers. An embracing financial environment is vital to RE financing and investment. As an example, international banking has been credited with the success of RE in the Bahamas [14]. There is need for efficient and effective investment for RE projects by both local and international institutions. The Inter-American Development Bank (IDB) and the International Finance Corporation (IFC) have both committed funds to help develop RE projects in island nations. The IDB has provided $70 million for the development of RE projects in that country. The IFC has partnered with BHD (the largest bank in Dominican Republic) to develop RE investment programs and lines of credit to finance RE projects.

Incentives are used to lower costs, reduce risk, and help create favorable market conditions as well as address concerns such as greenhouse gas emissions. The government of Trinidad and Tobago has provided incentives which include tax credits, tax allowances, wear and tear allowances, and exemptions from conditional duty and value added tax (VAT) for solar, wind and energy efficiency (see Table 7-4). Examples include the removal of tax on commercial vehicles which are manufactured to use compressed natural gas (CNG) and removal of customs duties on certain CNG kits with cylinders. Individuals using CNG kits and cylinders may be eligible for a tax credit of as much as $10,000. In the most recent budget, the government introduced the removal of import duties and value added tax (VAT)

Table 7-4. Incentives for solar and wind technologies.

Finance Act No. 13 of 2010 - Renewable Energy				
	Import Duty Exemption	0-Rated VAT	Tax Credit	Wear and Tear Allowance
Solar Water Heater (SWH) (usage)	From Caricom Sources	Fully assembled SWH	25% on the acquisition of SWH by H/Hold (Max. $2500.00)	on 150% of the cost of acquisition of SWHs
SWH (manufacture/ assembly)	SWH industry declared "Approved Industry" subject to conditional duty exemption	NA	NA	on 150 % of cost of plant, machinery, parts & materials for use in the manufacture of solar water heaters
Solar PV	PV cells, whether or not assembled in modules or made up into panels	PV cells, whether or not assembled in modules or made up into panels	NA	on 150% of the cost of acquisition of PV systems
Wind Energy	Wind turbines & supporting equipment	Wind turbines & supporting equipment	NA	on 150% of the cost of acquisition of wind turbines

on hybrid electric vehicles. Incentives were provided for energy efficiency improvements when certified energy services companies (ESCOs) acquire equipment to conduct energy assessments. Tax allowances are available to companies for the design and installation of energy saving systems.

Although financing remains a problem in the region, the cost of solar power has been falling rapidly [13]. The cost of solar photovoltaic (PV) plants has decreased by half since 2010 and installation costs have fallen by 65% [13]. Research and development into newer technologies will further decrease the cost for solar power in the future.

Market Barriers

Renewable energy technologies are often uncompetitive due to imperfect market conditions set by subsidized fuel and electricity. The two major barriers to commercialization are the lack of economies of scale and undeveloped infrastructure [15]. The Trinidad & Tobago RE market is in an early stage of development due to present policies and legislation. Grenada overcome this barrier when GrenSol, a private company, and GRENLEC, Grenada's sole utility company, agreed to new requirements for RE grid connections. This opened the market for RE in Grenada, allowing producers an electric generation limit of 5 MW [14]. T&T needs to allow grid interconnections in order for RE to be successful.

Technical Barriers

There is a lack of technical programs and specialized training programs available in renewable energy technologies. There are also limited certification schemes both for energy efficiency equipment and renewable energy technologies. In Trinidad and Tobago there is a lack of trained personnel and a shortage of technical and managerial expertise to effectively develop and implement RE technologies.

Within the last few years, the Trinidad and Tobago Bureau of Standards (TTBS) and the Ministry of Energy and Energy Affairs (MEEA) started working to establish the following standards [2]:

- Development of Technical Standards for Solar Water Heaters (SWHs) Standard TTS 106:2012, Solar water heater systems—design and installation requirements.
- Standard TTS/EN 12975-1:2012: Thermal solar systems and components—solar collectors.

However, these technical barriers need to be overcome in order for RE to be successful in Trinidad and Tobago. Technical skills must be developed for RE assessments, installation and system maintenance. Training should be provided from high school to tertiary levels.

Institutional Barriers

There is need for a coordinating agency under the Ministry of Energy for RE introduction and development. There are limitations in the amount of research and development for RE technologies and for the demonstration and implementation of RE projects. There have been successful small-scale solar and wind projects.

Since 2011, T&TEC has been researching the integration of renewable energy. Testing is being conducted at three renewable energy project sites. These projects include: the solar photovoltaic panels installed at the Stanley P. Ottley Building at Mount Hope, University of Trinidad and Tobago's O'Meara Campus, and the T&TEC/Powergen hybrid solar and wind systems located at the Islamic Children's Home in Gasparillo [16]. These RE projects are fully operational and interconnected with the T&TEC grid. They have an installed capacity of approximately 2.0-2.4 kW [16]. More RE projects like these should be encouraged by both public and private institutions.

Information and Awareness Barriers

In Trinidad and Tobago, there is a lack of awareness and education concerning renewable energy. There are varying levels of awareness of the benefits, costs and applications of renewable energy among policymakers, the local private sector, finance institutions and prospective customers.

Information regarding RE development in Trinidad and Tobago is not widely disseminated and is not readily available. Such information should be made available to consumers and include available technology options and expertise for installation and maintenance.

Capacity Barriers

In Trinidad and Tobago, capacity for building RE systems is undeveloped. Capacity building includes training, technical support and resource networking [11]. Capacity building is a long-term process in which all stakeholders need to participate. Examples of stakeholders that will need to be part of capacity building include T&TEC, the Ministry of Energy, the

Ministry of Finance and the Ministry of Trade.

Training is an important part of capacity building. There are many examples. Grenada implemented capacity building by training electricians and engineers in RE. Saint George's University in Grenada partnered with the Grenada Solar Power Company to assess the various types of solar modules for use in the Caribbean. The University of Trinidad and Tobago is offering courses in renewable energy. The University of the West Indies has a master's program in renewable energy technology. The Arthur Lok Jack Graduate School of Business is offering a MBA program in sustainable energy management. Similar programs should be introduced from secondary to tertiary level to help develop capacity building in RE.

Social Barriers

Society seems reluctant to adopt RE and energy efficiency (EE). Despite the minimal fiscal incentives introduced by the government since 2010, the wider community remains disinterested. Trinidad and Tobago has a dependency on oil, gas, electricity, and the revenues they generate to preserve living standards. It is difficult to change this dependency which has become normative.

Many people do not understand the need for alternative forms of energy. This may be because they do not believe oil and gas reserves will be soon depleted. Perhaps they believe climate change is a myth. Lack of interest in the RE market detracts investment. A greater challenge is achieving market acceptance of environmentally beneficial clean energy resources as they are typically more expensive to develop than fossil fuels.

Strategies and Recommendations

There are many strategies that help address the barriers discussed:

- Develop strong proactive governmental policies and agencies for RE and EE.
- Create more incentives (e.g., tax credits and rebates).
- Offer disincentives such as a consumption tax on electric water heaters.
- Create innovative financial schemes for purchasing RE equipment and systems.
- Conduct energy-savings assessments.
- Revise T&TEC, RIC and EID Acts to enable IPPs to sell surplus

electricity to T&TEC, allow access for net metering, and establish feed-in tariffs. Net metering will allow the IPPs to bank their power on the grid and generate revenue from the surplus electricity generated by RE.

- There should be RE interconnection standards for IPPs linking to the national grid. IPPs will need open access to the national electricity grid and this can only occur when the proper technical and legal requirements are met.
- RE targets should be set and mandated by law.
- Expand product and technical standards.
- Education and training should be offered including specialized training in RE technologies.
- Fuel subsidies should be reduced and gradually phased out.
- Offer subsidies for RE and EE technologies.
- Provide an enabling financial environment.
- Provide greater access to capital.
- Attract and expand private sector involvement.
- Further develop RE capacity building programs by training skilled workers in partnership with technical and vocational schools. Include RE and EE in education curricula of schools, hold workshops to engage personnel who would be directly engaged in the RE industry (e.g., technicians, electricians, T&TEC inspectorate and teachers).
- Provide more relevant RE consumer information in the public domain via information centers and by identifying available services. Establish awareness campaigns including communication fairs, workshops, media, and micro-marketing of incentives for RE and EE, seminars, workshops and exhibitions.
- Government needs to lead by example. The government might incorporate solar water heaters in government housing schemes and institutions such as schools and hospitals.
- Review and modify current electricity tariff structures.
- Require utilities like TTEC to purchase a portion of their energy from power producers using renewables.

- Emission quotas, emission limits and RE standards should all be clearly stated within policies so that measures can be taken to meet those targets.
- The government should facilitate renewable energy projects which qualify under the Clean Development Mechanism (CDM) for carbon credits by reducing carbon emissions [6].
- Subsidies and soft loans (loans at lower than market rates) encourage investment and can reduce the investment cost. Soft loans provide easier access to credit with the appropriate conditions for financing investment which can help to overcome high upfront costs. These soft loans can be used to purchase RE equipment and installation. Micro-financing should be made available to help provide loans for small to medium sized RE enterprises.

Trinidad and Tobago has great potential to utilize RE technologies such as wind and solar. There is opportunity to manufacture solar panels in T&T due to its vibrant economy, inexpensive energy (compared to other islands in the region), access to abundant raw materials and an established industrial supply chain. Trinidad and Tobago can generate revenue from these new industries and provide renewable energy at a competitive market price.

T&T is an ideal location for new plants in Point Lisas to manufacture polysilicon, metallurgic silicon, float glass and solar photovoltaics [17]. Trinidad and Tobago has the opportunity to be a leader in compressed natural gas (CNG) vehicle market penetration [6]. This will create jobs for contractors and local mechanics. The National Gas Company (NGC) has invested over $2 billion to assist in the conversion of CNG vehicles over the next five years [6].

However, for RE technologies to be successful in T&T, key barriers must be overcome. The technical, legislative, financial, awareness, capacity building, economic, market, institutional and social barriers must be addressed before RE can become a sizeable component of Trinidad and Tobago's energy mix.

References

[1] EurochamTT (2015). Summary. http://eurochamtt.org/summary.html,accessed 5 February 2015.

[2] Ministry of Energy and Energy Affairs (2015). Development of a renewable energy policy framework. http://www.energy.gov.tt/our-business/alternative-energy/develop-

ment-of-a-renewable-energy-policy-framework, accessed 6 March 2015.
[3] Jobity, R. (2013). Tertiary natural gas workshop July/August 2013. http://ngc.co.tt/wp-content/uploads/pdf/NGC_Webinar_The%20Structure_History_and_Role_of_the_Natural_Gas_Industry_2013-08-22.pdf, accessed 12 October 2014.
[4] Ministry of Energy and Energy Affairs (2011).
[5] International Energy Agency (2006). Carrots and sticks: taxing and subsidizing energy. http://www.iea.org/publications/freepublications/publication/carrots-and-sticks-taxing-and-subsidising-energy.html, accessed 1 March 2015.
[6] The Energy Chamber of Trinidad and Tobago (2014). A closer look at Trinidad and Tobago's fuel subsidy. http://www.energy.tt/index.php?categoryid=355&p2001_articleid=1233, accessed 4 March 2015.
[7] T&T Express Newspaper (2013, March). Ramnarine: T&T's fuel subsidy needs attention. http://www.trinidadexpress.com/news/Ramnarine-TTs-fuel-subsidy-needs-attention-200535471.html, accessed 5 March 2015.
[8] InvesTT Limited, World Bank (2013, November 21). Cheaper, faster electricity connections power Trinidad's economy—See more at: http://www.investt.co.tt/blog/investt-blog/2013/november/world-bank-cheaper-faster-electricity-connections-power-trinidads-economy#sthash.PMCV0IDs.dpuf. http://www.investt.co.tt/blog/investt-blog/2013/november/world-bank-cheaper-faster-electricity-connections-power-trinidads-economy, accessed 4 February 2015.
[9] NaturalGas.org (2013, September 20). Natural gas and the environment. http://naturalgas.org/environment/naturalgas, accessed 7 March 2015.
[10] Inter-American Development Bank (2014, November 14). The Caribbean has some of the world's highest energy costs—now is the time to transform the region's energy market. http://blogs.iadb.org/caribbean-dev-trends/2013/11/14/the-caribbean-has-some-of-the-worlds-highest-energy-costs-now-is-the-time-to-transform-the-regions-energy-market, accessed 5 February 2015.
[11] T&T Guardian Newspaper (2012, November 5). Solar energy is expensive but worth it. http://www.guardian.co.tt/lifestyle/2012-11-04/solar-energy-expensive-worth-it, accessed 6 March 2015.
[12] Emanuel, E., Dillon, A., and P. Willard (2013). An assessment of fiscal and regulatory barriers to the deployment of energy efficiency and renewable energies in Grenada. Project, Trinidad and Tobago: United Nations.
[13] Parnell, J. (2015, January 19). Solar leading falling renewable energy costs, says IRENA. http://www.pv-tech.org/news/solar_leading_falling_renewable_energy_costs_says_irena, accessed 3 March 2015.
[14] Schwerin, A. (2013, July). Analysis of the potential solar energy market in the Caribbean. http://www.credp.org/Data/Solar_Market_Analysis_Caribbean.pdf, accessed 1 February 2015.
[15] Union of Concerned Scientists (1999). Barriers to Renewable Energy Technologies. http://www.ucsusa.org/clean_energy/smart-energy-solutions/increase-renewables/barriers-to-renewable-energy.html#.VPfMD_ldWSo, accessed 7 March 2015.
[16] Trinidad and Tobago Electricity Commission (2012, September). Watts happening? https://ttec.co.tt/magazine/2012/Watts%20Happening%20July-Sept-2012.pdf, accessed 6 November 2014.
[17] InvesTT (2014, February 27). Study pegs energy-rich Trinidad and Tobago as ideal manufacturing locale for solar panels. http://www.investt.co.tt/blog/investt-blog/2014/february/study-pegs-energy-rich-trinidad-and-tobago-as-ideal-manufacturing-locale-for-solar-panels, accessed 20 February 2015.

Chapter 8

Relationship between Nuclear Energy Consumption and Economic Growth

Korhan Gokmenoglu and Mohamad Kaakeh

This study empirically observed the nexus between nuclear energy consumption and economic growth in the country of Spain. Unit root and stationarity tests, Johansen cointegration tests, vector error correction models (VECM), and Granger causality tests were applied to annual data for the years 1968 to 2014. Empirical results confirmed the existence of a long-term equilibrium relationship between two variables and showed that nuclear energy consumption using Granger causality tests causes economic growth, which leads to important policy implications.

Spain is an oil-importing country. As environmental concerns increase, policy makers should consider increasing the share of nuclear power in the country's energy portfolio since it provides energy with no carbon emissions. However, policy makers should consider the safety measures and the correct disposal methodologies for nuclear waste.

INTRODUCTION

The industrial revolution, which emerged in Europe in the early 19th century, intensified energy demand because newly automated production processes required vast amounts of energy. To meet the production system's energy needs and enhance economic growth, coal and oil were primarily used [1]. Because of its higher efficiency oil supplanted coal. Rapid population growth, urbanization and new products intensified the demand for oil. Oil became vital to transportation and electrical generation. It became the first choice to satisfy global energy and economic demands because of its abundance and relatively low cost compared to other energy sources.

While oil satisfied the growing energy needs of the industrial era and contributed to global economic growth, it brought new challenges. Dependence on oil makes oil-importing countries vulnerable to oil price fluctuations. In the 1970s, two oil price shocks caused many developed economies to fall into recession and suffer from skyrocketing inflation, ultimately creating global economic instability [2]. Afterwards, oil markets were heavily affected by the formation of oil cartels, the Iraqi invasion of Kuwait in 1990, sanctions against Iran, the Russian and Asian financial crises, the Iraq war and the Arab Spring revolts. These and other related events made oil price stability a major concern for policy makers. Oil price volatility and national energy security became issues because of the social and political instability in oil-exporting countries.

Political tension in the oil exporting countries of Iraq, Syria, Yemen and Libya, endangers oil supplies. Problems with organic energy sources include their finite quantities, environmental impacts and nonrenewable nature. Decreasing oil reserves might reduce global energy supplies in the next few decades. Issues concerning reliance on fossil fuels include their negative environmental impacts. Fossil fuels produce gases such as carbon dioxide (CO_2) that harm the environment, increase the greenhouse effect, and cause global warming. If greenhouse gas emissions continue to increase during the next 30 years, global atmospheric temperatures will increase, surpass historical levels, harm the Earth's ecosystem and affect humanity [3]. To alleviate these concerns, maintain energy security and decrease oil dependence, it is crucial to take steps toward finding alternative sources to enhance energy diversification. Our world needs clean energy sources to ensure sustainable growth and development. Despite using a type of fossil fuel, civil nuclear energy is considered to be an important solution to this problem since it does not emit greenhouse gas emissions. For this reason, it is considered to be a form of clean energy [4].

Nuclear energy began to be used for electrical energy generation during the late 20th century. Its deployment grew rapidly and now provides over 11% of the world's electricity [5]. This increase results from several desirable properties and its advantages over other fossil fuels. Nuclear energy is produced using uranium. Unlike oil and natural gas, uranium reserves are plentiful. In addition to vast availability, the geographic distribution of uranium guarantees cost and supply stability [6]. The relatively even dispersion of uranium reserves throughout the world secures an uninterrupted energy supply and alleviates political concerns. The price of uranium has not fluctuated as frequently as oil and natural gas. Moreover, nuclear energy offers lower greenhouse gas emissions than oil or natural gas power plants [7]. Nuclear energy has an economic advantage when compared to certain renewable energy sources such as offshore wind and large-scale hydroelectric power—a lower levelized cost [8].*

In the long term, energy demand will increase as a result of population growth and ongoing industrialization. Additional energy-generating sources will be needed. Echávarri suggested that nuclear electricity generation will increase and play a major role in the future, especially if governments adopt energy policies to decrease greenhouse gas emissions [6].

This research investigates the effects of nuclear energy consumption on economic growth, using Spain as a case study to create policy recommendations. The widespread use of nuclear energy to generate electricity in Spain makes the country an interesting place to study the nuclear energy consumption–economic growth nexus. More than 20% of Spain's electricity is generated by seven nuclear reactors. As electricity is a major component of production processes, nuclear energy plays a vital role in Spain's economic growth [9].

Nuclear energy consumption has been affected by governmental policies. The Spanish government adopted an energy policy that aims to diversify the energy supply by integrating different energy sources including nuclear energy. Although the government is currently considering greater use of nuclear energy, its policies toward it have changed several times. For example, a 2014 decision to reduce the country's share of nuclear energy was replaced by a more favorable approach in 2015 when the government decided that the lifespans of nuclear power plants are only limited by the Nuclear Safety Council's safety standards. In other words, a nuclear plant can operate if it complies with international safety standards [10]. To guide governmental policy makers, it is important to conduct an in-depth analysis of the effects of nuclear energy consumption on economic growth. This study aims to investigate this relationship by employing several econometric tests. The Johansen co-integration test is applied to investigate the existence of the long-term relationship between selected variables. The vector error correction model is implemented to estimate the long-term coefficient, and the Granger causality test is used to discern the existence of any causal

*Geothermal, biomass, and land-based wind power have a lower levelized cost than uranium (*Annual Energy Outlook*, 2015).

relationships among the variables.

This chapter provides a literature review, describes the methodology, discusses the empirical results and summarizes our conclusions.

LITERATURE REVIEW

Researchers have proposed four hypotheses regarding the causal relationships between economic growth and energy consumption [11,13]. The growth hypothesis suggests that energy consumption affects economic growth directly or indirectly through capital and labor, as energy consumption plays a vital role in production processes [12-22]. The conservative hypothesis asserts that conservative energy policies that reduce energy consumption will not have an effect on economic growth [12-17,20,22-24]. The feedback hypothesis suggests that energy consumption and economic growth complement one another and are interrelated [11,13,17,20,22,23,25,26]. The neutrality hypothesis states that energy consumption has minimal to no effect on economic growth [13,15-17,22-24,27].

Kraft and Kraft conducted the first empirical research on the nexus between energy consumption and economic growth [28]. Using gross national product (GNP) as a proxy for economic growth, they determined that there is a long-term relationship between these variables and that there exists a causal relationship between GNP and energy consumption. Following the work of Kraft and Kraft, many researchers have investigated the energy consumption–economic growth nexus, using traditional energy consumption data, such as oil or natural gas consumption, as proxies for energy consumption. Yoo investigated the causal relationship between oil consumption and economic growth in Korea using the Granger causality method [29]. His findings suggest that increasing oil consumption would increase economic growth. Ziramba studied the same relationship in South Africa [30]. Using Toda-Yamamoto Granger causality test, his results showed that there is a causal relationship between oil consumption and economic growth. Using an innovative accounting approach for Granger causality, Shahbaz, Lean, and Farooq demonstrated that natural gas consumption creates economic growth and that conservative energy policies may slow down Pakistan's economy [31]. Lach studied the relationship between both natural gas consumption and oil consumption with economic growth [32]. His findings demonstrated that in the short term, both forms of energy consumption create economic growth. In the long term, it is the opposite. To have a better understanding regarding the relationship between energy use and economic growth, some researchers have used electricity consumption, coal consumption or total energy usage as proxies for energy consumption [20,33,34].

With recent advances in renewable energy, their effects on economic growth have been enlightening [35,36]. Sustainable development is one of the primary reasons many researchers have become interested in this relationship. As an important clean energy source, nuclear energy has also gained considerable attention. Al-mulali included a sample of 30 countries in a balanced panel for the years from 1990–2010 [25]. He found that there is a short-term causal relationship between nuclear energy consumption and economic growth. Apregis and Payne found the same results for the long term. In the short term, the relationship was bidirectional [26]. Applying Toda-Yamamoto Granger causality test to time series data, Payne and Taylor found that there was no causal relationship between nuclear energy consumption and economic growth in the United States [37,38]. Wolde-Rufael also applied Toda-Yamamoto Granger causality test on time series data from India and found that nuclear energy consumption causes economic growth [19,37]. The literature on the relationship between nuclear energy and economic growth appears limited. Researchers have not yet reached a consensus on the nature and direction of the linkage between these variables.

Empirical studies inspecting the relationship between nuclear energy consumption and economic growth in Spain are limited and lack consensus. The literature referencing Spain can be divided into two groups–panel and time series studies. Many of the most recent studies belong to the first group. Al-Mulali included Spain in his panel and found that there is a short-term bidirectional relationship between economic growth and nuclear energy consumption [25]. They concluded that nuclear energy consumption can increase economic growth without increasing CO_2 emissions. Apergis and Payne studied a panel that had 16 countries including Spain [26]. The authors found the same results as did Al-Mulali for the short term. However, they observed that for the long term there is a unidirectional relationship between nuclear energy consumption and GDP. Nazlioglu et al. examined a set of panel data containing Spain and concluded that nuclear energy consumption (using a Granger causality test) creates economic growth [16]. One of the main problems with the panel approach is that pooling the data might cloud the characteristics of each cross section and does not allow the individual properties of cross sections to be investigated in detail. If the sample size is large enough, it will be better to use time series analysis to better understand country-specific properties. Also, definitions of long term vary.

To our best knowledge, the number of time series analyses on nuclear energy consumption and economic growth in Spain is quite limited and yield contradictory results. Wolde-Rufael and Menyah investigated the relationship between nuclear energy consumption and economic growth in Spain using time series data [20]. They observed that the relationship between nuclear energy consumption and economic growth was bidirectional and followed the feedback hypothesis. Yoo and Ku analyzed the relationship between economic growth and nuclear energy consumption in Spain, but did not conduct a causality test on their sample [22]. However, the latest observations for both studies are from 2005.

There have been unprecedented developments in the energy market over the past decade. The global financial crisis of 2007–2009 affected energy markets, especially the market for oil. Before, during and after this crisis oil prices exhibited very high volatility. These prices increased to record high levels before the financial crisis, crashed during the financial crisis and fluctuated afterwards [39]. Uncertainties in the supply of and demand for energy and price fluctuations delayed some of the planned nuclear energy and infrastructure projects in the short term [6]. As the effects of the financial crisis faded, the demand for energy began to increase, once again increasing Spain's nuclear energy capacity. The Fukushima Daiichi disaster in 2011, initiated by the tsunami following the Tōhoku earthquake, highlighted environmental issues with nuclear plants. Given these observations, there is a need to examine the relationships between nuclear energy consumption and economic growth using post-crisis period data. The aim of our study is to fill this gap and obtain robust findings on the nuclear energy consumption–economic growth nexus in Spain.

Documenting literature regarding the causal relationships between nuclear energy consumption and economic growth was part of our research.

DATA AND METHODOLOGY

Data

To investigate the relationship between nuclear energy consumption and economic growth, data covering the period from 1968 to 2014 are used. Real GDP in constant prices of 2005 denominated in U.S. dollars were obtained from the World Bank and used as a proxy for economic growth. The nuclear energy consumption series was obtained from the BP Statistical Review of World Energy and is expressed in terms of terawatt-hours (TWh). The logarithmic forms of both variables were used.

Descriptive Statistics and Correlation Matrix

Table 8-1 shows the descriptive statistics of the series used in this study. Range and standard deviations of the nuclear energy consumption series are noticeable. This series varies from 0.079 (small numbers belonging to the early years of the data set) terawatt per hour to 63.71 terawatt per hour, which shows the growth in nuclear energy consumption. The correlation matrix shows that there is a positive relationship between nuclear energy consumption and economic growth.

Table 8-1. Descriptive statistics.

Statistic	NC	GDP
Mean	37.99	789,044**
Median	55.10	762,149**
Maximum	63.71	126,498**
Minimum	0.079	343,586**
Standard Deviation	24.69	2,864,191**
Skewness	-0.469	0.274
Kurtosis	1.385	1.713
NC*	1.000	0.857
GDP*	0.857	1.000

* *denotes correlation,* ** *denotes billions of dollars.*

Methodology

This study utilized time series econometrics techniques to examine the relationship between nuclear energy consumption and economic growth. To choose the most suitable econometric methods to investigate the long-term and causal relationships between the variables, stochastic properties of the variables should be investigated first. To determine if the series have a unit root, we used the Dickey and Fuller unit root test and the Kwiatkowski, Phillips, Schmidt, and Shinn stationarity test for confirmation [40,41]. Because all series have a unit root, the Johansen cointegration test was used to examine the existence of a long-term relationship among the variables [42]. Vector error correction models were used to estimate the long-term coefficients. The Granger causality test was applied to find the direction of the causal relationship.

Unit root tests. Prior to any econometric analysis, the order of integration of the variables must be investigated using unit root tests. To investigate the level of integration of the series, two tests were used based on the work of Dickey and Fuller and Kwiatkowski et al. [40,41]. The Augmented Dicky-Fuller (ADF) test has three different models which are represented by equations. The first does not account for intercept or slope (Equation 8-1). The second takes intercept into consideration (Equation 8-2), and the third includes both intercept and slope (Equation 8-3). Equation 8-1 is unlikely and the most restrictive, so the results of Equations 8-2 and 8-3 are seen as more reliable.

$$\Delta Y_t = \beta_1 + \gamma Y_{t-1} + \alpha \sum \Delta Y_{t-1} + \varepsilon_t \tag{8-1}$$

$$\Delta Y_t = \beta_1 + \gamma Y_{t-1} + \alpha \sum \Delta Y_{t-1} + \varepsilon_t \tag{8-2}$$

$$\Delta Y_t = \beta_1 + \beta_2 t + \gamma Y_{t-1} + \alpha \sum \Delta Y_{t-1} + \varepsilon_t \tag{8-3}$$

For these equations, where Yt is the tested variable, Δ is the change in the variable, t is time, and εt is the error term. $\beta 1$, $\beta 2$, γ, and α, are the coefficients to be estimated. The augmentation is conducted by adding the lagged values of the dependent variable ΔYt to confirm that the errors are not correlated with lag terms.

The null hypothesis of the ADF test is that a series has a unit root, and if rejected, the alternative hypothesis is accepted, indicating that the series is stationary.

H_0: $\gamma = 0$ (Y_t has a unit root)

H_1: $\gamma < 0$ (Y_t is stationary)

The most important criticism of the ADF test is that its power is low if the process is highly persistent but stationary. This problem becomes more severe if the sample size is small. A failure to reject the null hypothesis means that either the null hypothesis is true or there is insufficient information. For highly persistent data-generating mechanisms, the ADF test tends to not reject the null hypothesis. To solve this problem, the Kwiatkowski–Phillips–Schmidt–Shin (KPSS) test can be used to conduct a confirmation analysis. The null hypothesis of this test assumes stationarity, and in the test the data will appear stationary by default if there is little information about the sample. The null and alternative hypotheses of the KPSS test are the following:

$$H_o: \sigma^2\varepsilon = 0$$

$$H_1: \sigma^2\varepsilon > 0$$

The KPSS test has the same assumptions as the ADF test, but does not use t statistics. Instead, it uses the Lagrangian multiplier to test whether the error term has a constant variance. The regression models of the KPSS test and the multiplier are given respectively by:

$$Y_t = \beta_0 D_t + \mu_t + \varepsilon_t$$

$$\mu_t = \mu_{t-1} + \varepsilon_t (T^{-2} \sum^{T}_{t=1} \hat{S}^2_t) / \lambda \text{ and } S_t = \sum^{T}_{t=1} \varepsilon_j \tag{8-4}$$

Cointegration and error-correction model. If each variable has a unit root, a cointegration test should be conducted to investigate the existence of a long-term relationship between these variables. Because both of our series are nonstationary, the Johansen cointegration test, which uses the maximum likelihood approach, can be used to investigate the long-term relationships among series [42]. The Johansen cointegration test is based on the error correction representation of the vector autoregressive model and its examination.

$$Y_t = a_1 Y_{t-1} + a_2 Y_{t-2} + \dots\dots + a_k Y_{t-k} + \varepsilon_t \tag{8-5}$$

The cointegration test examines the existence of a long-term equilibrium between the variables, but does not give a numerical value for either the long- or short-term coefficients. To estimate the coefficients, error correction models are used. The error correction model is represented in the following equation:

$$\Delta Y_t = \beta_0 + \beta_1 \Delta X_{t-\pi} \hat{e}_{t-1} + e_t \tag{8-6}$$

$\beta 1$ shows the result of the speed of adjustment from the short-run disequilibrium to the long-run equilibrium, and $\hat{e}\ t\!-\!1$ is the error correction term.

Granger causality test. Granger was the first to study the direction of the causal relationships among variables [43]. The main logic behind the Granger causality test is that if the past values of variable *X* precede the value of variable *Y*, then there is causality from *X* to *Y*. The Granger causality test estimates the following equations to capture the causal relationship between the variables:

$$LnNC_t = \sum^{n}_{t=1} \alpha_i LnGDP_{t-1} + \sum^{n}_{j=1} \beta_i LnNC_{t-j} + \varepsilon_{1t} \quad (8\text{-}7)$$

$$LnGDP_t = \sum^{n}_{t=1} \lambda_i LnGDP_{t-1} + \sum^{n}_{j=1} \delta_i LnNC_{t-j} + \varepsilon_{2t} \quad (8\text{-}8)$$

The output of these equations will set the causal relationship between variables LnNC and LnGDP. There are four possible outcomes: If the coefficients of the lagged value of LnGDP are significant in Equation 8-7 and the coefficients of the lagged value of LnNC are not significant in Equation 8-8, the relationship would be unidirectional from LnGDP to LnNC. The second scenario is the opposite, with the unidirectional relationship direction from LnNC to LnGDP. If both coefficients of LnGDP in Equation 8-7 and LnNC in Equation 8-8 are significant, the causal relationship will be bidirectional. The last possible outcome is the nonexistence of any significant lagged value, indicating no causal relationship between LnNC and LnGDP.

EMPIRICAL FINDINGS

Unit Root and Stationary Test Results

The results of the unit root test and stationarity test are summarized in Table 8-2. Test findings revealed that both series are not stationary at their levels. The ADF test clearly shows that both variables are integrated of order one. The KPSS test results showed that both variables have unit roots at their level form; however, the first difference of these series is stationary, which indicates that both variables have no unit root. The general conclusion is that both series are integrated of order one. After confirming that the data are not stationary, the cointegration test was conducted to investigate the long-run relationship between the variables.

Table 8-2. Stationarity and unit root results.

Statistics (Level)	lnNC	Lag	lnGDP	Lag
τT (ADF)	-1.254	3	-2.033	1
τμ (ADF)	-1.290	3	-1.280	1
τ (ADF)	0.391	3	2.036	1
τT (KPSS)	0.216*	5	0.244***	4
τμ (KPSS)	0.760*	5	0.880*	5
Statistics (First Difference)	**lnNC**	**Lag**	**lnGDP**	**Lag**
τT (ADF)	-4.349*	6	-3.462***	0
τμ (ADF)	-4.507*	2	-3.392**	0
τ (ADF)	-4.692*	2	-2.665*	0
τT (KPSS)	0.112	4	0.092	4
τμ (KPSS)	0.458*	4	0.328	4

Note: NC represents nuclear energy consumption; GDP represents the gross domestic product. τT represents the intercept and trend model; τμ represents intercept model; τ represents the model without intercept and trend. Lag lengths used in ADF test to remove autocorrelation in the error terms are indicated under "lag" label.
, ** and * denote rejection of the null hypothesis at the 1%, 5%, and 10% levels respectively*

Cointegration Results

The Johansen cointegration method was applied to the nonstationary series to investigate the existence of a long-term equilibrium relationship between the variables. As Johansen suggested, Akaike's information criterion was used to choose lag lengths, and Pantula's method was used to select the optimal model [42]. The findings supported the existence of one cointegrating equation at 5% level of alpha, which means that nuclear energy consumption and gross domestic product have the tendency to move together in the long-term. The results for the Johansen cointegration test are indicated in Table 8-3.

Table 8-3. Cointegration test results between nuclear energy consumption and economic growth.

Null hypothesis	Eigenvalue	Trace Statistic	5% Critical Value	1% Critical Value
H_0: r = 0	0.280	23.824**	19.96	24.60
H_0: r ≤ 1	0.0181	8.994	9.24	12.97

(**) denotes rejection of the hypothesis at the 1%(5%) levels

Vector Error Correction Model Results

Based on the results of the Johansen cointegration test, a long-term relationship exists between nuclear energy consumption and economic growth. At this stage, the long- and short-term coefficients can be estimated by the vector error correction model (VECM). However, the VECM is sensitive to the number of lags included, so we must first determine optimal lag length. By using Schwarz and Hannan-Quinn information criteria, we found 2 for the optimal lag, as indicated in Table 8-4.

Table 8-4. VAR lag order selection criteria results.

Lag	AIC	SC	HQ
0	2.161	2.243	2.191
1	-5.238	-4.993	-5.148
2	-5.651	**-5.241***	**-5.500***
3	-5.549	-4.975	-5.337

*Note: * denotes lag order selected by the criterion*

Table 8-5 presents the VECM results. The speed-of-adjustment term is significant at the 5% level of significance. As indicated by the results in Table 8-6, lnNC converges to its long-term equilibrium level by 10.7% speed of adjustment every period by the contribution of lnGDP. The long-term coefficient is −1.607 and is significant at the 10% level of significance. No significant short-term relationship can be observed in the results.

Granger Causality Test Results

Granger causality test results are shown in Table 8-6. As the results indicate, the null hypothesis of noncausality was rejected. More specifically,

Table 8-5. Results on VECM output.

Result	Variable	Coefficient	Standard Error	t-Statistic
Speed of adjustment	ΔlnNC	-0.107	0.048	-2.210**
Short-run relationship	ΔlnGDP(-1)	1.354	1.717	0.788
	ΔlnGDP(-2)	-1.626	1.688	-0.963
Long-run relationship	lnGDP(-1)	-1.607	0.861	-1.866***

Note: *(**) (***) *denotes rejection of the hypothesis at the 1% (5%) (10%) level*

the results show that nuclear energy consumption Granger causes economic growth, indicating that it follows the growth hypothesis, which in turn points out that energy consumption (nuclear in this case) affects economic growth.

Table 8-6. Results of the Wald test.

Null hypothesis	Chi-square	df	Prob value
lnGDP Does not Granger cause lnNC	0.975	2	0.614
lnNC Does not Granger cause lnGDP	7.186	2	0.027*

Note: *(**) *denotes rejection of the hypothesis at the 1%(5%) level*

CONCLUDING REMARKS

Spain imports 95% of the oil it requires. The country is sensitive to nonrenewable energy sources because of volatility in oil and gas prices, political risks in oil exporting countries, environmental concerns associated with increased carbon emissions, and dependency on imported oil. Authorities should consider other sustainable and reliable energy resources to overcome these problems including nuclear energy. It is a reliable energy source with plentiful raw material (uranium) that lacks the price volatility associated with oil and natural gas. Using nuclear power can reduce greenhouse gas emissions, decrease the cost of generated electricity, and minimize reliance on imported oil.

The purpose of this study was to investigate the long-term and causal relationship between economic growth and nuclear energy consumption in Spain to determine policy implications. Information from 1968 to 2014 was analyzed using time series techniques. First, the data were confirmed to be integrated of order one using the ADF unit root test and the KPSS stationarity test. The Johansen cointegration test revealed that there was one cointegrating vector among variables that confirmed the existence of a long-term relationship. This long-term relationship confirms that economic growth is affected by the consumption of nuclear energy. VECM was used to estimate the long- and short-term coefficients. Lastly, the Granger causality test was conducted to determine the direction of the causal relationship. As the Granger results indicated, nuclear energy consumption causes economic growth, and the series follows the growth hypothesis, which confirms that any change in nuclear energy consumption directly affects the growth of Spain's economy. In light of this result, more attention needs to be paid to nuclear energy consumption policies, as slight changes in policies can have consequences on economic growth.

Our results show that there is a strong relationship between nuclear energy consumption and economic growth in Spain. For an energy-dependent country such as Spain, policy makers might seriously consider nuclear energy as a replacement for imported fossil fuels. To this aim, the Spanish government should increase the number of infrastructure projects and introduce policies that encourage private sector businesses, such as tax incentives, subsidies, and sales tax on energy produced by using nuclear power. These policies will enhance Spain's growth potential. At the same time, it will help to ensure a stable energy source, achieve lower greenhouse gas emissions, and create energy independence and diversification.

The government needs to be mindful that each energy resource has its own particular threats. The costs of nuclear energy development are not declining as has been the case with some forms of renewable energy, such as wind power, geothermal and solar energy, which have fewer negative externalities. To minimize the negative externalities associated with nuclear energy, such as water consumption, health concerns and environmental impacts, precautionary measures should be undertaken. Government research institutes and private organizations should support the development of more efficient technologies to minimize the cost of nuclear energy and nuclear waste disposal. The government should regulate resource extraction procedures, establish appropriate site selection procedures, verify designs for reactors and mandate operational safety. Proper long-term disposal and containment of nuclear wastes must be guaranteed.

It is important for Spain's future to balance clean energy, economic growth, and nuclear reliability so that its natural resources can be efficiently used.

References

[1] Wrigley, E. (2010). *Energy and the English industrial revolution.* Cambridge University Press.

[2] Kilian, L. and L. Lewis. (2011). Does the Fed respond to oil price shocks? *The Economic Journal,* 121(555), pages 1,047-1,072.

[3] Mora, C., Frazier, A., Longman, R., Dacks, R., Walton, M., Tong, E. and C. Ambrosino (2013). The projected timing of climate departure from recent variability. *Nature,* 502(7470), pages 183-187.

[4] Mallah, S. (2011). Nuclear energy option for energy security and sustainable development in India. *Annals of Nuclear Energy,* 38(2), pages 331-336.

[5] International Atomic Energy Agency (2014). IAEA annual report. www.iaea.org, accessed October 20, 2016.

[6] Echavarri, L. (2009). The nuclear energy option: How will the financial crisis affect nuclear energy? *OECD Observer,* 273, pages 52-53.

[7] Jewell, J. (2011). Ready for nuclear energy? An assessment of capacities and motivations for launching new national nuclear power programs. *Energy Policy,* 39(3), pages 1,041-1,055.

[8] Annual Energy Outlook (2015). www.eia.gov, accessed October 20, 2016.

[9] Chien, H. (2014). Crisis and essence of choice: explaining post-Fukushima nuclear energy policy making. *Risk, Hazards and Crisis in Public Policy,* 5(4), pages 385-404.

[10] International Energy Agency (2015). www.iea.org, accessed October 20, 2016.

[11] Apergis, N. and J. Payne (2010). Renewable energy consumption and economic growth: evidence from a panel of OECD countries. *Energy Policy,* 38(1), pages 656-660.

[12] Akhmat, G. and K. Zaman (2013). Nuclear energy consumption, commercial energy consumption and economic growth in South Asia: bootstrap panel causality test. *Renewable and Sustainable Energy Reviews,* 25, pages 552-559.

[13] Chang, T., Gatwabuyege, F., Gupta, R., Inglesi-Lotz, R., Manjezi, N. and B. Simo-Kengne (2014). Causal relationship between nuclear energy consumption and economic growth in G6 countries: evidence from panel Granger causality tests. *Progress in Nuclear Energy,* 77, pages 187-193.

[14] Aslan, A. and S. Çam (2013). Alternative and nuclear energy consumption-economic growth nexus for Israel: evidence based on bootstrap-corrected causality tests. *Progress in Nuclear Energy,* 62, pages 50-53.

[15] Chu, H. and T. Chang (2012). Nuclear energy consumption, oil consumption and economic growth in G-6 countries: bootstrap panel causality test. *Energy Policy,* 48, pages 762-769.

[16] Nazlioglu, S., Lebe, F. and S. Kayhan (2011). Nuclear energy consumption and economic growth in OECD countries: cross-sectionally dependent heterogeneous panel causality analysis. *Energy Policy,* 39(10), pages 6,615-6,621.

[17] Omri, A., Mabrouk, N. and A. Sassi-Tmar (2015). Modeling the causal linkages between nuclear energy, renewable energy and economic growth in developed and developing countries. *Renewable and Sustainable Energy Reviews,* 42, pages 1,012-1,022.

[18] Wolde-Rufael, Y. (2010). Coal consumption and economic growth revisited. *Applied Energy,* 87(1), pages 160-167.

[19] Wolde-Rufael, Y. (2010). Bounds test approach to cointegration and causality between nuclear energy consumption and economic growth in India. *Energy Policy,* 38(1), pages 52-58.
[20] Wolde-Rufael, Y. and K. Menyah. (2010). Nuclear energy consumption and economic growth in nine developed countries. *Energy Economics,* 32(3), pages 550-556.
[21] Yoo, S. and K. Jung (2005). Nuclear energy consumption and economic growth in Korea. *Progress in Nuclear Energy,* 46(2), pages 101-109.
[22] Yoo, S. and S. Ku (2009). Causal relationship between nuclear energy consumption and economic growth: a multi-country analysis. *Energy Policy,* 37(5), pages 1,905-1,913.
[23] Lee, C. and Y. Chiu (2011). Nuclear energy consumption, oil prices, and economic growth: evidence from highly industrialized countries. *Energy Economics,* 33(2), pages 236-248.
[24] Lee, C. and Y. Chiu (2011). Oil prices, nuclear energy consumption, and economic growth: new evidence using a heterogeneous panel analysis. *Energy Policy,* 39(4), pages 2,111-2,120.
[25] Al-Mulali, U. (2014). Investigating the impact of nuclear energy consumption on GDP growth and CO_2 emission: a panel data analysis. *Progress in Nuclear Energy,* 73, pages 172-178.
[26] Apergis, N., Payne, J., Menyah, K. and Y. Wolde-Rufael (2010). On the causal dynamics between emissions, nuclear energy, renewable energy and economic growth. *Ecological Economics,* 69(11), pages 2,255-2,260.
[27] Menyah, K. and Y. Wolde-Rufael (2010). CO_2 emissions, nuclear energy, renewable energy and economic growth in the U.S. *Energy Policy,* 38(6), pages 2,911-2,915.
[28] Kraft, J. and A. Kraft (1978). Relationship between energy and GNP. *Journal of Energy and Development* (U.S.), 3(2), pages 401-403.
[29] Yoo, S. (2006). Oil consumption and economic growth: evidence from Korea. *Energy Sources,* 1(3), pages 235-243.
[30] Ziramba, E. (2015). Causal dynamics between oil consumption and economic growth in South Africa. *Energy Sources, Part B: Economics, Planning, and Policy,* 10(3), pages 250-256.
[31] Shahbaz, M., Lean, H. and A. Farooq (2013). Natural gas consumption and economic growth in Pakistan. *Renewable and Sustainable Energy Reviews,* 18, pages 87-94.
[32] Lach, L. (2015). Oil usage, gas consumption, and economic growth: evidence from Poland. *Energy Sources, Part B: Economics, Planning and Policy,* 10(3), pages 223-232.
[33] Wolde-Rufael, Y. (2006). Electricity consumption and economic growth: a time series experience for 17 African countries. *Energy Policy,* 34(10), pages 1,106-1,114.
[34] Hondroyiannis, G., Lolos, S. and E. Papapetrou (2002). Energy consumption and economic growth: assessing the evidence from Greece. *Energy Economics,* 24(4), pages 319-336.
[35] Bobinaite, V., Juozapaviciene, A. and I. Konstantinaviciute (2015). Assessment of causality relationship between renewable energy consumption and economic growth in Lithuania. *Engineering Economics,* 22(5), pages 510-518.
[36] Inglesi-Lotz, R. (2016). The impact of renewable energy consumption to economic growth: a panel data application. *Energy Economics,* 53, pages 58-63.
[37] Toda, H. and T. Yamamoto (1995). Statistical inference in vector autoregressions with possibly integrated processes. *Journal of Econometrics,* 66(1), pages 225-250.
[38] Payne, J. and J. Taylor (2010). Nuclear energy consumption and economic growth in the U.S.: an empirical note. *Energy Sources, Part B: Economics, Planning, and Policy,* 5(3), pages 301-307.
[39] Bhar, R. and A. Malliaris (2011). Oil prices and the impact of the financial crisis of 2007–2009. *Energy Economics,* 33(6), pages 1,049-1,054.

[40] Dickey, D. and W. Fuller (1981). Likelihood ratio statistics for autoregressive time series with a unit root. *Econometrica: Journal of the Econometric Society,* pages 1,057-1,072.

[41] Kwiatkowski, D., Phillips, P., Schmidt, P. and Y. Shin (1992). Testing the null hypothesis of stationarity against the alternative of a unit root: How sure are we that economic time series have a unit root? *Journal of Econometrics,* 54(1), pages 159-178.

[42] Johansen, S. and K. Juselius (1990). Maximum likelihood estimation and inference on cointegration-with applications to the demand for money. *Oxford Bulletin of Economics and Statistics,* 52(2), pages 169-210.

[43] Granger, C. (1969). Investigating causal relations by econometric models and cross-spectral methods. *Econometrica: Journal of the Econometric Society,* pages 424-438.

[44] Zhang, X. and X. Cheng (2009). Energy consumption, carbon emissions and economic growth in China. *Ecological Economics,* 68(10), pages 2,706-2,712.

Chapter 9

Renewable Energy Potentials in Regions of Hungary

Bálint Hartmann, Ph.D., Attila Talamon, Ph.D., Viktória Sugár

The national "Energy map (E-map)" project (KEOP 7.9.0/12-2013-0017) created an interactive energy map for Hungary to serve as the basis for financial resource use planning and implementation from 2014 to 2020. The map incorporated all available regional energy data in sectoral components, using statistical databases and graphical information system (GIS) methodologies.

This chapter introduces the methodology and selected results of the evaluation process. In the first part, definitions of theoretical, technological and economic potentials of renewable energy sources are created, using the National Renewable Action Plan of Hungary. The assessment considers hydropower, wind energy, solar, geothermal and biomass energy potentials. It discusses how the regional potential of each energy source was determined, using available databases. The second part of the chapter provides an overview of the results. Finally, we offer an update on renewable energy policy changes that have occurred in Hungary.

INTRODUCTION

Studies that evaluate the utilization possibilities of renewable energy sources often estimate the potential of each energy source. These studies typically offer a very broad range of estimates. This occurs due to the varying ecological approaches used, resulting in differing definitions of potential concepts and methodologies. For this reason, before estimating Hungary's renewable energy potential, the definitions used in this chapter are detailed.

The initial estimates of the potential of renewable energy sources are usually similar in terms of methodology, since they are based on measurable physical parameters (e.g., irradiation, wind speed, flow rate or geothermal gradient). The physically available volume of energy is usually considered to be the *theoretical potential*. Proper determination of the theoretical potential is essential since it provides input data for other quantities or subsets.

One of these subsets is the *ecologically sustainable potential*, whose definition is closely related to the purposes of the study. This potential volume is used to estimate existing and future project possibilities. The methodology for determining the ecologically sustainable potential can only be created separately for each energy source. Literature shows wide variances in such cases. There are three general processes: bottom-up assessments, top-down assessments by creating regulatory and other boundary conditions, and estimates applying international best practices and analogies. The efficiency of these methods is largely dependent on the type of renewable energy source: bottom-up assessments are often used to examine biomass while top-down assessments are commonly used for wind farms.

Another widely used subset of theoretical potential is the *economic potential*, which assesses the volume of economically usable renewable energy sources. This estimation of potentials is complex; its volume is affected by available technologies, energy demand, existing infrastructure, regulatory aspects and possibly existing support schemes. For this reason, it is reasonable to examine the economic potential only for concrete projects, where locally or regionally available energy sources are well known.

The combination of the two subsets, economic and ecologically sustainable potentials, is also called the *sustainable potential* which requires long term social and environmental sustainability.

The definition of *convertible potential* refers to the volume of energy demand that can be supplied by currently existing technologies. For such estimates, knowledge is required of the main parameters of energy generation technologies which may include the conversion efficiencies and annual full load operating hours.

For our purposes, the National Renewable Energy Action Plan (NREAP) was used including its estimates of potentials, thus its definitions also required interpretation [1]. For this study, sustainable potential (i.e., the sustainable potential coordinated with technical, economic, social and ecological aspects) was calculated using either the long-term (until 2030) realizable volume or the medium-term (until 2020) realizable volume chosen as targets. We emphasize that such a refined distinction is necessary since the thematic order of long-term plans usually provides estimates for a distant horizon, while concrete targets and action plans are deduced from shorter time periods. Such methodologies are useful in case of Hungary, since the targets defined by the NREAP are below the mandatory European Union (EU) targets. Thus, possible increases should be evaluated not only until 2020 but also beyond. The potential definitions, used in the background study of the NREAP are shown in Figure 9-1.

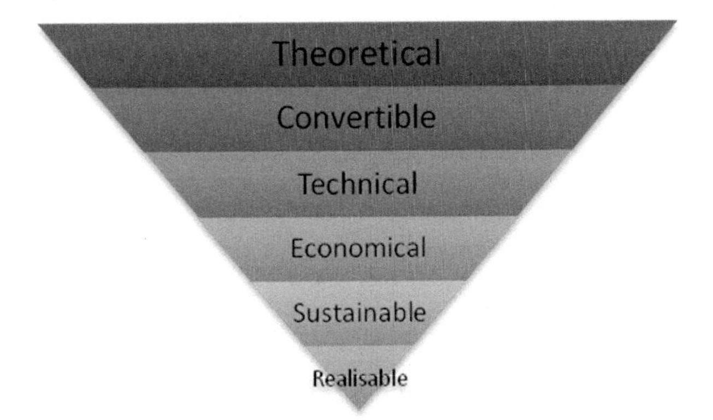

Figure 9-1. Hierarchical position of renewable energy potential concepts (NREAP).

- *Theoretical potential*—the physically modifiable energy source volume and structure, without accounting for constraints (elements: solar, hydropower, biomass, wind and geothermal energy).
- *Convertible potential*—the long-term total of usable volumes, taking into account constraints (mostly financial and regulatory).
- *Technical potential*—the volume of energy sources that can be utilized optimally by available, up-to-date technologies.
- *Economic potential*—the long-term (through 2030) volume of energy sources that can be utilized in an economically feasible manner under certain circumstances.
- *Sustainable potential*—the long term (through 2030) volume of suggested energy source volume and structure, coordinated with technical-economic-social-ecological aspects, consisting of optimal technologies.

In the background study, the sum of these potential concepts results in the quantities provided in Table 9-1.

The long-term technical potential of renewable energy sources can be estimated through 2030. Their combined volume is approximately 500 PJ/a. A slightly smaller number (425 PJ/a) represents the economic potential. The study, performed by the authors, was primarily based on these categories, noted in bold in Table 9-1. The long-term sustainable potential is estimated to be 350 PJ/a; it can be seen that this is the first column where sources other than biomass play a relevant role in future plans. The medium-term sustainable potential resulted as 237.5 PJ/a.

ASSESSMENT OF POTENTIALS

Hydropower Energy

The theoretical, technical and economic hydropower energy potential of Hungary is estimated by the literature as 100, 20 and 10 PJ/a, respectively.

Estimating hydropower energy potential is relatively simple compared to estimates for other sources since only a limited number of possible installation locations need to be examined. Numerous measurements are available from hydropower sites. Approximately 90% of total Hungarian hydropower energy potential can be exploited from three rivers (Danube, Tisza and Dráva), while the remaining 10% is distributed among twelve others. Only the shares of Rába, Hernád and Sajó rivers exceed 1% each, and the majority of micro hydro plants are installed on the first two (see Table 9-2).

Long-term plans for hydropower in Hungary can be considered in four major groups: micro hydro plants, existing hydro plants and dams, small hydropower plants and plants on the Danube. Annual estimated GWh production quantities are provided for each site.

The first group are micro hydro plants, installed on small rivers, with installed capacities of a few MWs. According to the literature, twelve units can be constructed: five each on the Hernád (4.8 GWh total) and Sajó (5.2 GWh total) rivers with one each on the Körös (10 GWh) and Maros (12 GWh) rivers.

Another 85 GWh of production is expected from the development of existing hydro plants and dams. This group includes the power plants of Békésszentandrás (12.5 GWh), Nick (5 GWh), Tass (3.1 GWh), Dunakiliti (28.4 GWh), Kisköre (26 GWh) and Tiszalök (12 GWh).

Small hydro plants are planned for the Tisza river. According to hydrological assessments, three units should also be constructed in the areas of

Table 9-1. Theoretical, convertible, technical, economic, sustainable and realizable potential of renewable energy sources, as used in the NREAP.

Potential concept	Theoretical potential [PJ/a]	Convertible potential to 2050 [PJ/a]	Technical potential to 2030 [PJ/a]	Economic potential to 2030 [PJ/a]	Sustainable potential to 2030 [PJ/a]	Sustainable potential (existing and installable) to 2020 [PJ/a]
Solar heat	417,600	103	75	65	50	15
Solar electricity		1,749	50	25	15	7
Hydro electricity	100	27	20	15	5	2.3
Wind electricity	36,000	532	30	25	20	15.5
Wind combined head and electricity						
Biomass heat	420-500	203-328	150	180	180	150
Biomass electricity						
Biomass fermentation biogas			50	30	20	13.2
Other biomass (waste)			15	10	8	4.3
Geothermal electricity	102,180,000	343,000	20	15	12	6.1
Geothermal heat plant			30	25	20	13.2
Heat pump			35	30	15	10
Other combined		100	25	15	5	1
Total		**345,839**	**500**	**425**	**350**	**237.5**

Table 9-2. Theoretical hydropower energy potential of Hungarian rivers.

River	Duna	Tisza	Dráva, Mura	Rába	Hernád	Other
Average theoretical potential [GWh/a]	5,348	708	756	187	139	308
Share [%]	72	9.5	10	2.5	1.9	4.1

Dombrád (100 GWh), Vásárosnamény (90 GWh) and Csongrád (90 GWh).

Long-term plans for hydropower projects primarily focus on the Danube's potential. The smallest of these units is the hydropower plant in The Hague which has allocated another 1,000 GWh of the Gabcikovo hydropower plant for use by Hungary.

The present small utilization is reflected by the totals: realization of the previous plans would represent surplus electricity totaling 2,872 GWh compared to the present annual volume of 200 GWh.

These generation potentials can be assigned to administrative units (districts) since the plant locations are in most cases known. The exceptions are rivers Sajó and Hernád, where potentials are assigned to districts in proportion to river lengths. In Equations 9-1 and 9-2, E represents the potential, $rkmj$ and $rkmi$ are the river's distance from the first and last settlements in the examined district.

$$Ehydro,Sajó,i = (rkmj - rkmirkm) \times E\ hydro,Sajó \quad (9\text{-}1)$$

$$Ehydro,Hernád,i = (rkmj - rkmirkm) \times E\ hydro,Hernád \quad (9\text{-}2)$$

In the case of the Gabcikovo power plan, the potentials cannot be assigned to local use, since most of the volume is generated in Slovakia rather than Hungary. Therefore, it was assigned to the nearest district, Mosonmagyaróvár.

In total, hydro energy potentials were assigned to 27 of 178 districts, but only the districts of Adony, Kalocsa and Mosonmagyaróvár represent volumes greater than one PJ/a.

Wind Energy

The theoretical, technical and economic wind energy potentials for Hungary are estimated by the literature as 36,000, 30 and 25 PJ/a, respectively.

Wind characteristics have a key role when estimating a region's wind energy potential. A location's average wind speeds are internationally accepted parameters. Locations where wind speeds are less than 5 m/s are generally not considered to be feasible for development. Using international benchmarks, locations worthy of consideration are typically those with average wind speed levels reaching 5-7 m/s at 50 meters above ground level. In practice, investors usually have higher thresholds and choose the best available locations. In Hungary, some of these have already been developed. Investment thresholds and development locations are also affected by the local price of electricity and the availability of subsidies.

To assign energy potentials to districts, wind speed data for 50, 100 and 200 meters above ground level from the Global Wind Atlas (Technical University of Denmark) were used. Local wind speed values of distinct geographic locations are determined by microscale modeling, using WAsP (Wind Atlas Analysis and Application Program). WAsP considers the terrain of the ground which is needed to properly calculate the Hellmann Coefficients. The Atlas was created with 250 m spatial resolution, while available data uses one kilometer resolution which suited our purposes. From the available levels, data for 100 m above ground level was used as it is approximately equal to the hub height of many modern wind turbines.

To assess regional potentials, wind speeds of the district seats were recorded, while potential volumes were assigned in proportion to the geographic areas of the districts. In Equation 9-3, E represents the potential, v the wind speed and A is the area of the examined district.

$$E_{wind,i} = \frac{v_i \times A_i}{\sum^{175}_{j=1} v_j \times \sum^{175}_{j=1} A_j} \times E_{wind,country} \qquad (9\text{-}3)$$

The best wind speeds in Hungary are found in Győr-Moson-Sopron and Komárom-Esztergom counties. Wind characteristics are also favorable in Transdanubia. Detailed wind potential distributions are shown in Figure 9-2.

Solar Energy

Theoretical, technical, and economic solar energy potentials for Hungary are estimated by the literature to be 417,600, 125 and 90 PJ/a, respectively. For technical potential, the proportion of heat energy to electricity is

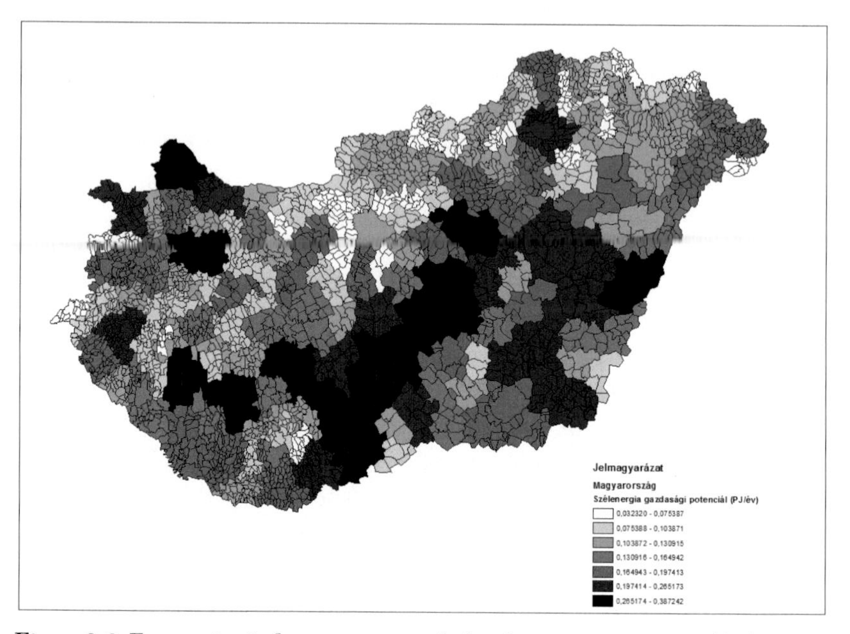

Figure 9-2. Economic wind energy potentials for electricity generation (darker areas represent better potentials, ranging from 0.0323 and 0.387 PJ/a).

75:50 PJ, while the ratio for economic potential is 65:25 PJ.

For regional assessment of potentials, the Global Solar Dataset of 3TIER was used. This provides annual average power (W/m^2) of global horizontal irradiation with a spatial resolution of three kilometers. This dataset provides a decade of data using analysis results from satellite images taken in 30-minute intervals. The satellites recorded the visible light spectrum and the images were processed by 3TIER.

When assigning potential volumes for solar energy we assumed that both the technical and economic potentials correlate with the geographic areas of districts. The reason for this assumption is that for both solar collectors and photovoltaic panels, residential use and small installed capacities are expected to dominate the market. Furthermore, the solar irradiation values of Hungarian regions vary only slightly and installations are not expected to be limited to only a few districts. Each district's potential was calculated using the solar irradiation levels recorded at the district seats, while potential volumes were assigned in proportion to geographic areas of each district. In Equations 9-4 and 9-5, E represents the potential, I the irradiation and A the total area of the examined district.

$$E_{solar,electricity,i} = \frac{I_i \text{ x } A_i}{\Sigma^{175}{}_{j=1}\, I_j \text{x } \Sigma^{175}{}_{j=1} A_i} \text{ x } E_{solar,elecrtricity,country} \tag{9-4}$$

$$E_{solar,heat,i} = \frac{I_i \text{ x } A_i}{\Sigma^{175}{}_{j=1}\, I_j \text{x } \Sigma^{175}{}_{j=1} A_i} \text{ x } E_{solar,heat,country} \tag{9-5}$$

The distribution of regional potentials is correlated with Hungary's solar irradiation map, thus significant utilization of solar energy is anticipated in districts located in the southern great plain region. The gap between national maximum and minimum is far less than for wind energy. Figure 9-3 shows the geographical distribution of economic potential for solar photovoltaics.

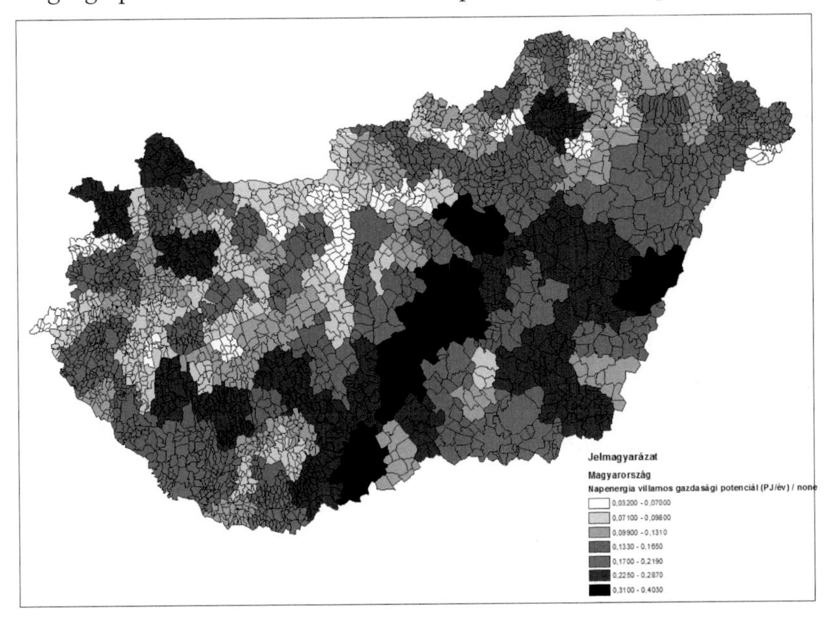

Figure 9-3. Economic solar energy potentials for electricity generation (darker areas represent better potentials, ranging from 0.0320 to 0.4030 PJ/a).

Geothermal Energy

Theoretical, technical and economic geothermal energy potential for Hungary is estimated by the literature as 102,180,000, 85 and 80 PJ/a, respectively. For technical and economic potential, the proportion of power plant, heating plant, and heat pump utilization is 20:30:35 and 15:25:30 PJ/a, respectively.

For geothermal energy, the higher temperature of the Earth's upper layers is utilized, as the energy of steam or water is drawn to generate electricity and/or heat. The best utilization of the temperature sources are given by the Lindal-diagram. For electricity generation, high-temperature sources (above 180°C) are best, since these can directly drive a power plant's turbine and provide base load generation. The low-temperature geothermal resources are suitable for heat pumps and cool tubes. Large-scale heat energy and electricity production requires resources reaching 40°C (104°F) and 120°C (248°F), respectively. The scale of utilization varies since heat pumps are predominantly used for residential and small commercial applications while heating plants and electrical power plants have large installed capacities.

To assign the potential volumes for each region, the GeoElec information service of the European Geothermal Energy Council was used. The GeoElec project operated between 2011 and 2013 as a consortium of ten partners to support geothermal energy use in Europe. This project assessed the potential of geothermal energy within a 20 km spatial and 250 m vertical resolution (below ground). GeoElec's website provides theoretical potential volumes and technical potentials depending on certain economic boundary conditions and life cycle costs.

The distribution of regional potentials was performed using the potentials of district capitals, while potential volumes were assigned in proportion to the geographic areas of the districts. In Equations 9-6 and 9-7, E represents the potential, Pg is the per area power of geothermal resources and A is the area of the examined district.

$$E_{geothermal,electricity,i} = \frac{Pg_i \times A_i}{\Sigma^{175}{}_{j=1}\, Pg_{ij} \times \Sigma^{175}{}_{j=1} A_i} \times E_{geothermal,electricity,country} \quad (9\text{-}6)$$

$$E_{geothermal,heat,i} = \frac{Pg_i \times A_i}{\Sigma^{175}{}_{j=1}\, Pg_{ij} \times \Sigma^{175}{}_{j=1} A_i} \times E_{geothermal,heat,country} \quad (9\text{-}7)$$

Based on the results, the best locations for geothermal energy in Hungary are in Hajdú-Bihar, Tolna and Bács-Kiskun counties and their neighboring districts. The distribution of economic potential for electricity generation is shown on Figure 9-4.

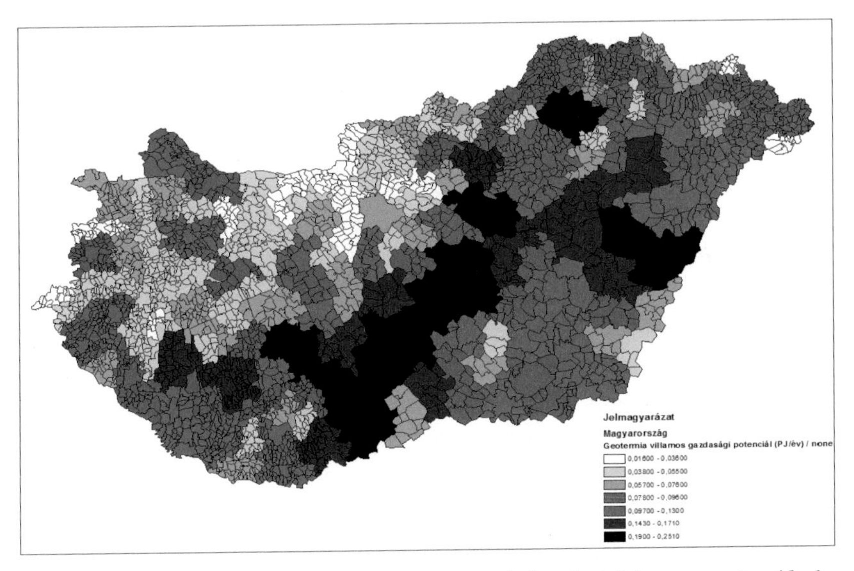

Figure 9-4. Economic geothermal energy potentials for electricity generation (darker areas represent better potentials, ranging between 0.0160 and 0.2510 PJ/a).

Bioenergy

Among all renewable energy resources, biomass assessments are the most complex. The reasons for this include the wide variation of possible combustible fuels, the large palette of energy generation technologies and higher levels of uncertainty using available data. We made several assumptions which are detailed below.

Studies estimating Hungary's biomass potential typically fail to separately identify biomass from biogas resources. However, technical and economic biomass potentials are separable while heat energy and electricity production can be allocated.

To assess regional biomass potentials, both top-down and bottom-up approaches were used. Since information was unavailable to allocate each fuel's proportion of generated heat and electricity production using the NREAP (and its background studies), we used our own assumptions. Our methodology was based on seven categories of potentially used biomass:

1. Wooden biomass
2. Energy crops

3. Agricultural by-products
4. Waste incineration
5. Biogas from organic material (manure)
6. Biogas from organic waste
7. Biogas from sewage

Since the NREAP handles category 4 separately, no further division was necessary. For waste incineration, combined heat and power (CHP) production was assumed. Typical efficiencies using CHP require 100 units of primary energy to generate 45 units of heat and 34 units of electricity (i.e., the ratio of latter secondary products is 57:43). All CHP producers were assumed to have the same efficiencies.

Biogas potentials were estimated for the gross volumes, without allocation between categories 5, 6 and 7. To assess the potentials of all sources, proportions found in the literature were used. Based on these, biogas generated from organic material (category 5) represents 69.7%, biogas from organic waste (category 6) represents 15.8%, and the remaining 14.5% is from sewage (category 7). For all three categories gas engines and CHP production were the assumed technologies. Considering the three remaining categories (1, 2 and 3), a literature review was performed based partly on the information of the NREAP.

After comparing these data sources, the respective proportions of wooden biomass, energy crops and agricultural by-products were 25%, 41% and 34%. Additional subgroups could have been created, based on the technologies used. As this would not have significantly affected the results, CHP generation was assumed for all fuels.

To summarize, three homogeneous groups were created (1-2-3, 4, 5-6-7). If 100% share is assumed, the technical and economic potentials for biomass and biogas can be divided based on fuel source. This division is shown in Table 9-3, highlighting the homogenous subgroups with different backgrounds.

Separate handling of the subgroups was important throughout the examinations, since the assignment of potential volumes to districts is performed based on the methodologies. These methodologies are discussed next.

Using the proportions of Table 9-3, exact potential volumes can be determined, as shown in Tables 9-4. and 9-5.

Table 9-3.
Fuels and secondary utilization based grouping of biomass potential.

	Electricity	**Heat energy**
Biomass		
1: Wooden biomass	10.76%	14.24%
2: Energy crops	17.65%	23.35%
3: Agricultural by-products	14.63%	19.37%
4: Waste incineration	57%	43%
Biogas		
5: Biogas from organic material (manure)	29.97%	39.73%
6: Biogas from organic waste	6.79%	9.01%
7: Biogas from sewage	6.28%	8.32%

Table 9-4.
Fuels and secondary utilization based grouping of technical biomass potential.

	Electricity	**Heat energy**
Biomass		
1: Wooden biomass	10.76%	14.24%
2: Energy crops	17.65%	23.35%
3: Agricultural by-products	14.63%	19.37%
4: Waste incineration	57%	43%
Biogas		
5: Biogas from organic material (manure)	29.97%	39.73%
6: Biogas from organic waste	6.79%	9.01%
7: Biogas from sewage	6.28%	8.32%

Table 9-5.
Fuels and secondary utilization based grouping of economic biomass potential.

	Electricity	**Heat energy**
Biomass		
1: Wooden biomass	19.37 PJ/a	25.63 PJ/a
2: Energy crops	31.77 PJ/a	42.03 PJ/a
3: Agricultural by-products	26.33 PJ/a	34.87 PJ/a
4: Waste incineration	5.70 PJ/a	4.30 PJ/a
Biogas		
5: Biogas from organic material (manure)	8.99 PJ/a	11.92 PJ/a
6: Biogas from organic waste	2.04 PJ/a	2.70 PJ/a
7: Biogas from sewage	1.88 PJ/a	2.50 PJ/a

Wooden Biomass

To assess wooden biomass potentials, the volume of annual national forestry production was allocated in proportion to the geographic areas of the districts. The calorific value of wood was calculated using the produced volume of different species as 9.7 GJ/m^3, while industrial statistics publish the share of firewood as 52.6%. A district's share of can be calculated as:

$$E_{biomass1,share,i} = [(A_{forest,county} \text{ x } A_i/A_{forest,county})/A_{forest,country} \text{ x } V_{wood,country}] \text{ x } q_{production} \text{ x } q_{firewood} \text{ x } C_{wood} \text{ x } (1/10^3) \quad (9\text{-}8)$$

Annual potential volume of a district is calculated as:

$$E_{biomass1,electricity,i} = \frac{E_{biomass1,share,i}}{\sum_{j=1}^{175} E_{biomass1,share,j}} \text{ x } 10.76\% \text{ x } E_{biomass1\text{-}2\text{-}3,country} \quad (9\text{-}9)$$

In Equations 9-8 and 9-9, E is the potential, V is the volume of produced wood and A is the geographical area. The two coefficients ($q_{production}$ and $q_{firewood}$) represent the share of total production compared to growth volume and the share of firewood from production respectively. The magnitudes of these coefficients are 0.607 and 0.562 respectively, based on data of the Hungarian Central Statistical Office. C_{wood} is the estimated average calorific value of total produced wood. To determine this the following calorific values were used: 10.5 GJ/m^3 for oak, tern, beech, hornbeam, acacia and other hardwoods, 9.0 GJ/m^3 for pine species and other wood, 8.0 GJ/m^3 for poplar species and willow. These values average 9.7 GJ/m^3. A 10.8% multiplier provides the assessed potential that can be used as the potential for the examined fuel subgroup to produce electricity (see Table 9-3). Heat energy potentials can be calculated in a similar manner.

Energy Crops

To assess the biomass potential from energy crops, the primary data needed is the area of unused land (withdrawn from cultivation) as such land can be used to grow energy crops. District potentials were determined in proportion to geographic areas. To estimate the yield potentials, an average was calculated from the per area yield of typical plants, resulting in 12.5 t/

ha. The calorific value (16 MJ/kg) was determined according to the conditions of mixed flora. A district's potential share can be calculated as:

$$E_{biomass2,share,i} = [(A_{unused,county} \times A_i/A_{county})/A_{unused,county}] \times q_{yield} \times C_{energy} \times (1/10^3) \quad (9\text{-}10)$$

Annual potential volume of a district is calculated as:

$$E_{biomass2,electricity,i} = \frac{E_{biomass2,share,i}}{\sum_{j=1}^{175} E_{biomass2,share,j}} \times 17.65\% \times E_{biomass1\text{-}2\text{-}3,country} \quad (9\text{-}11)$$

For Equations 9-10 and 9-11, E is the potential, A is the land area, and q_{yield} is the estimated yield of energy crop plantations (approximately 12.5 t/ha for mixed woody and herbaceous tillage). $C_{energycrop}$ is the estimated average calorific value (16 MJ/kg) of produced biomass assuming both woody and herbaceous plants. The 17.65% multiplier indicates the proportion of known total potential that can be assessed as the potential for the specific fuel subgroup to produce electricity (see Table 9-3). Heat energy potentials can be calculated in a similar manner.

Agricultural By-products

Statistical data for several agricultural products were chosen to assess the volume of agricultural waste and by-products (i.e., wheat, maize, barley, oat and triticale). The size and the yield of arable land were determined for each county's crops. The first parameter was divided in proportion of the area of districts, while the second was handled as a homogeneous value for each county. The volume of utilizable by-products collected from the fields is assumed to reach 68% of the product volume; its calorific value was 14 MJ/kg. The potential share of a district can be calculated as:

$$E_{biomass3,share,i} = [\Sigma^{n}_{i=1} (A_{cropi,county} \times A_i/A_{county}) \times q_{cropi,yield,county}] \times q_{by\text{-}product} \times C_{agr.by\text{-}product} \times (1/10^6) \quad (9\text{-}12)$$

Annual potential volume of a district is calculated as:

$$E_{biomass3,electricity,i} = \frac{E_{biomass3,share,i}}{\sum^{175}_{j=1} E_{biomass3,share,j}} \times 14.63\% \tag{9-13}$$

$$\times E_{biomass1-2-3,country}$$

For Equations 9-12 and 9-13, E is the potential, A is the land area, q_{yield} is the estimated yield of crops in the examined county, and $q_{by\text{-}product}$ is the volume of by-products compared to the product (68%). $C_{agr.by\text{-}product}$ is the estimated average calorific value (14 MJ/kg) of biomass. The 14.63% multiplier indicates the proportion of known assessable potential for the examined fuel subgroup to produce electricity (see Table 9-3). Heat energy potentials can be calculated in a similar manner.

Waste Incineration

Incineration potential of solid municipal waste can be assessed by the volume of combusted waste (which is the currently utilized part of the potential), available for all counties, and the volume of landfill waste (which sets an upper limit). Both volumes were divided among districts in proportion to their population, assuming that no significant difference can be observed among the waste production of the regions. The calorific value of municipal solid waste was calculated as 16 MJ/kg. A district's potential share can be calculated as:

$$E_{biomass4,share,i} = [(Q_{incinerated,county} \times L_i/L_{county}) + (Q_{lerakott,county} \times L_i/L_{county})] \times C_{waste} \times (1/10^6) \tag{9-14}$$

Annual potential volume of a district is calculated as:

$$E_{biomass4,electricity,i} = \frac{E_{biomass4,share,i}}{\sum^{175}_{j=1} E_{biomass4,share,j}} \times 43\% \times E_{biomass4,country} \tag{9-15}$$

For Equations 9-14 and 9-15, E is the potential, L is the population of the district, and m is the mass of the waste. C_{waste} is the estimated average calorific value (16 MJ/kg) of biomass. The 43% multiplier indicates the proportion of known potential that can be assessed as a potential for the examined fuel subgroup for electricity production (see Table 9-3). Heat energy potentials can be calculated in a similar manner.

Biogas from Organic Materials

Biogas raw material from organic materials is dominantly manure and small amounts of industrial organic waste. Our assessment emphasized wastes from the first group since there was limited information on the second. Among Hungarian livestock, cattle, swine and poultry were considered. Each county's waste volumes from these animals were divided among the districts in proportion to their geographical area. Manure of different species not only differs in volume but also in composition, affecting the quality of bedding. Using estimates from research, per unit biogas yield was set as 160, 26.7 and 0.3255 m^3 for cattle, swine and poultry, respectively. The calorific value of biogas was estimates to be 22 MJ/m^3. Potential share of a district can be calculated as:

$$E_{biomass5,share,i} = [(Q_{cattle,county} \times A_i/A_{county} \times q_{cattle,yield}) + (Q_{swine,county} \times A_i/A_{count} \times q_{swine,yield}) + (Q_{poultry,county} \times A_i/A_{county} \times q_{poultry,yield})] \times C_{biogas} \times (1/10^6) \quad (9\text{-}16)$$

Annual potential volume of a district is calculated as:

$$E_{biomass5,electricity,i} = \frac{E_{biomass5,share,i}}{\sum_{j=1}^{175} E_{biomass5,share,j}} \times 29.97\% \times E_{biomass5\text{-}6\text{-}7,country} \quad (9\text{-}17)$$

For Equations 9-16 and 9-17, E is the potential, A is the area of the district, and Q is number of animals. The coefficient q_{yield} represents biogas yield per animal while C_{biogas} is the estimated average calorific value (22 MJ/m^3) of biogas. The 29.97% multiplier indicates the proportion of known potential that can be assessed as a potential for the examined fuel subgroup for electricity production (see Table 9-3). Heat energy potentials can be calculated in a similar manner.

Biogas from Organic Wastes

Biogas from organic waste can be handled similarly to municipal waste, since the raw material is the landfill waste in both cases. Biogas yield of municipal solid waste is approximately 100 m^3/t. The potential share of a district can be calculated as:

$$E_{biomass6,share,i} = \frac{m_{landfill,county} \times L_i}{L_{county}} \times q_{yield} \times C_{biogas} \times (1/10^9) \quad (9\text{-}18)$$

Annual potential volume of a district is calculated as:

$$E_{biomass6,electricity,} \frac{E_{biomass6,share,i}}{\sum_{j=1}^{175} E_{biomass6,share,j}} \times 6.79\% \times E_{biomass5-6-7,country} \quad (9\text{-}19)$$

For Equations 9-18 and 9-19, E is the potential, L is the population of the examined district, and m is the mass of the waste. The 6.79% multiplier indicates the proportion of known potential that can be assessed as a potential for the examined fuel subgroup for electricity production (see Table 9-3). Heat energy potentials can be calculated in a similar way.

Biogas from Sewage

To assess the potential of biogas from sewage, each county's volume of treated sewage was divided among the districts in proportion to their populations. Biogas yield of sewage was estimated using the data of Budapest Sewage Works Private Limited as 0.2/m^3. The potential share of a district can be calculated as:

$$E_{biomass7,electricity,} \frac{V_{sewage,county}}{L_{county}} \times L_i \times q_{yield} \times C_{biogas} \times (1/10^9) \quad (9\text{-}20)$$

Annual potential volume of a district is calculated as:

$$E_{biomass7,electricity,} \frac{E_{biomass7,share,i}}{\sum_{j=1}^{175} E_{biomass7,share,j}} \times 6.28\% \times E_{biomass5-6-7,country} \quad (9\text{-}21)$$

For Equations 9-20 and 9-21, E is the potential, L is the population of the examined district, and V is the volume of sewage. The 6.28% multiplier indicates the proportion of known potential that can be assessed as a potential for the examined fuel subgroup for electricity production (see Table

9-3). Heat energy potentials can be calculated in a similar way.

After assessing the technical and economic potential of all fuels, subgroups 1-4 and 5-7 were cumulated for both heat energy and electricity production. The results are shown in Figures 9-5 through 9-8. Cleary, traditional agricultural areas offer high potentials. County seats are also performing well and the large volumes of municipal waste and sewage could also be used as fuel.

CONCLUSIONS

This chapter introduced a methodology to assess regional potentials of various renewable energy sources, and selected results of the evaluation process. If the potential of all renewable energy sources is cumulated, we obtain the distribution shown in Figure 9-9. This offers a theoretical cumulative potential, since in practice some renewable technologies directly compete. The three best performing districts are Mosonmagyaróvár, Kalocsa and

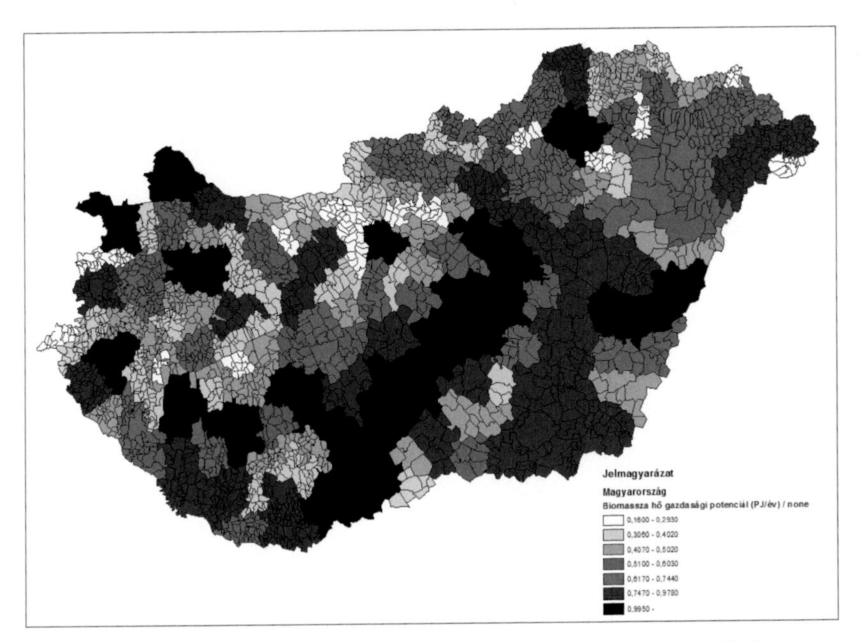

Figure 9-5. Economic biomass energy potentials for heat generation (darker areas represent better potentials, ranging between 0.160 and 0.995 PJ/a).

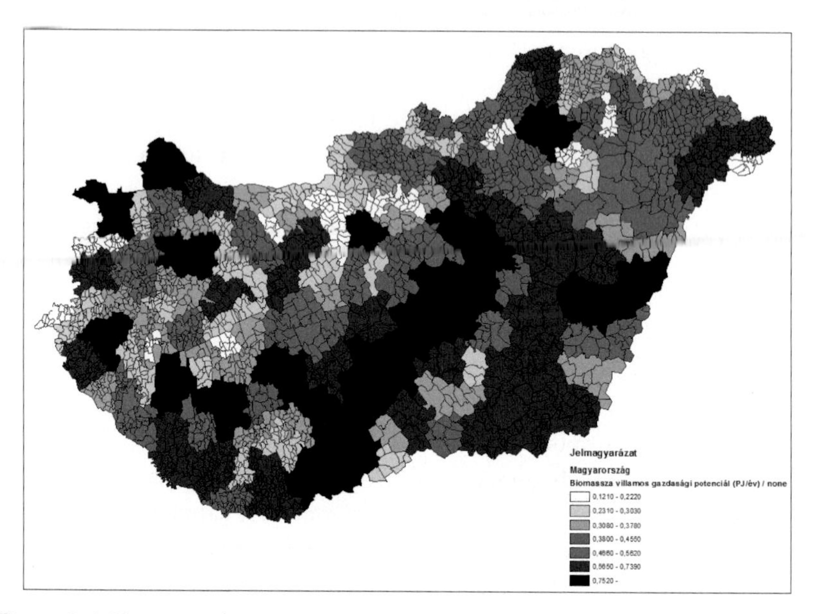

Figure 9-6. Economic biomass energy potentials for electricity generation (darker areas represent better potentials, ranging between 0.121 and 0.752 PJ/a).

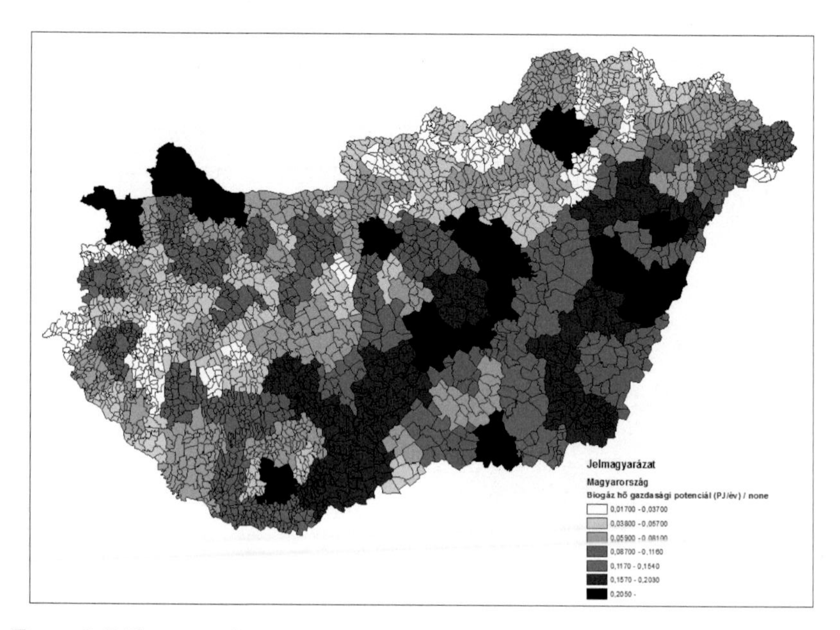

Figure 9-7. Economic biogas energy potentials for heat generation (darker areas represent better potentials, ranging between 0.017 and 0.205 PJ/a).

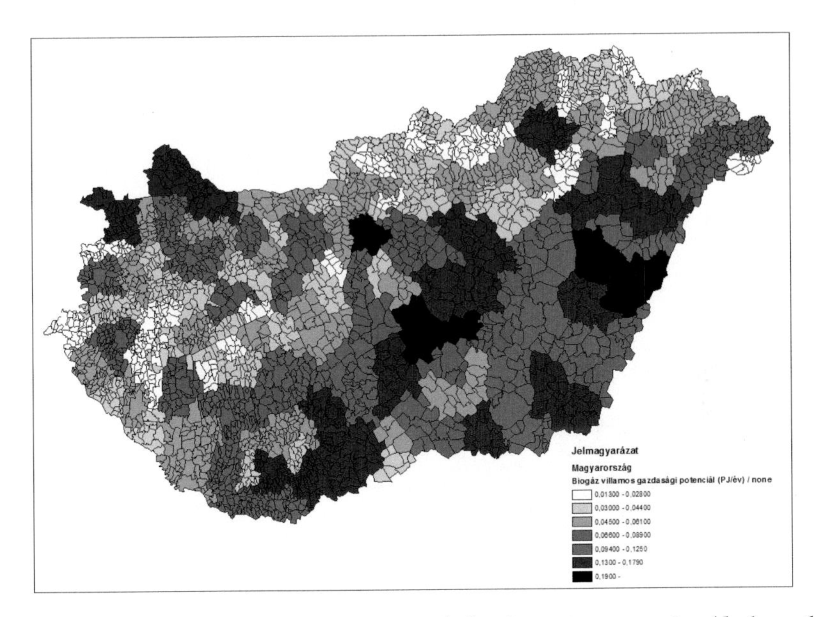

Figure 9-8. Economic biogas energy potentials for electricity generation (darker colors represent better potentials, ranges between 0.013 and 0.190 PJ/a).

Budapest. For the first two we conclude that they have performed relatively well in all aspects. For Budapest, the performance partly results from city size (approximately two million inhabitants) since waste materials and sewage utilization was emphasized in NREAP background studies for biomass.

Since the European Commission has published guidelines for state aid for environmental protection and energy 2014-2020 (2014/C 200/01) in 2014, all EU member states are working towards new support mechanisms for renewable energy sources. The main ideas of the guidelines were that future support for renewable energy producers should take the form of premiums rather than fixed support schemes and that mandatory tenders for new capacities should be developed beginning in 2017.

Several policy changes recently have been approved by the Hungarian government. It announced the Act CXXXVIII of 2016 which modified several climate policy and green industry development policies. Governmental Decree 393/2016 (XII. 5.) modified previous support schemes for renewable based electricity production.

Decrees 17/2016 (XII. 21.) and 5/2016 (XII. 21.) approved by the Hungarian Energy and Public Utility Regulatory Authority and the Min-

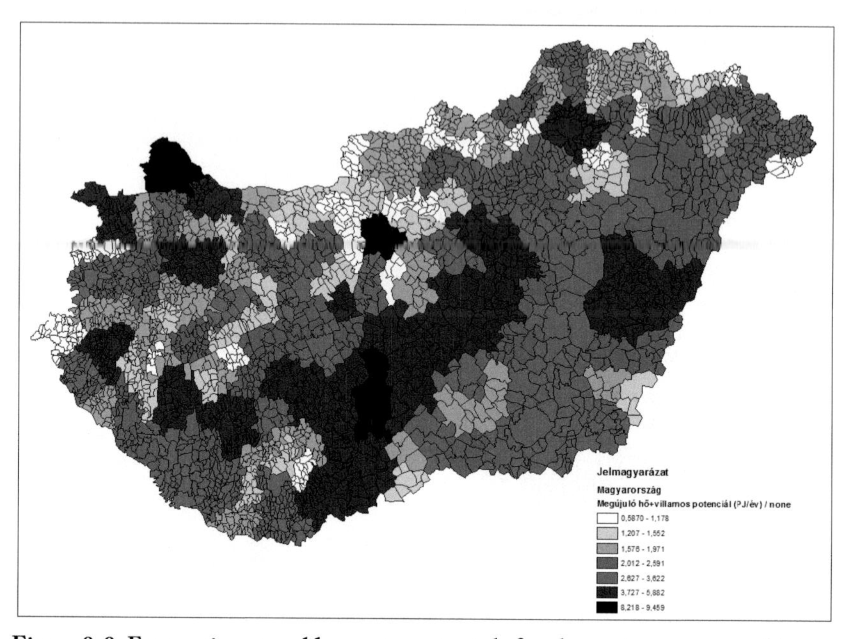

Figure 9-9. Economic renewable energy potentials for electricity generation (darker areas represent better potentials, ranging between 0.587 and 9.459 PJ/a).

istry for National Development (MND), respectively quantified the new premiums and the technical requirements for power plants. Details of the tendering and settlement processes were announced in MND Decrees 62/2016 and 63/2016 (XII. 28.). Thus, newly introduced legislation has largely transformed the former support schemes for renewables.

The new policies redefine green and brown premiums. New electrical generation facilities with over 0.5 MW of installed capacity are eligible for green premiums. Plants with generation capacities below 1 MW do not require tender offerings (with the exception of wind power plants), and receive reference market prices and administrative premiums. The reference market price is the one-day average of prices on the Hungarian power exchange (HUPX), which is weighted by the actual production of wind and solar plants. The period and volume of subsidized electricity is determined by the regulator. If the capacity is above 1 MW, the premiums are determined in a tender offering.

The brown premium was designed for existing power plants (dominantly biomass and biogas units after pay-off), so that they can remain in operation after the expiration of previous subsidy periods. In case of the *normal brown premium*, the subsidized price is set to cover operation costs and is revised annually. For biomass co-firing plants, the premium is determined by the share of fuels. In some circumstances, power plants may sell electricity above the reference maker price and provide auxiliary services. The second case is called the *alternative brown premium*, which is designed to compensate for the competitive disadvantage of biomass co-firing. This premium is revised annually and calculated as the difference between the operation costs of fossil fuel and biomass firing. It also accounts for any additional costs (e.g., CO_2 prices). Brown premiums must be claimed and permissions are given for extendable five year periods.

Since the new legislation will begin in 2017, concerns have arisen among professionals regarding the tendering process. The annual tendered volume is determined by the minister for national development for five year periods which are revised annually. Other limitations include the limitations of annual support (supported price multiplied by annual volume), technologies (minimal and maximal allocable capacity), distribution utility areas and maximal offers. The intensity of competition will largely be determined by the premiums and quantities of cost-efficient renewable sources. Since we have aimed to highlight the best options for renewable energy utilization, we believe that the energy map tool will be able to support future investors and decision makers in their efforts to reach Hungary's renewable energy targets.

An important goal of our research is to find ways to maximize the use of locally available resources. This will be reflected in our future efforts, part of which focuses on the synchronization of the database with the National Building Energy Performance Strategy, and the definitions of the energy performance of buildings directive. The latter defines a nearly zero building as one "that has a very high energy performance, as determined in accordance with Annex I. The nearly zero or very low amount of energy required should be covered to a very significant extent by energy from renewable sources, including energy from renewable sources produced on-site or nearby." As more buildings are constructed in accordance with the new directive, greater use of locally available renewable energy sources is inevitable. The developed energy potential map will be a useful tool to this end. We aim to improve the database by performing a sensitivity analysis of the assessment, and developing decision support systems for municipalities and districts.

References

[1] National Renewable Energy Utilization Plan of Hungary 2010-2020 (2011). In Hungarian: Magyarország megújuló energia hasznosítási nemzeti cselekvési terve 2010-2020. Ministry of National Development.

[2] Delegation for the National Renewable Energy Action Plan, Volumes A-H (2009). In Hangarian: Nemzeti Megújuló Energiahasznosítási Cselekvési Terv háttértanulmánya, A-H kötetek, Pylon Kft.

[3] Comparison of renewable energy based energy generation technologies, based on economic and environmental aspects (2014). In Hungarian: Megújuló energiaforrást fölhasználó energiafejlesztési technológiák gazdasági és környezeti szempontok alapján történő összehasonlítása. Hungarian Academy of Sciences, Centre for Energy Research.

[4] Renewable energy potential assessment methodologies (2015). In Hungarian: Megújuló energiapotenciál felmérési módszertanok. Hungarian Academy of Sciences, Centre for Energy Research.

Chapter 10

Forecasting Domestic Energy Consumption for the Nordic Countries

Samad Ranjbar Ardakani, Seyed Mohsen Hossein, and Alireza Aslani

Residences account for a large share of total annual energy use in the Nordic countries due to the extremely cold climates and high household heating demand. Most domestic energy in the Nordic countries is used for space heating and hot water. The purpose of our study was to forecast the annual energy consumption of the Nordic residential sectors by 2020 as a function of socio-economic and environmental factors, and to offer a framework for the predictors in each country.

Our research models the domestic energy use in Nordic countries based on social, economic and environmental factors. Applying multiple linear regression (MLR), multivariate adaptive regression splines (MARS), and the artificial neural network (ANN) analysis methodologies, three models have been generated for each country in the Nordic region. Using these models, we forecasted the Nordic countries domestic energy use by 2020 and assessed the causal links between energy consumption and the investigated predictors. The results showed that the ANN models have a superior capability of forecasting domestic energy use and specifying the importance of predictors compared to the regression models. The models revealed that changes in population, unemployment rate, work force, urban population, and the amount of CO_2 emissions from the residential sectors can cause significant variations in Nordic domestic sector energy use.

INTRODUCTION

Europe's Nordic countries include Denmark, Finland, Norway, Iceland and Sweden. Due to social perspectives and the desire for secure energy supplies, renewable energy utilization has an important role in the development of energy policies. According to the International Energy Agency (IEA), the Nordic countries were responsible for 186.2 megatons of annual CO_2 emissions during the early 1990s which was reduced 17% by 2014. Despite European Union (EU) efforts to decrease the CO_2 emissions of the Nordic countries, their share has increased by 0.3%, due mainly to increased emissions by Norway and Iceland. After the EU established goals to reduce greenhouse gas emissions that contribute to global warming, the countries approved treaties and adopted the EU's domestic sector policies on sustainable development.

EU policies recognize that renewable energy and fuel substitution are key to lowering fossil fuel dependency and meeting their short-term and long-term goals [1-4]. As a consequence of these efforts, IEA and World Bank data indicate that the share of CO_2 emissions from residential buildings and commercial and public services declined from 13.6% in 1990 to 4.2% in 2014. Though fossil fuel prices have declined in the recent years, the Nordic countries (except Iceland and Norway) have decreased emissions. The Nord Pool, the region's primary electricity market operator, facilitated regional de-carbonization. Renewable energy technologies (e.g., wind power in Denmark and hydropower in Norway and Sweden) provide abundant supplies of electricity for the Nord market. By balancing the market's generated power, the Nord Pool has simplified regional electricity accessibility. The Nordic countries, except Denmark, invested heavily in energy-intensive industries. During the past two decades they have substantially increased their shares of electricity and biomass, reduced their shares of industrial and residential sector fossil fuel use, and enhanced their economic and energy security [5].

Denmark is one of the world's leaders in energy efficiency. Based on IEA reports, the country has decoupled gross domestic product (GDP) from energy consumption and CO_2 emissions [1]. Danish wind power has a key role in the Nordic electricity market. The country invested heavily in electrification due to dependency on renewables, particularly wind power. In 2015, Denmark's share of wind generated electricity was 42%—the world's highest [5]. Denmark consumed 538.8 PJ of energy in 2014, and the domestic sector was responsible for 31% of total final consumption (TFC) [6].

Mean building energy use per person rose by 0.2% per year between 1990 and 2014 in the Nordic countries. IEA experts explained the increase due to growth in the services sector. After approving the EU's Energy Performance of Buildings Directive (EPBD), the Nordic governments revised their energy consumption policies for buildings and developed new standards and requirements for their residential sectors [7].

Denmark's share of residential and commercial sector total CO_2 emissions declined from 12% in 1990 to 8% in 2014. During this period, the shares of renewables and wastes in residential energy consumption increased from 8.4% to 20.2%. Many Danish households are not simply power consumers, but also produce electricity using household wind power and solar panels [8]. Due to their climatic conditions, Denmark's building energy use has high demand for heating energy [9]. The Danish, Finnish, and Swedish governments provide energy system flexibility by using combined heat and power (CHP) systems with heat storage, a strategy adopted just after the 1973 oil crisis [10]. At that time, 90% of Denmark's energy demand was satisfied by imported fossil fuels despite the country having substantial North Sea oil and gas reserves. Using district heating and implementing energy efficiency measures allowed Denmark to become a net oil exporter [11].

Finland's total energy consumption in 2014 was 1,027.7 PJ. Finland has second highest per capita Nordic energy use after Iceland, which accounts for almost a quarter of the region's energy consumption. Among IEA countries, Finland consumes a median amount of energy [2]. Finland's industrial sector (particularly the energy-intensive pulp and paper industries) accounts for the largest share of its energy use. Finland's domestic energy sector was responsible for 20.6% of TFC in 2014 [12]. Between 1990 and 2014, Finnish building floor area increased 40%, energy use per m^2 increased by 127%, and the county's per capita total building final energy consumption increased by 60% [5]. Finland's residential sector share of total CO_2 emissions decreased from 12.2% in 1990 to 4% in 2014, while the share of renewables in its TCF increased 8%.

In an effort to comply with carbon-neutral district heating policies, a third of Finland's residential buildings were using district heating by 2012. The government also encouraged families to live in apartments rather than detached houses by pricing the cost of district heating for detached houses much higher than for apartments [13]. Compactly designed apartments and structures consume less heating energy since they have lower conductive heat transfer [14]. Finally, increasing urban density facilitates the utilization of district heating for residential buildings which enhances energy efficiency.

Iceland has the lowest population among the Nordic countries and the lowest total annual energy consumption. The country has the world's highest per capita electricity use due to its low population density, high electricity use by its aluminum industry and the extremely cold climate. The total energy use of Iceland in 2014 was 114.6 PJ. The shares of its industrial and residential sectors were 51.4% and 13.7% respectively [15]. Geothermal energy is used to meet the county's high demand for space heating and electricity generation. Geothermal power plants generate more than a quarter of Iceland's electricity and supply almost two-thirds of its primary energy use. About 85% of Iceland's total primary energy supply in 2014 was from sindigenous renewable resources [16]. Iceland does not belong to the Nord Pool due to the distances between Iceland and Norway or Denmark [17]. However, the Nord Pool supported the Icelandic Electricity Grid (Landsnet) in establishing a market for electricity based on the use of the Elbas system and continuous trading [18].

The TFC of Norway in 2014 was 842.1 PJ, and the share of its residential sector energy use was 19.1% [19]. Due to energy-intensive industries and high building heating demand, the country has a high per capita electricity use second only to Iceland [20]. While Norway's government attempted to decrease the share of buildings in total CO_2 emissions by 6.3%, the share of renewables in the TCF was relatively constant from 1990 to 2013 and declined in 2014. In Norway, 81% of residential energy use in 2013 was from electricity. The share of district heating in Norwegian energy consumption in 2013 was only 2%. Lacking district heating availability, in 2012 the primary heating source for more than 70% of Norwegian households was electricity. There are large variations in housing energy composition in Norway. While only 5% of households living in Oslo used heat pumps in 2012, the corresponding value for Hedmark/Oppland was about 40% [21,22].

The total energy consumption of Sweden was 1,335 PJ in 2014 and the share of its residential sector was 20.8% [23]. The economy of Sweden relies heavily on energy-intensive industries, including steel manufacturing, pulp, paper, and heavy vehicle production. The country substantially decreased its industrial and residential sector fossil fuel usage while increasing the use of renewables such as biomass and solar thermal. The Swedish government has also invested heavily in nuclear energy. The share of its domestic buildings in total CO_2 emissions declined by 14.7% between 1990 and 2014. The EU's present policy is to decrease the share of nuclear power for electricity generation in Nordic countries from 22% in 2013 to 6% by 2050, with Sweden developing low-carbon energy solutions other than nuclear power after 2030 [5]. A large part of final energy consumption in Sweden is used for space heating in buildings. In 2013, the energy use for space heating and hot water in residential buildings was 11% less than in 2000. This reduction was due to the improved energy efficiency of heating appliances, particularly heat pumps, and district heating energy intensity [24].

By 2014, over 21% of total energy consumption in the Nordic countries was consumed by the residential sectors. However, the domestic sector in the EU was responsible for 24% of TFC. Because of the high energy demand for space heating in the Nordic countries, the corresponding governments were partially successful in applying their domestic sector energy efficiency policies [5].

Assessing residential sector energy trends is more difficult when compared to sectors such as transportation and industry. The issues are mainly due to the variety of buildings with different characteristics, the variable behavior of building occupants, and the lack of comprehensive data sets to model domestic energy use [25]. Understanding the detailed characteristics of the domestic sector is essential to clarify the interrelated and complex features of end-use energy consumption in the Nordic countries. The purpose of our study was to forecast annual energy consumption of the Nordic residential sector by 2020 as a function of socio-economic and environmental factors, and to provide a framework for specifying the importance of each country's predictors. Determining factors that influence residential energy consumption is essential for policy-makers to adopt the best ways to lower energy use and CO_2 emissions. Hence, we discuss the determinants of domestic energy use chosen in this study to model residential consumption. We next introduce statistical methodologies for modeling, including MLR, MARS concept and the ANN approach [26]. Based on these methods, we generated three models for each country and compared their potential for predicting energy use and understanding the importance of investigated predictors.

DETERMINANTS OF DOMESTIC ENERGY CONSUMPTION

Research shows the importance of assessing the factors that affect domestic sector energy use and evaluating their interactions [27,28]. Policy-makers have recognized many parameters influencing domestic end-use energy consumption, which are classified as socio-economic factors, dwelling factors and appliance factors. Jones et al. [29] in a comprehensive literature review introduced these factors (see Figure 10-1).

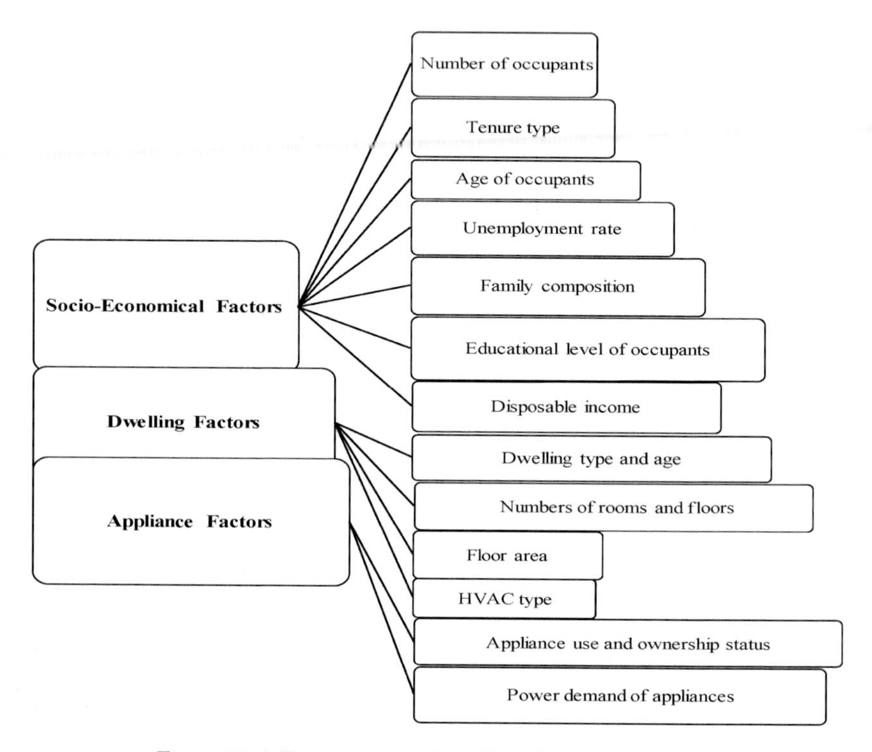

Figure 10-1. Determinants of residential sector energy use.

Previous studies revealed that there is a direct relationship between energy use, especially electricity use, and residential sector size [30-37]. The larger the population, the higher the energy consumption. Researchers showed that single occupancy dwellings in Ireland consume 19% less electricity weekly compared to two person households [36]. In another study, Zhou and Teng found that for every additional household member in China, electricity consumption increases by 8% [38].

In our study, we used the annual population of Nordic countries as a parameter demonstrating the size of the domestic sector. This choice was mainly due to the lack of reliable data on the actual number of residents in some of the investigated countries. Using annual dwelling stock data helped overcome this problem. The population of the Nordic countries has risen by 13% (about 3 million) since 1990. The Nordic Council reported that during this period the populations of Iceland and Norway increased by 28% and 21% respectively. Births overtaking deaths and immigration outstripping emigration are the main causes of population growth during the last two decades [39]. Table 10-1 contains the population of Nordic countries between 1990 and 2015.

Table 10-1. The population of Nordic countries between 1990 and 2015. Source: Worldbank.org.

Country	Population 1990	Population 1995	Population 2000	Population 2005	Population 2010	Population 2015/16
Sweden	8,558,835	8,826,939	8,872,109	9,029,572	9,378,126	9,798,871
Denmark	5,140,939	5,233,373	5,339,616	5,419,432	5,547,683	5,676,002
Finland	4,986,431	5,107,790	5,176,209	5,246,096	5,363,352	5,482,013
Norway	4,241,473	4,359,184	4,490,967	4,623,291	4,889,252	5,195,921
Iceland	254,826	267,468	281,205	296,734	318,041	330,823

Another important social parameter influencing energy consumption is the age structure of the population. Some researchers claimed that this factor influences both the macro and micro level energy use, especially in the transportation and residential sectors [40-46]. Other studies evaluated the impact of age structure on total annual CO_2 emissions and residential sector shares [47-51]. These studies emphasized the importance of age structure as one of the determinants of energy consumption. We evaluated the impact of this parameter on the Nordic region's domestic sector end-use energy consumption. We considered that people of different ages have different disposable income levels and accordingly variable rates of energy consumption. Children and older persons use less energy in comparison to mature adults because of differences in their activities. The percentage of people aged between 15 and 64 was used as an age structure factor in our study. Table 10-2 shows these data from 1990 to 2015. It is important to understand the effects of this parameter on the Nordic residential energy use. The Nordic Council declared that the ratio of old to young people is increasing due to lower birth rates and longer life spans. Nordic countries have among the highest birth rates among European countries. Finland has the oldest age structure and the county's life expectancy has risen to 84 over the last two decades. Forecasts predict that more than 8% of the Nordic peoples will be over the age of 80 by 2040 [39].

Table 10-2. The share of population aged between 15 and 64 (work force). Source: Worldbank.org.

Country	% of total labor force 1990	% of total labor force 1995	% of total labor force 2000	% of total labor force 2005	% of total labor force 2010	% of total labor force 2015/16
Sweden	64.3	63.7	64.3	65.3	65.3	62.8
Denmark	67.4	67.4	66.7	66.1	65.4	64.2
Finland	67.3	66.8	66.9	66.7	66.4	63.2
Norway	64.8	64.6	64.8	65.6	66.2	65.7
Iceland	64.4	64.3	65.1	66.2	66.9	66.0

Disposable income influences domestic energy consumption. Previous studies showed that disposable income directly impacts electricity consumption [33,35,36,52,53]. Leahy and Lyons found that Ireland's income elasticity compared to electricity consumption by appliances was 4% [36]. Though Nordic countries commonly use district heating systems in urban areas, the increase in disposable income may enhance the utilization of individual heating facilities in the domestic sector and increase the TFC. For Norway and Iceland, this likely results in wider use of electricity for heating [5]. Therefore, we used the annual growth rate of household disposable income as a determinant of residential energy consumption.

The level of urbanization is an important determinant of domestic energy usage, effectively influencing policies regarding socio-economic progress [43,54-58]. As people migrate from rural to urban areas, they engage in urban activities and work in the industrial, commercial or service sectors. This initially enhances industrial production and GDP, since the migrated people often require new infrastructure which increases development. This increases urban energy consumption and CO_2 emissions. Increasing energy use can contribute to urban heat island (UHI) effects if offsetting energy efficiency measures are not implemented. Muye et al. showed that the migration of rural residents to urban areas in China increases emissions since use of biofuels, electricity, coal, and liquefied petroleum also increases [59].

People living Nordic countries use district heating and geothermal energy in urban areas which has a positive impact on the region's domestic energy use. We evaluated the impacts of the percentage of urban population on Nordic residential energy consumption. According to the Nordic Council, the population growth in the urban and suburban areas was greater than elsewhere. During the past two decades, many peripheral areas lost population to the cities increasing urbanization [39]. Table 10-3 provides urban population percentages for the Nordic countries from 1990 to 2015.

Among social factors, there is less study on the causes and effects of educational levels on residential energy consumption. Aixiang assessed the links between energy consumption, the number of technological scientists, the number of people studying in the tertiary education level, and the amount of the research and development (R&D) funding in China [60]. The results indicated that improving energy efficiency in China highly depends on education levels and technology use in Jiangsu province society. Given the dramatic increase in the percentage of Nordic people with tertiary education, the study evaluates the impact on the residential energy consumption. Along with the impressive improvement in educational levels and population increases over the last two decades, the domestic per capita energy use of the Nordic countries was relatively constant. This indicates the importance and effectiveness of educational development and spending in R&D in these countries. All of the Nordic countries, invested more than the 2% EU average in R&D with the exception of Norway (investing 1.7%)

Table 10-3. Percentages of urban population in Nordic countries (1990 to 2015/16). Source: Worldbank.org.

Country	% of urban population 1990	% of urban population 1995	% of urban population 2000	% of urban population 2005	% of urban population 2010	% of urban population 2015/16
Sweden	83.1	83.8	84.0	84.3	85.1	85.8
Denmark	84.8	85.0	85.1	85.9	86.8	87.7
Finland	79.4	81.0	82.2	82.9	83.6	84.2
Norway	72.0	73.8	76.1	77.5	79.1	80.5
Iceland	90.8	91.6	92.4	93.0	93.6	94.1

[61]. The Nordic Council estimated that 35% to 50% of Nordic people aged between 15 and 74 studied at the secondary education level and 22% to 34% at the tertiary educational level. Among the Nordic countries, Norway and Finland have the most population with tertiary education and Denmark has the least [62].

Resident employment status influences the end-use energy consumption of households and affects the level of disposable income [29]. Unemployed people with low incomes lack potential to enhance their energy consumption. Studies on the relationships between energy consumption and rates of unemployment consistently report that there is no significant link between these two factors [63-65]. Nevertheless, we considered the parameter as one of our independent variables. One of the EU targets for labor is an employment rate of 75% for the working age group of 15 to 64 by 2020. It appears that the Nordic countries with the exception of Finland can partially meet this target [66]. Table 10-4 shows the unemployment rates between 1991 and 2014.

The last two determinants of domestic energy consumption considered in our study involved environmental issues. The link between energy

Table 10-4. The unemployment rate in the Nordic countries (share of total labor forces). Source: Worldbank.org.

Country	% unemployed 1991	% of unemployed 1995	% of unemployed 2000	% of unemployed 2005	% of unemployed 2010	% of unemployed 2014/16
Sweden	3.3	9.3	5.9	7.8	8.7	8.0
Denmark	9.1	7	4.5	4.8	7.5	6.6
Finland	6.5	15.3	9.7	8.4	8.4	8.6
Norway	5.4	4.9	3.4	4.6	3.6	3.4
Iceland	2.5	4.9	2.3	2.6	7.6	5.0

Source: Worldbank.org

use and the first parameter, CO_2 emissions, was investigated in previous studies which indicated that there is a direct relationship between them [56,67,68]. Table 10-5 shows the percentages of CO_2 emissions by the residential sector between 1990 and 2013. As the last parameter, this study analyses the link between the annual percentage of renewables and waste in household energy consumption to clarify how the enhancement of the parameter over the last two decades has influenced end-use energy consumption.

The work's inspiration comes from Fumo and Biswas, who used regression analysis to predict domestic sector energy use [69]. They claimed that for bottom-up approaches to developing a model for domestic energy use, the statistical methods are more simple and useful in comparison to engineering approaches. Among the statistical methodologies, regression analysis showed a promising potential to model sector energy use. Using statistical approaches, we first predict domestic end-use by 2020, and then discuss the importance of the predictors for Nordic countries. The number of models and types of methods in our study vary. We compare the capa-

Table 10-5. Percentage of CO_2 emissions from the Nordic countries residential sector. Source: Worldbank.org.

Country	% of residential CO2 emissions 1990	% of residential CO2 emissions 1995	% of residential CO2 emissions 2000	% of residential CO2 emissions 2005	% of residential CO2 emissions 2010	% of residential CO2 emissions 2013/16
Sweden	17.7	15.5	12.1	7.3	5.5	3.6
Denmark	12.4	10.2	9.4	9.2	8.4	8.3
Finland	12.2	10.9	6.4	6.1	4.6	4.4
Norway	9.0	5.9	4.7	4.1	3.7	3.0
Iceland	2.6	2.0	1.4	0.9	0.5	0.5

Source: Worldbank.org

bilities of traditional and advanced statistical methodologies to model residential sector energy demand in the Nordic countries mostly based on the social factors.

METHODOLOGIES

The statistical approaches, including regression, conditional demand analysis, and neural network are capable of linking a response variable with one or more predictor variables. The response variable in this study is residential energy consumption in Nordic countries. All of the previously mentioned predictors are defined as continuous variables.

Using this model, we can estimate domestic sector energy use for the Nordic countries. We generated three models for each country based on the MLR analysis, MARS concept and the ANN approach [70,71]. Forecasting with these methods requires a prediction set, provided with another methodology. For this purpose, the additive Holt-Winter (HW) algorithm is used.

Additive Holt-Winter

The additive HW method is a version of the HW algorithm which is capable of forecasting with historical data despite trendy behavior [72]. The series of formulas are:

$$\hat{y}_t + h_t = l_t + h_{bt} + s_{t-m} + h_m +$$
$$l_t = \alpha(\gamma_t - s_{t-m}) + (1-\alpha)\,(l_{t-1} + b_{t-1})$$
$$b_t = \beta \text{ x } (l_t - l_{t-1}) + (1-\beta) \text{ x } b_{t-1}$$
$$s_t = \gamma(\gamma_t - l_{t-1} - b_{t-1}) + (1-\gamma)s_{t-m}$$

The error correction equations are:

$$l_t = l_{t-1} + b_{t-1} + \alpha e_t$$
$$b_t = b_{t-1} + \alpha\beta \text{ x } e_t$$
$$s_t = s_{t-m} + \gamma e_t$$
$$e_t = \gamma_t - (l_{t-1} + b_{t-1} + s_{t-m}) = \gamma_t - \hat{y}_{t|t-1}$$

Where l is level, b is trend, s is the seasonal component, α, β, and γ are smoothing parameters, m is the period of seasonality, e is error, and y is the observation [73].

Multiple Linear Regression

The multiple linear regression (MLR) methodology is one of the traditional modeling approaches capable of explaining the variations in a response variable using the change in predictors. The MLR models a dependent variable as a function of independent factors and estimates the regression coefficient for each predictor [70]. The regression coefficients represent the value at which the response variable changes when the independent variables change. The MLR model has the following form:

$$Y = b_1x_1 + b_2x_2 + \ldots + b_nx_n + c$$

In this equation, b_i is the regression coefficient. The less the degree of variability of the residual values in relation to the overall variability, the greater the prediction accuracy model. This ratio is referred as the coefficient of determination, named R^2, and represents the capability of the model in fitting the input data as a function of the target variable. The importance value of determinants for the MLR is the t-statistic corresponding to the regression coefficient estimate of each independent variable [74].

Multivariate Adaptive Regression Spline

The nonparametric regression technique, MARS, is a form of the stepwise linear regression developed by Jerome Friedman in 1991 [75]. The model is much simpler in comparison to other approaches such as neural network and random forest. However, the MARS was derived from linear regression methodology; it can organize nonlinear links between the target and predictors. This methodology has been used by other researchers to predict building energy efficiency.

The MARS is popular for its flexible models and automatically adjusting the models to propose the most reasonable interaction between the response and dependents. It uses linear basis functions such as $(x-t)_+$ and $(t-x)_-$ in combination to propose the final equation as a polynomial function. The sign + means the MARS only considers the positive part of defined linear functions. Therefore,

$$(t-x)_- = \{x-t, \Delta x<t \; 0, \textit{otherwise}$$
$$(x-t)_+ = \{x-t, \Delta x>t \; 0, \textit{otherwise}$$

The variable t represents the knot location of predictor x. The general form of a MARS model is,

$$\gamma = f(x) = \beta_o + \Sigma^1{}_{i=1}\ \beta_i\ h\ (x_i)$$

In this equation, $h_i(x)$ is a function from set C, containing the candidate reflected pairs function. β_i is the coefficient estimated with standard linear regression. MARS considers the values as weight representing the importance of the dependent variables. During forward stepwise, the MARS chooses those basis functions from set C which effectively reduces the residual error in each step. Then the methodology applies a backward procedure to prune the model by eliminating basis functions with the smallest increasing effect in the least squares. It uses a generalized cross validation (GCV) error function to evaluate the goodness of fit, considering the residual error and model complexity [74]. The formula for the GCV function is:

$$\mathrm{GCV} = [\Sigma^1{}_{i=1}\ (\gamma_i - f(x_i))^2] \div [1\text{-}\ (1+cd)/I]\ ^2$$

For this formula, I is the number of cases in the dataset, d is the degree of freedom, and c is the penalty factor for choosing the basis functions from set C. For all MARS models in our study, the degree of interaction is 6, the penalty is 3, and the threshold value is zero. To determine the variable importance, we computed the reduction in GCV for each dependent variable added to the model [74]. Then the reductions were summed for each corresponding continuous predictor to determine their importance. Using this methodology, the dependent variables not included in the pruned final model have a zero importance.

Artificial Neural Network

The artificial neural network (ANN) is a simple modeling approach often used by policy-makers. An ANN system comprises highly interconnected nodes called neurons [71]. For generated optimal networks, ANN tools effectively map patterns of input parameters onto the patterns of corresponding output variables by training the nodes and making them suitable for an alternative series of patterns. When the model is generated, the ANN applies the model to a new input pattern to forecast the proper output pattern. Among the various types, this study uses a feed forward ANN with a backward propagation learning algorithm for training the model. The formula is:

$b_j = f(\Sigma^1{}_{i=1}\, w_{ij}\, \alpha_i - T_j)$

For this formula, b_j is the output vector, a_i is the input vector, w_{ij} is the weighting between neurons i and j, T_j is the internal threshold, and f is a hyperbolic tangent transfer function. To improve the performance of the model, the ANN uses the following equations to adjust the threshold values and the weight factors [76].

$$T_j^{new} = T_j^{old} + \eta(\Sigma\delta_{pj}) + \alpha\Delta T_j^{old}$$
$$w_{ij}^{new} = w_{ij}^{old} + \eta(\Sigma\delta_{pj} O_{pi}) + \alpha\Delta w_{ij}^{old}$$

In these equations, η is the learning rate in the model, α is the momentum coefficient of the backward propagation learning algorithm, while ΔT and Δw are the previous change of threshold values and weight factors. O and p are the outputs and respective patterns. We normalized all of each pattern's data with values between -1 and 1, generated the initial weight factor with a random selection between -0.2 and 0.2, and set threshold values to zero.

Understanding the importance of variables using ANN is complicated compared to the previous approaches. One of the best solutions to this problem is to perform a sensitivity analysis of the model. We used the weight method sensitivity analysis, proposed by Garson, and applied by Montaño and Palmer to feed forward neural networks [77,78]. The formula for this methodology is:

$$Q_{ik} = [\Sigma^L{}_{j=1}((w_{ij}/\Sigma^N{}_{r=1} w_{rj}) v_{jk})] \div [\Sigma^N{}_{j=1}(\Sigma^L{}_{j=1}\,((w_{ij}/\Sigma^N{}_{r=1} w_{rj}) v_{jk}))]$$

In this equation, w_{ij} is the connection weight between the input neuron i and the hidden neuron j, and v_{jk} is the connection weight between the hidden neuron j and the output neuron k. Q_{ik} is the variable importance.

The overall performance of each model is measured using the coefficient of determination (R^2), calculated by:

$$R^2 = 1 - [\Sigma^N{}_{i-1}(y_i - f(x_i))]^2 \div [\Sigma^N{}_{i-1}(y_i - \bar{y})]^2$$

For this equation, f is the estimated function, y is the observations, and N is their quantity.

Denmark

Performing the forecasting model process, three models were generated for Denmark. The ANN model showed superior performance in comparison to the other regression methodologies. The R^2 of ANN, MARS, and MLR was 0.99, 0.8 and 0.61 respectively. Although the ANN model estimated that Denmark's residential energy use will experience a parabolic trend between 2015 and 2020, the regression models showed an approximately constant trend. However, there is a large difference between the estimated values. Considering that the final energy consumption of Denmark's domestic sector in 2014 was 165,630 Tj, the ANN model provides a more reasonable estimate of 2015 energy consumption. Figure 10-2 shows the forecasted trend of energy use between 2015 and 2020.

Research by Williams and Gomes indicated that predictors are highly affected by the type of model [74]. Figure 10-3 shows the weighted mean normalized importance of predictors. We used the R^2 of each model to determine the weighted mean of each predictor. The sensitivity analysis of Denmark's neural network model revealed that household disposable income, share of population aged 15 to 64, unemployment rate, and share of CO_2 emissions in total annual emissions have the greatest effects on energy consumption. For the MARS model, the population growth rate had the highest importance; the share of CO_2 emissions and the other significant predictors of ANN model ranked in the mid-range. The MLR model showed that the growth of population and household disposable income have the greatest effects on energy use. Hence, the three models agreed that the growth of disposable income, unemployment, population growth, and the percentage of people aged from 15 to 64, have significant effects on Denmark's residential energy use. An elasticity analysis indicted that household disposable income, gross enrollment ratio, and the percentage of the population aged from 15 to 64 are more effective in comparison to the other parameters (see Figure 10-4).

Finland

For Finland, all three models have an acceptable R^2. The coefficient of determination for the ANN, MARS, and MLR models was 0.99, 0.6, and 0.66 respectively. The difference between the R-value of the ANN model and the other two models is surprising. The 2014 energy consumption of Finland's domestic sector was 212,020 Tj.

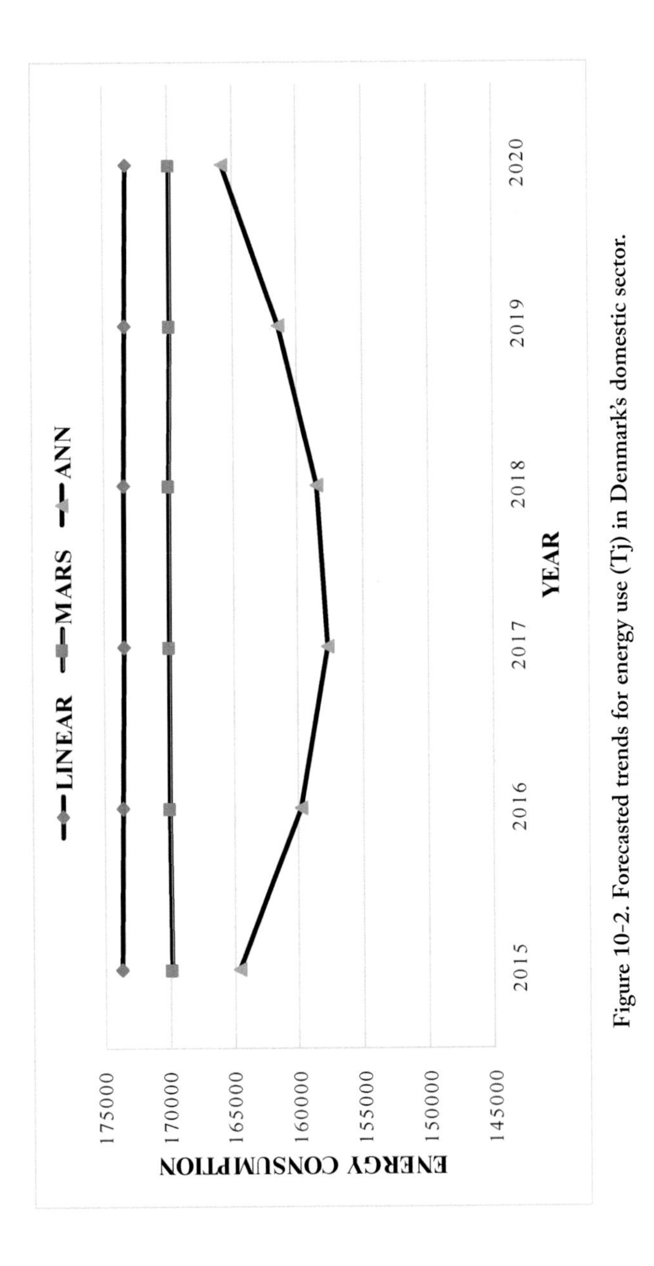

Figure 10-2. Forecasted trends for energy use (Tj) in Denmark's domestic sector.

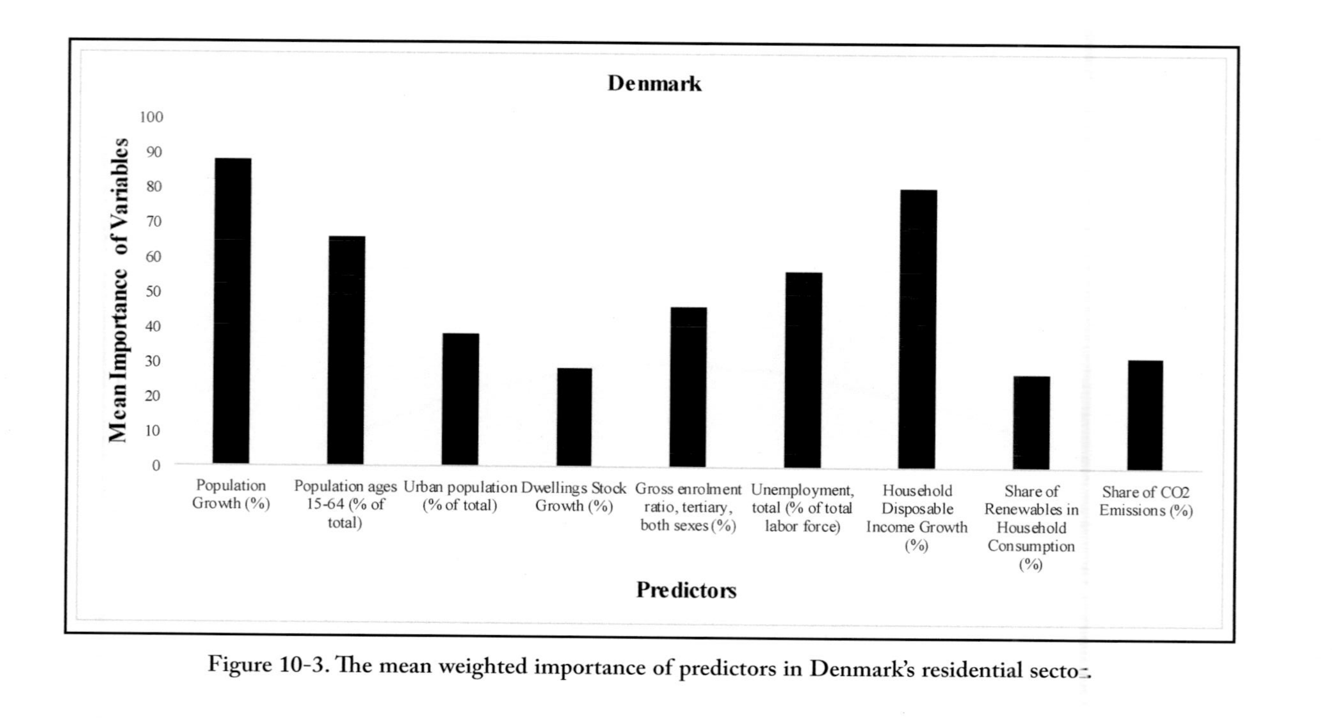

Figure 10-3. The mean weighted importance of predictors in Denmark's residential secto.

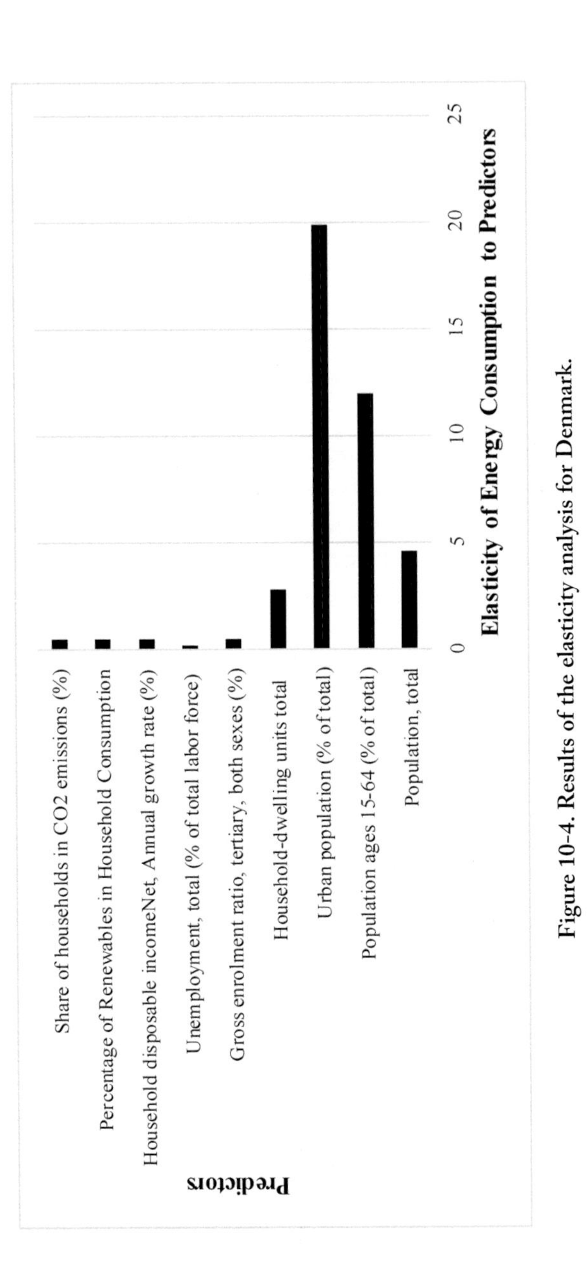

Figure 10-4. Results of the elasticity analysis for Denmark.

Among the models, the MLR had the most reasonable estimation for 2015, and the MARS model predicted the worst one. Based on the ANN and MLR forecasts, the energy use in the domestic sector will decline between 2015 and 2020. This considers that energy use declined between 1991 and 2014. However, the MARS estimated that the energy consumption would increase slightly during this period. Hence, the MARS methodology was unable to efficiently forecast Finland's energy use. Figure 10-5 shows the estimated values for energy use between 2015 and 2020 for Finland's residential sector.

The MARS performed poorly in regard to determining predictor importance. The model revealed that changes in the share of renewables has the most effect on domestic sector energy consumption. The ANN and MLR models presented conclusive evidence that the parameter has an insignificant effect. The two models achieved a consensus on the effectiveness of the unemployment rate, the percentage of population aged 15 to 64, and population growth. The elasticity analysis also showed that urban population, age structure, and population growth have the most importance.

Finland's residential sector energy consumption fluctuated widely between 1991 and 2014. While energy use dropped 3.9% in 1993, it increased 3.6% in 1994. A similar situation also occurred in 2000 and 2001. Most economists believe that the global economic crisis in 2008 deeply affected Finland's economy in 2009 and 2010. During this period, Finland's residential energy use increased 15% while declining 12.5% in 2011. These fluctuations affected the performance of the regression models. Alternative-

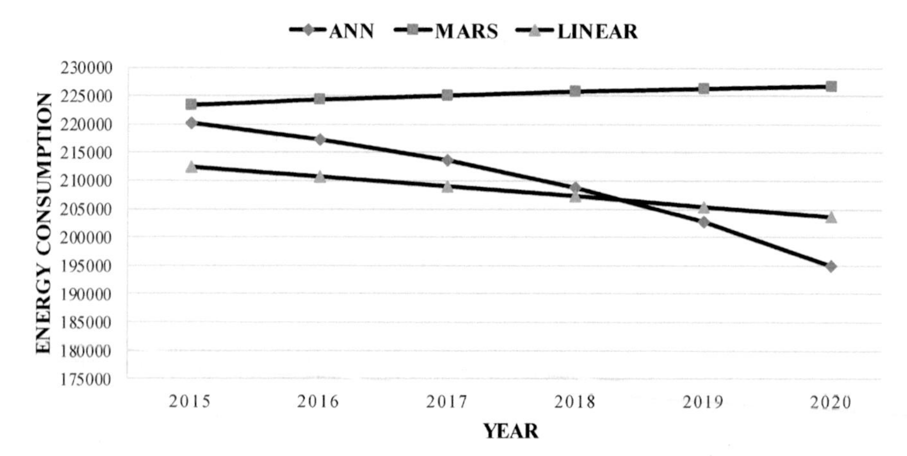

Figure 10-5. Forecasted trends for energy use (Tj) in Finland's domestic sector.

ly, the elasticity analysis results seem to contrast, since domestic energy use experienced sharp fluctuations, and some of the predictors changed gradually over the last two decades. The proportion of people aged 15 to 64, as the active people in Finland, has gradually declined between 1991 and 2014. At the same time, the total population and the urban population increased by 9.5% and 5.9% respectively. Consequently, the age structure in Finland positively effects energy consumption, and surprisingly, the growth of total population and urban population had a negative effect.

There is evidence that in some of the heavily urbanized countries (e.g., Ireland, Denmark, and the UK), electricity usage from apartments and semi-detached houses is much less than for detached houses [36,79,80]. The transformation of Finland's social structure by urbanization between 1990 and 2015 created opportunities for people to live in more compact quarters such as apartments. During this period, the number of attached houses and blocks of flats increased by 248% and 34% respectively. With the country's low population density (17 inhabitants per km^2) and its high demand for space heating, greater urbanization combined with increased biomass use in CHP systems substantially improved domestic sector energy efficiency. Figure 10-6 shows the weighted average importance of the dependent variables. Figure 10-7 shows the elasticity of energy use to the predictors for Finland's domestic sector.

Iceland

The three methodologies efficiently modeled the energy consumption in Iceland's residential sector. The ANN model and MARS model had R-values of 0.99, and the value for the MRL model was 0.96. While all three models had an equal estimation for energy use in 2015, the MARS model predicted an increasing trend for the following years, and the ANN and MLR models predicted a nearly constant distribution for energy use in the following years. The historical data showed that domestic energy use increased 46% between 1991 and 2014. Based on the MARS forecasting model, energy consumption increases 6.1% from 2015 to 2016. This difference between the forecasted values after 2017 occurred because the MARS model was mistaken in choosing the most suitable predictors. Figure 10-8 shows the forecasted domestic sector energy use in Iceland through 2020.

All of the models agreed on the effectiveness of the percentage of urban population and population aged 15 to 64. Figure 10-9 shows the average normalized weighted importance of the dependent variables for Iceland.

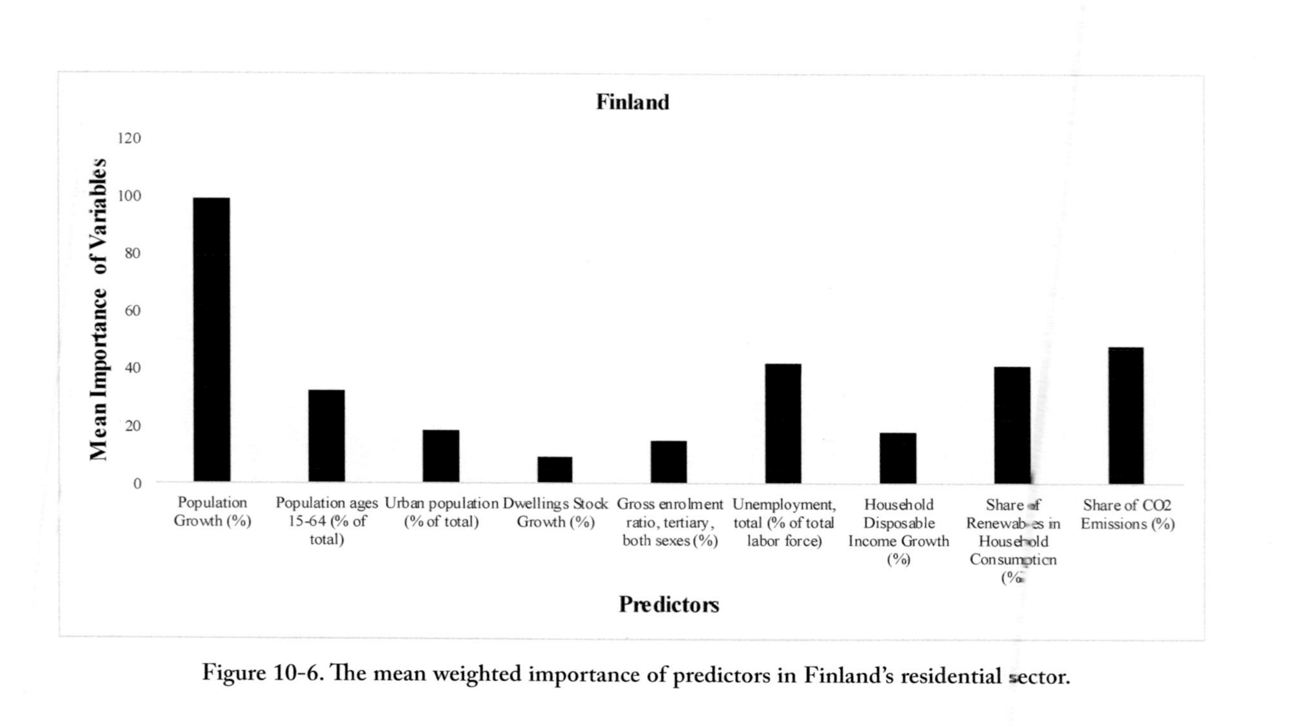

Figure 10-6. The mean weighted importance of predictors in Finland's residential sector.

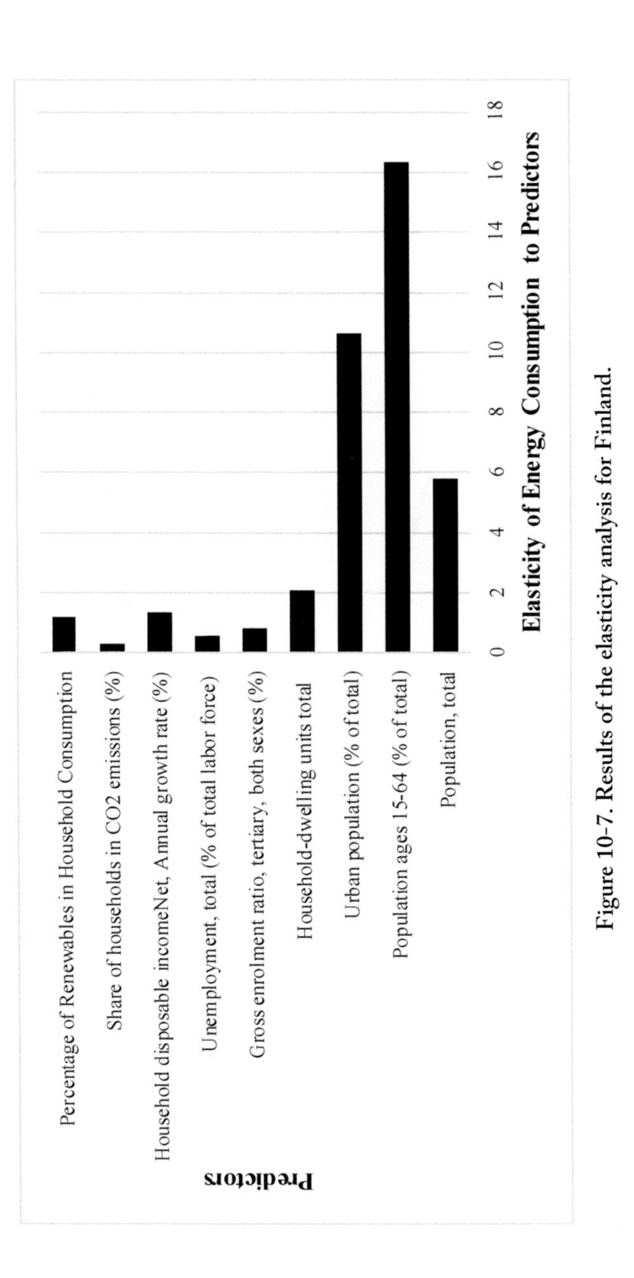

Figure 10-7. Results of the elasticity analysis for Finland.

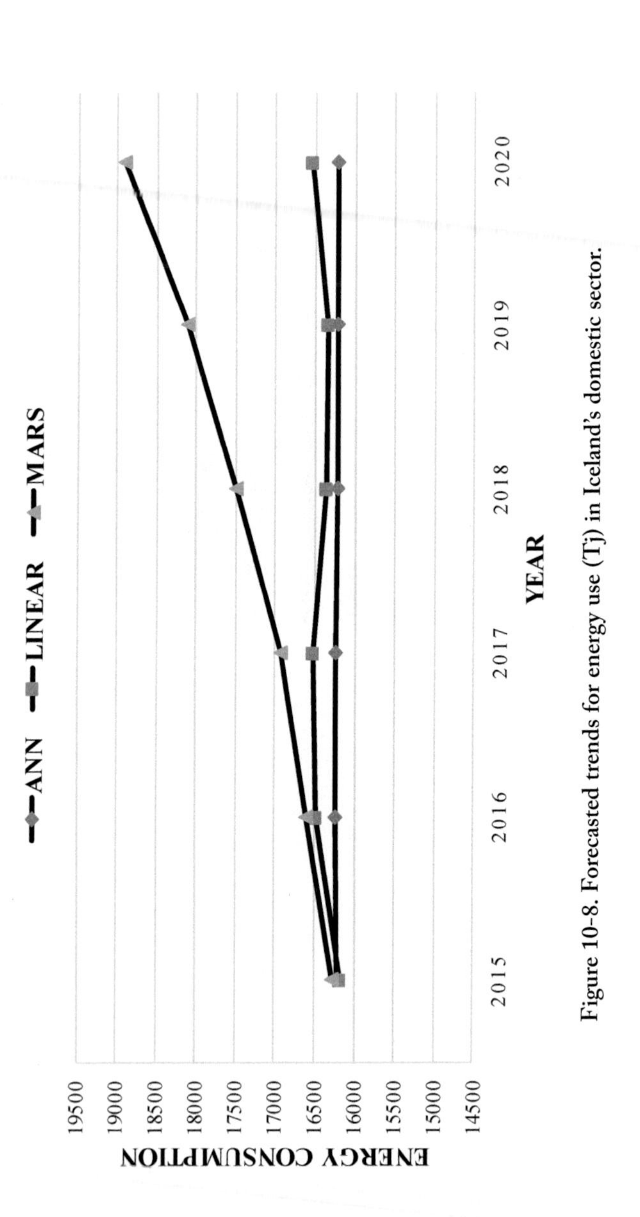

Figure 10-8. Forecasted trends for energy use (Tj) in Iceland's domestic sector.

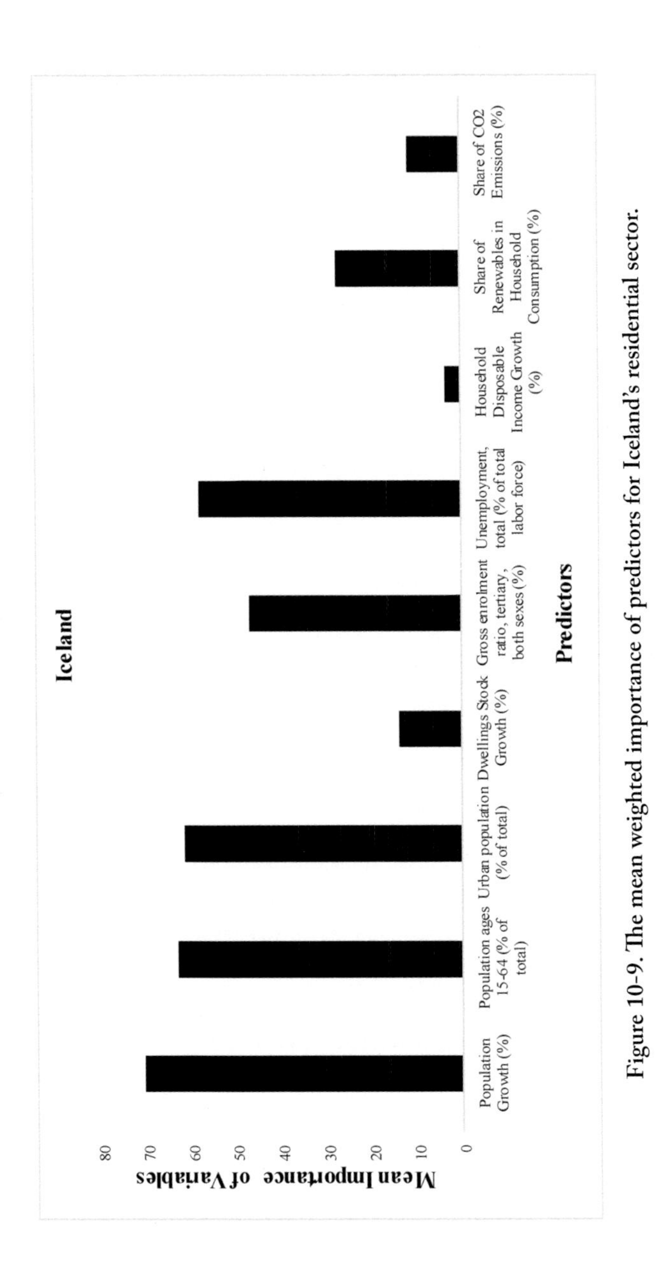

Figure 10-9. The mean weighted importance of predictors for Iceland's residential sector.

The elasticity analysis (see Figure 10-10) shows that the gross enrollment ratio and unemployment rate are insignificant determinants of Iceland's domestic energy use; yet based on the MARS model the two parameters had the most importance. The MLR and ANN models indicated that the residential sector share of total CO_2 emissions and share of renewables in the final energy consumption were significant. However, the two parameters were not used in the MARS proposed model.

Sweden

Although the proposed variables efficiently modeled Sweden's residential energy use, the forecasted values for energy use by 2020 offer entirely different trends (see Figure 10-11). The R-value of the ANN, MARS, and MLR models was 0.99, 0.9, and 0.76 respectively. The ANN model forecasted declining energy use while the MLR model predicted no change for Sweden's residential end-use energy consumption between 2015 and 2020. The MARS model predicted a slightly increasing trend for this period.

There has been some fluctuation in level of Sweden's domestic energy consumption during the last two decades. The biggest changes occurred in 1996 and 2010, when unemployment rates were very high. In places with high unemployment rates and poor weather conditions, people tend to spend more time indoors which increases energy use. Figure 10-12 shows the weighted normal importance of the predictors and Figure 10-13 shows the elasticity of energy use to those predictors. All three models, particularly the MARS, indicated the importance of unemployment as a determinant of energy consumption. Nonetheless, the sector set a record low for energy use in 2014, consuming 277,710 Tj, the least amount in the past two decades. Hence, the fluctuations in historical data adversely affected the models, and forecasted predictors with the AHW algorithm. This caused Sweden's forecasts to be incorrect. The ANN model predicted that Sweden's residential energy consumption would decline 61% by 2020, the MARS model forecasted that it would increase 20%, and the MLR forecasted that it will be constant. The models represent high R^2 values, making judgments about their forecasting capabilities seemingly impossible. Regardless, the models achieved consensus on the effectiveness of some parameters—age structure, unemployment rate, dwelling stock and CO_2 emissions from residential buildings—and for the others there were differences between their magnitudes and ranks.

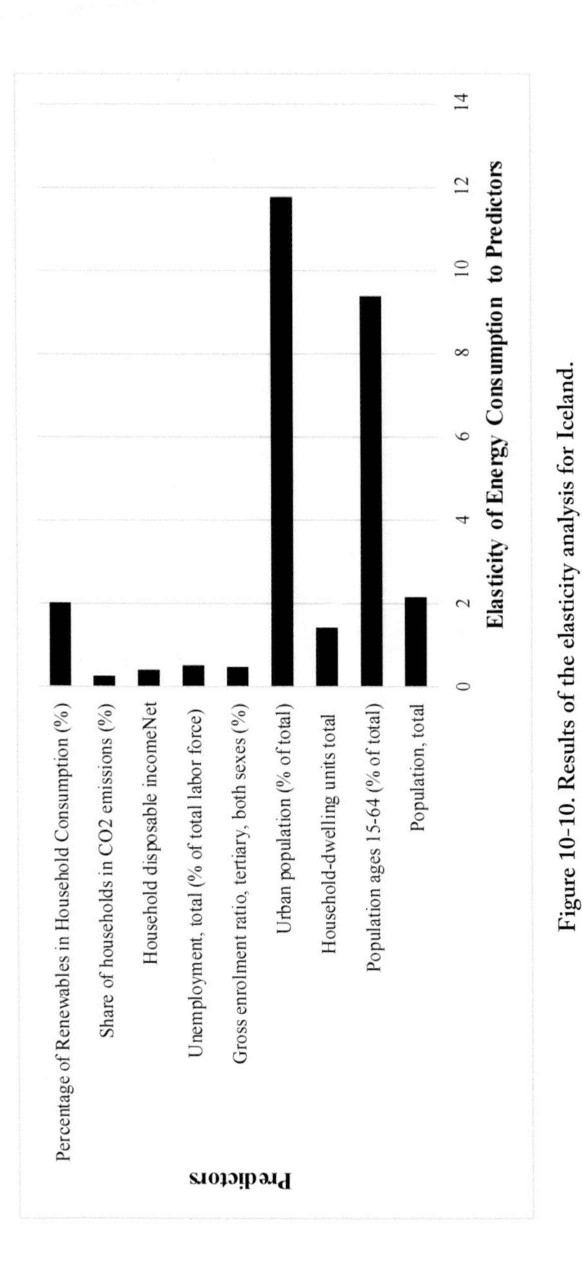

Figure 10-10. Results of the elasticity analysis for Iceland.

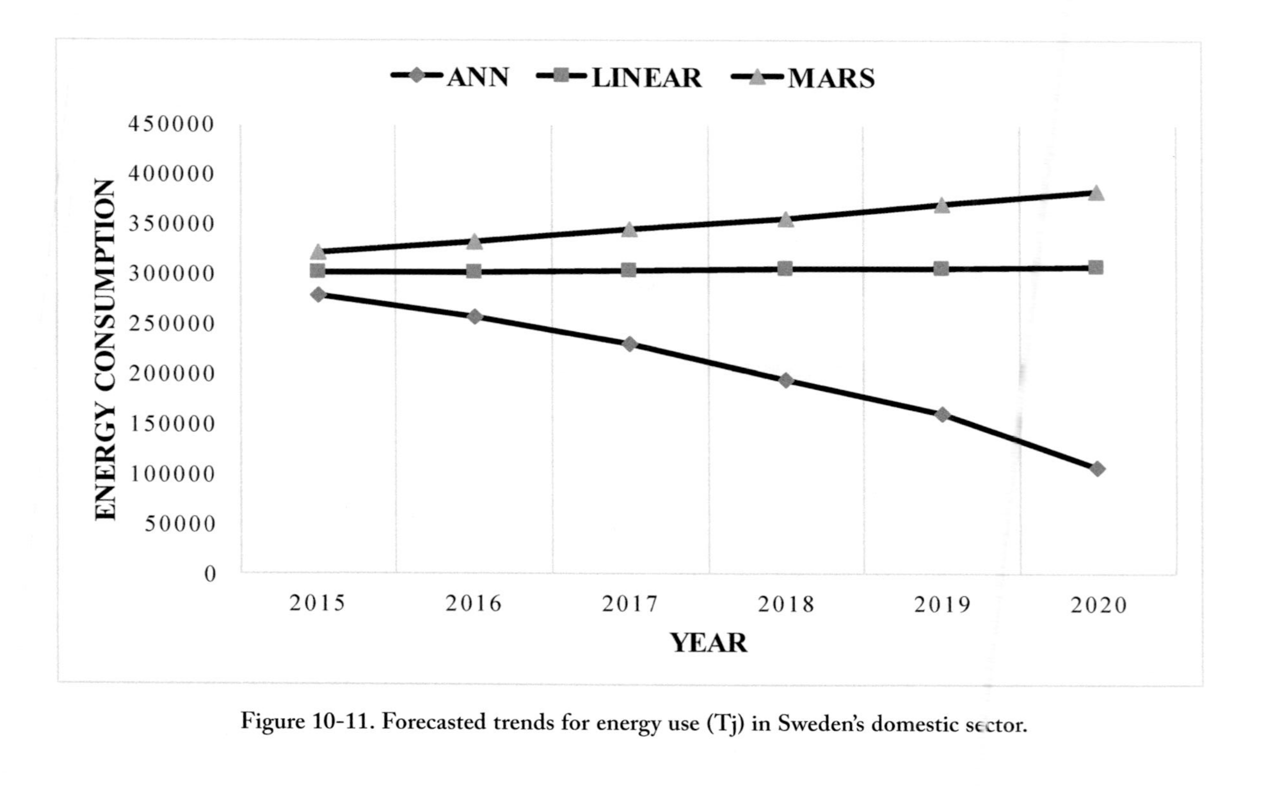

Figure 10-11. Forecasted trends for energy use (Tj) in Sweden's domestic sector.

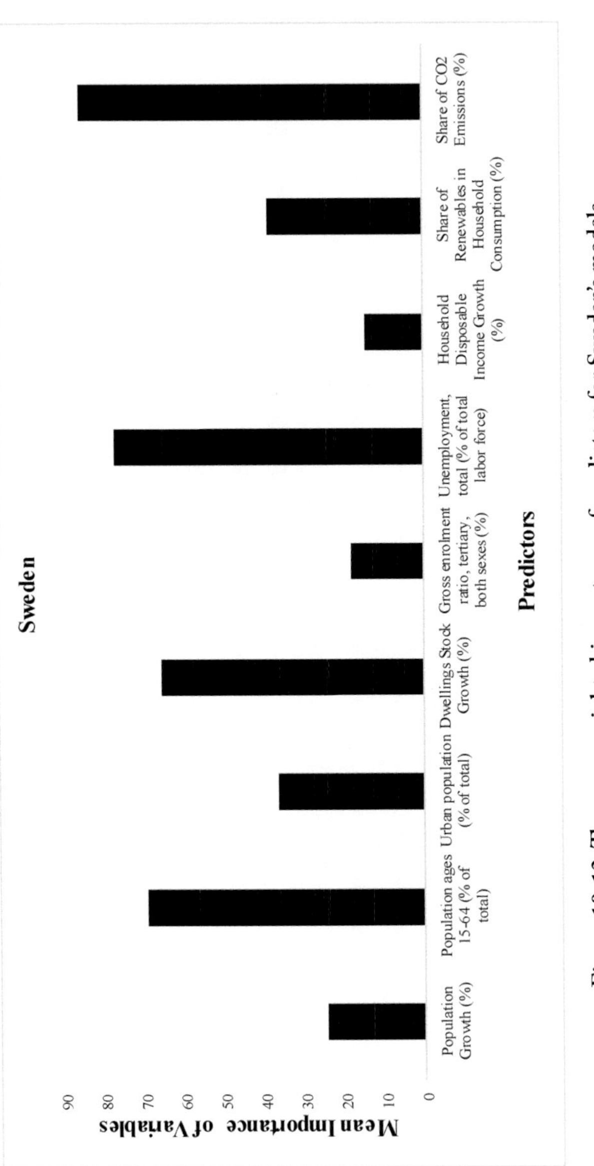

Figure 10-12. The mean weighted importance of predictors for Sweden's models.

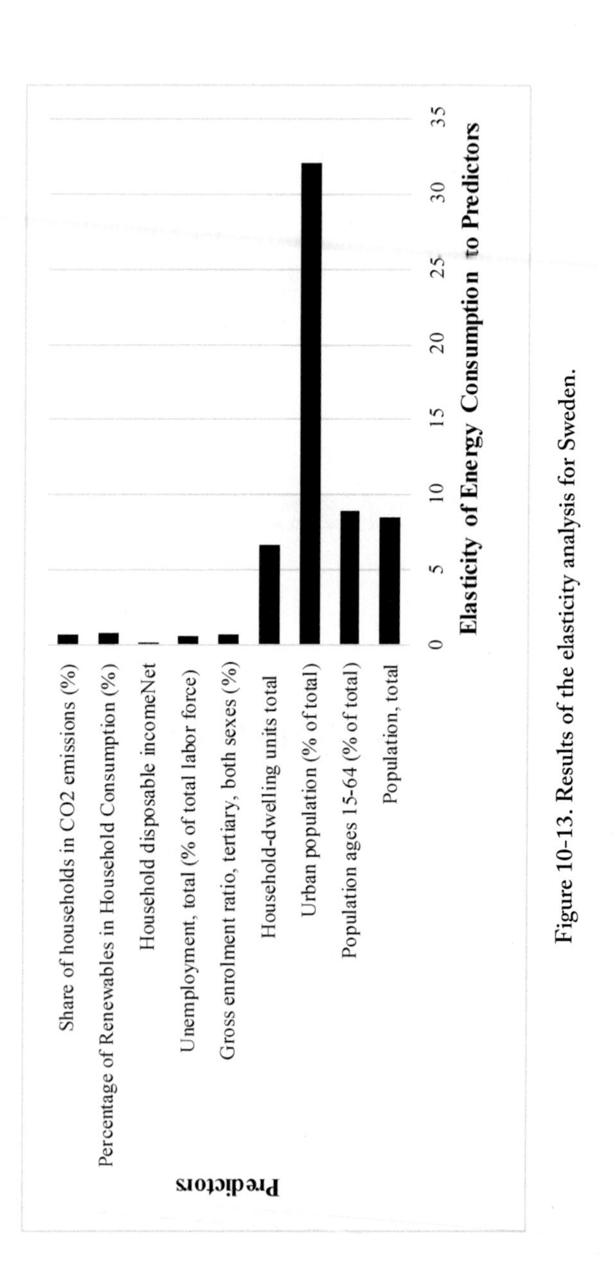

Figure 10-13. Results of the elasticity analysis for Sweden.

Norway

Similar to the previous models, the ANN model had the highest performance considering the R^2 while the MLR model had the lowest. The R-values for the ANN, MARS and MLR models were 0.91, 0.8, and 0.58 respectively. The final 2014 energy use of Norway's domestic sector was 160,606 Tj. Hence, the ANN model, followed by the MARS model provided the most reasonable estimate of 2015 energy use. Both forecasted similar slightly decreasing trends for energy consumption between 2015 and 2020. Figure 10-14 shows the estimated values of energy consumption in Norway's domestic sector. The historical data also show a downward trend in domestic sector per capita energy consumption. However, the total energy use of households increased slightly between 1991 and 2014, particularly after 2008. Despite the effects of the global economic crisis on energy consumption, the number of detached houses and farmhouses equipped with heat pumps increased by 18% between 2009 and 2014. The increase in the number of detached houses and farmhouses during this period adversely affected residential sector energy use. The Norway statistics data show that residents living in these houses invested in heat pump technologies.

The proportion of the population living in Norway's urban areas increased 8% between 1991 and 2014. According to the ANN and MARS models, variations in urban population have the greatest impact on residential energy use (see Figure 10-15). The elasticity analysis revealed that small changes in the age structure, urban population, population growth, gross enrollment ratio, and unemployment rates have considerable impacts on residential energy consumption. The elasticity values of energy consumption to the predictors are provided in Figure 10-16. These parameters are significant in the ANN and MARS models. The two methods efficiently modeled Norway's residential energy use and similarly forecasted energy consumption to 2020. Despite neglecting the share of renewables, the predictors have a close importance value in both models. Norway's models were generated with eight predictors, since valid data were unavailable for the growth rate of dwelling stocks.

Discussion

Three different methodologies were used to model the energy use of the Nordic domestic sectors. Among them, the ANN method efficiently modeled domestic energy use for all of the countries based on the investigated predictors.

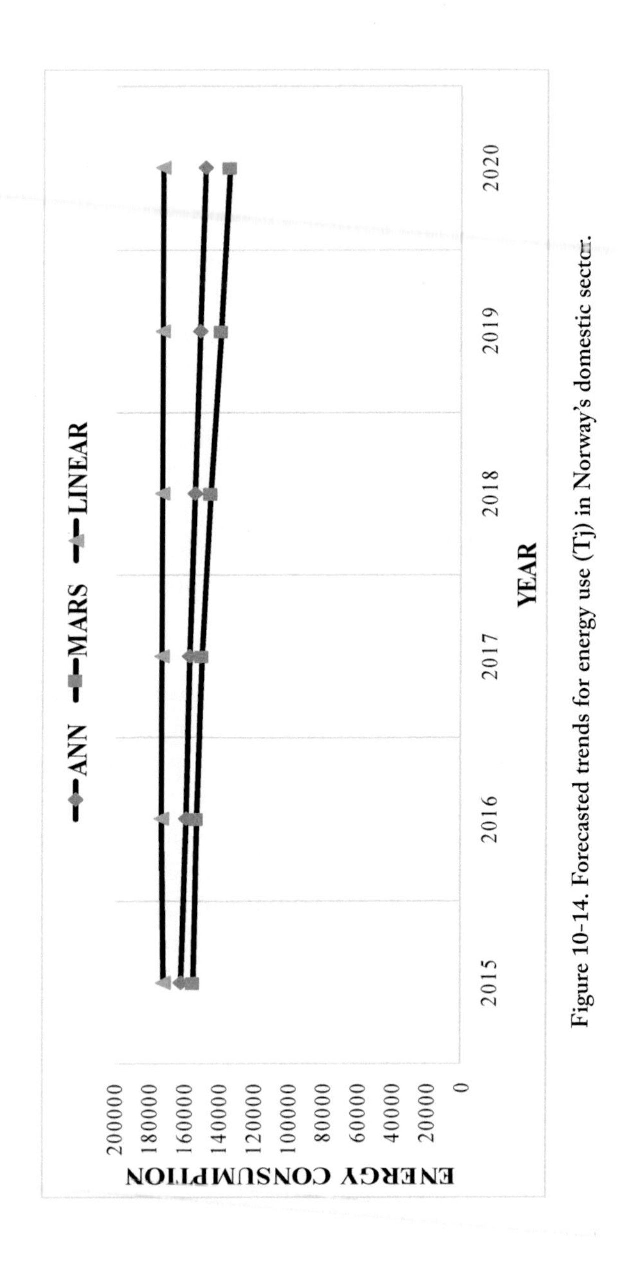

Figure 10-14. Forecasted trends for energy use (Tj) in Norway's domestic sector.

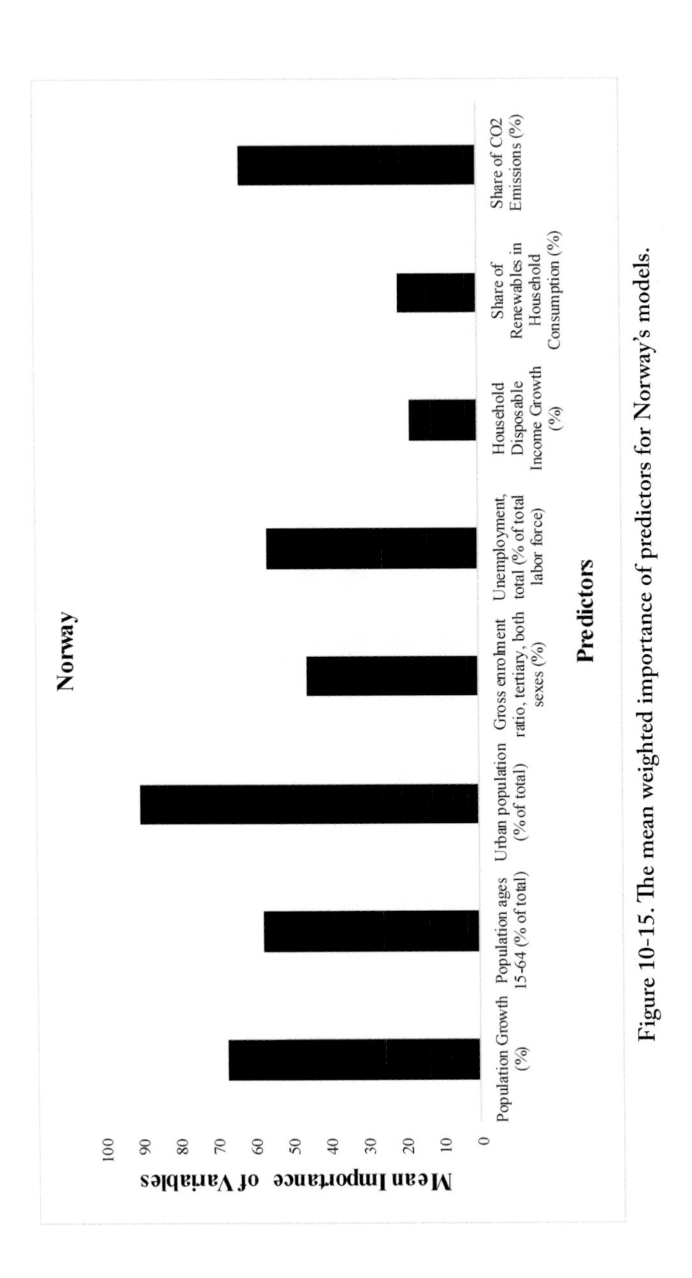

Figure 10-15. The mean weighted importance of predictors for Norway's models.

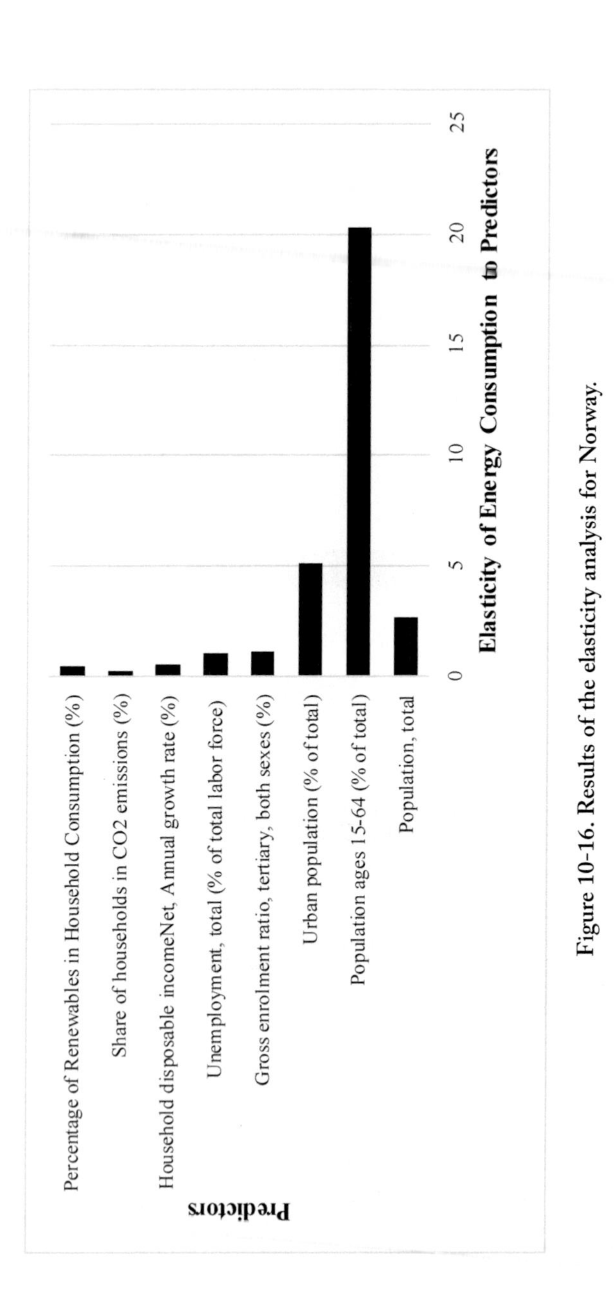

Figure 10-16. Results of the elasticity analysis for Norway.

Forecasting energy use by 2020, the ANN and MLR models offered similar trends for all of the residential sectors, except for Sweden. While showing acceptable performance, the MARS model could not reasonably forecast residential sector energy use by 2020, except for Norway's domestic sector. The study indicated that the MARS is mainly suitable for high input dimensional problems. There is a big disadvantage to using these approaches for forecasting energy use. They require an additional method (i.e., the AHW) to forecast historical data for future years which affects forecasting precision. Precision forecasting needs to consider shorter interval data.

The main purpose of our study was to determine the effects of social and environmental factors on energy use. To this end, we conducted the forecasting analysis mainly to validate the results of models and evaluate their potential to model residential sector energy consumption. The models revealed that changes in some of the investigated factors (total population growth rate, unemployment rate, age structure, urban population, and the share of CO_2 emissions from the residential buildings) has significant effects on annual residential energy use. Over the past two decades the population of the Nordic countries increased by 13.3%. Most of this growth occurred in urban areas. The governments further developed urban areas, enhancing dwelling stocks, especially apartments and attached houses. Since the Nordic countries commonly use CHP systems to meet their heating requirements, urban development policies effectively impact changes in residential sector energy use. Compact apartments and attached houses using CHP and geothermal systems are typically more energy efficient than detached houses.

In the past, the Nordic countries were heavily dependent on imported fossil fuels for their transportation systems and energy intensive industries. Their share of fossil fuels in residential end-use energy consumption was relatively high. This was a direct cause of energy supply problems faced by the countries after the oil crisis in 1973. Afterwards, they heavily invested in renewable energy technologies. Sweden and Denmark have established leadership roles in sustainable energy systems. According to the IEA in 2013, the share of renewables in the final energy consumption of Sweden was greater than 33%. Historical data offer the following for the period between 1991 and 2014:

- Denmark's residential sector relies heavily on solid biofuels and solar thermal energy; their shares have increased.

- Finland's domestic sector uses resources similar to Denmark. Since the Danes live in lower latitudes and have greater solar resources, they have 8.4 times as much solar thermal infrastructure.
- Swedish households also use primary solid biofuels, biogases and solar thermal.
- The share of renewables in the domestic sectors of Iceland and Norway have declined during this period. Iceland uses geothermal resources for primary energy which generated consistent power until 2013 and declined afterwards. Norway's domestic sector relies on renewable resources including solid biofuels and hydro-electricity for heating demand. Energy supply systems must rely on multiple resources to maintain resiliency. Norway and Iceland need new renewable resources to respond to their large residential sector energy demand.

Promoting policies for more compact cities with carbon-free energy systems supports the objectives of improved energy security, health, economic development and sustainability in Nordic countries. There are strong arguments to phase out nuclear power plants, particularly in Sweden. The overall mean weighted normalized importance of population growth and share of urban population indicate the impacts of these relationships.

While past studies showed that unemployment rates are not significant, ours shows that this parameter effectively and indirectly influences residential energy consumption in some Nordic countries [63-65]. We also found a close relationship between energy use in residential energy consumption and age structures in Nordic countries. This means Nordic countries with young populations need to invest more in residential sector energy efficiency and renewable energy. The rapid growth of young populations affects unemployment rates by increasing the size of the labor force. The results showed that the other independent variables are less important in the proposed models. Table 10-6 provides a summary of the importance of variables in different categories.

CONCLUSIONS

This study forecasted domestic sector end-use energy consumption in Nordic countries by 2020, investigating the causal links among energy consumption, total population growth, urban population growth, age structure,

Table 10-6. Importance of investigated predictors in different models and countries.

Countries		Population Growth (%)	Population ages 15-64 (% of total)	Urban population (% of total)	Dwellings Stock Growth (%)	Gross enrolment ratio, tertiary, both sexes (%)	Unemployment, total (% of total labor force)	Household Disposable Income Growth (%)	Share of Renewables in Household Consumption (%)	Share of CO_2 Emissions (%)
Denmark	ANN	69.4	93.1	49.4	25.3	37	86.4	100	56.3	72.2
	MARS	100	54.8	54.8	43.0	48.2	43.0	45.9	0	0
	MLR	100	40.6	0	14.2	58.2	33.7	96.2	20.3	13.4
	Mean	88.0	66.0	38.2	28.5	46.5	56.7	80.3	27.4	31.8
Finland	ANN	100	54.6	19.3	23.4	14.6	58.4	6.7	15.9	54.1
	MARS	99.7	0	0	0	4.0	4.0	0.65	100	2.0
	MLR	97.7	39.8	41.2	0	29.2	65.7	56.0	0.9	100
	Mean	99.3	31.8	18.4	9.1	14.8	41.5	17.6	41.0	48.2
Iceland	ANN	38.4	69.2	100	20.8	7.1	21.6	9.0	57.0	8.2
	MARS	85.0	85.0	24.9	0	100	85.0	0	0	0
	MLR	100	25.5	52.7	22.2	36.6	76.1	0	19.8	32.5
	Mean	70.7	63.2	61.6	14.0	47.0	57.9	3.51	27.5	11.8
Sweden	ANN	48.7	65.9	48.5	100	28.9	56.7	37.0	74.2	63.9
	MARS	0	60.9	36.0	21.0	0	100	0	15.4	98.9
	MLR	20.6	85.6	19.5	72.8	26.7	78.3	0	16.1	100
	Mean	24.5	69.4	36.5	65.5	18.3	77.4	14.46	38.6	85.5
Norway	ANN	91.9	26.7	100		67.1	70.3	34	17.3	51.9
	MARS	50.58	59.9	100		47.9	83.3	0	0	83.3
	MLR	52.2	100	62.8		10.6	0	19.3	54.1	54.1
	Mean	67.32	57.5	90.2		45.6	56.3	18.4	21.0	63.3
Overall Importance		69.4	58.3	49.2	24.5	35.3	60.4	25.5	29.1	48.8

education level, unemployment rate, dwelling stock, share of CO_2 emission, and the share of renewable energies in domestic sectors. Time variations of these parameters is mainly due to changes in the human development index (HDI), developing the economy and urban areas, and the level of investment in renewable and sustainable energies.

The three models offer good performance when considering their R^2, led by the ANN, followed by MARS, and finally the MLR model. Recently developed models like the ANN and MARS provided better performance than the traditional MLR. Although Fumo and Biswas [69] indicated on the performance of statistical approaches, our study revealed that by using these methods, it is difficult to obtain accurate predictions of Nordic domestic sector energy consumption because of the following reasons:

- The MLR approach only looks for the linear relationship between the dependent variable and predictors. Using quantile regression might provide a better solution for this problem in future studies.
- Another factor limiting the performance of the MLR is the assumption of the predictors being completely independent variables. While there are close relationships between most of the predictors in this study, the results also showed that using multilevel models like the ANN may enhance prediction accuracy.
- Though the MARS method makes no assumptions about the types of relationships between domestic energy use and predictors, this method (like the MLR) assumes no link between predictors in the corresponding model.
- The ANN method is a data-driven approach and needs no assumptions regarding the type and form of model. The methodology has disadvantages. Despite its impressive performance for prediction, it is difficult to interpret the generated model. The model is susceptible to over-training, which may cause instability in the obtained results. Accurate forecasting with the ANN requires a thorough understanding of the model. Using artificial intelligence networks is recommended for future studies.

Despite these drawbacks, our study revealed interesting information about the determinants of domestic energy use in Nordic countries. Despite the view of previous works about the effects of unemployment rates on domestic energy use, the sensitivity analysis showed that the parameter had a large effect on Nordic residential energy consumption between 1990 and 2014. The other important determinant is the share of residential sector CO_2 emissions. Except for Denmark, this factor has a key role in the models generated for Nordic countries. The other three major predictors are population growth, proportion of population in urban areas, and the share of people aged from 15 to 64 (work force). In most of the generated models, the work force parameter and the unemployment rates had similar normalized importance, led by the second, which demonstrates a close relationship between the two parameters.

The proportion of populations living in urban areas is a primary determinant of domestic energy use in Nordic countries. High regional heating demand and the use of district heating and geothermal energy enhances the importance of concentrating populations in Nordic countries. Using energy-efficient boilers in combination with inexpensive carbon-free energy in district heating systems has improved energy efficiency in the Nordic domestic sector during the last two decades.

The Nordic governments should heavily invest in residential infrastructure to accommodate an additional 3.9 million residents by 2050, plus further develop urban areas, district heating systems, and geothermal technologies to maintain or improve levels of energy. The issue raises an alarm for country's like Norway with little investment in district heating systems during the last two decades. Electricity is not the best future energy source for heating in Norway. A better option is for these countries to invest in small and medium CHP systems, especially in rural areas. Larger cities should enhance the capacity of large CHP and district heating systems, since these systems are more energy-efficient and environmentally friendly for Nordic climates.

Acknowledgements

The data and projections used in our study have been excerpted from authentic sources. The authors appreciate the assistance provided by following institutions:

International Energy Agency, www.iea.org.

The Worldbank group, www.Worldbank.org.

Danish Statistics Agency, www.statbank.dk.

Danish Energy Agency—Energistyrelsen, www.ens.dk.

Finnish Statistics Agency—Tilastokeskus, www.stat.fi.

Icelandic Statistics Agency, www.statice.is.
Norwegian Statistics Agency, www.ssb.no.
Swedish Statistics Agency, www.scb.se.
Nordic Council of Ministers and the Nordic Council, www.norden.org.

References

[1] Energy Policies of IEA Countries - Denmark (2011). Review https://www.iea.org.
[2] Energy Policies of IEA Countries - Finland (2013). Review. https://www.iea.org.
[3] Energy Policies of IEA Countries - Norway (2011). Review. https://www.iea.org.
[4] Energy Policies of IEA Countries - Sweden (2013). Review. https://www.iea.org.
[5] Karlsson, K., Münster, M., Skytte, K., Pérez, C., Venturini, G. and R. Salvucci (2016). Nordic energy technology perspectives.
[6] International Energy Agency (2014). Statistics IE. Balances, Denmark. https://www.iea.org.
[7] Railio, J. (2005). Energy Performance of Buildings Directive. Influences on European standardization and on ventilation and air-conditioning industry, update and follow-up. Page 3.
[8] Kitzing, L., Katz, J., Schröder, S., Morthorst, P. and F. Andersen The residential electricity sector in Denmark: a description of current conditions. Working paper, Technical Univrsity of Denmark, Kgs. Lyngby. http://orbit.dtu.
[9] The Danish Energy Agency. Energy efficiency trends and policies in Denmark. www.odyssee-mure.eu.
[10] Barriers for flexibility in the district heating-electricity interface (2016). www.lsta.com.
[11] Whitehead, F. (2014). Lessons from Denmark: how district heating could improve energy security. *The Guardian*.
[12] International Energy Agency (2014). Statistics IE. Balances, Finland. https://www.iea.org.
[13] Paiho, S. and F. Reda. (2016). Towards next generation district heating in Finland. *Renewable and Sustainable Energy Reviews*, 65, pages 915-24.
[14] Znouda, E., Ghrab-Morcos, N. and A. Hadj-Alouane (2007). Optimization of Mediterranean building design using genetic algorithms. *Energy and Buildings*, 39(2), pages 148-53.
[15] International Energy Agency (2014). Statistics IE. Balances. https://www.iea.org.
[16] National Energy Authority of Iceland (2016). Geothermal. http://www.nea.is/geothermal.
[17] Mäntysaari, P. (2015). *E.U. electricity trade law: the legal tools of electricity producers in the internal electricity market*. Springer.
[18] Pool, N. (2007). Annual Report. www. nordpoolspot. com/about.
[19] International Energy Agency (2014). Statistics IE. Balances. Norway. https://www.iea.org.
[20] Gebremedhin, A. (2012). Introducing district heating in a Norwegian town–potential for reduced local and global emissions. *Applied Energy*, 95, pages 300-4.
[21] Statistics Norway (2012). Energy consumption in households. https://www.ssb.no/en/husenergi.
[22] Eva Rosenberg Institute for Energy Technology (2015). Energy efficiency trends and policies in Norway. www.odyssee-mure.eu.
[23] International Energy Agency (2014). Statistics IE. Balances. Sweden. https://www.iea.org.

[24] International Energy Agency (2016). The IEA CHP and DHC Collaborative CHP/DHC Scorecard: Sweden. http://www.iea.org.

[25] Swan, L. and Ugursal, VI. (2009). Modeling of end-use energy consumption in the residential sector: a review of modeling techniques. *Renewable and Sustainable Energy Reviews*, 13(8), pages 1,819-35.

[26] Friedman, J. (1991). Multivariate adaptive regression splines. The annals of statistics. Pages 1-67.

[27] Lomas, K. (2010). *Carbon reduction in existing buildings: a transdisciplinary approach*. Taylor and Francis.

[28] Oreszczyn, T. and R. Lowe (2010). Challenges for energy and buildings research: objectives, methods and funding mechanisms. *Building Research and Information*, 38(1), pages 107-22.

[29] Jones, R., Fuertes, A. and K. Lomas (2015). The socio-economic, dwelling and appliance related factors affecting electricity consumption in domestic buildings. *Renewable and Sustainable Energy Reviews*, 43, pages 901-17.

[30] Bedir, M., Hasselaar, E. and L. Itard (2013). Determinants of electricity consumption in Dutch dwellings. *Energy and Buildings*, 58, pages 94-207.

[31] Tso, G. and K. Yau (2007). Predicting electricity energy consumption: a comparison of regression analysis, decision tree and neural networks. *Energy,* 32(9), pages 1,761-8.

[32] Wiesmann, D., Azevedo, I., Ferrão, P. and J. Fernández (2011). Residential electricity consumption in Portugal: findings from top-down and bottom-up models. *Energy Policy*, 39(5), pages 2,772-9.

[33] Druckman, A. and T. Jackson (2008). Household energy consumption in the UK: a highly geographically and socio-economically disaggregated model. *Energy Policy*, 36(8), pages 3,177-92.

[34] Kavousian, A., Rajagopal, R. and M. Fischer (2013). Determinants of residential electricity consumption: using smart meter data to examine the effect of climate, building characteristics, appliance stock, and occupants' behavior. *Energy*, 55, pages 184-94.

[35] Brounen, D., Kok, N. and J. Quigley. (2012). Residential energy use and conservation: economics and demographics. *European Economic Review*, 56(5), pages 31-45.

[36] Leahy, E. and S. Lyons (2010). Energy use and appliance ownership in Ireland. *Energy Policy*, 38(8), pages 4,265-79.

[37] Tso, G. and K. Yau (2003). A study of domestic energy usage patterns in Hong Kong. *Energy*, 28(15), pages 1,671-82.

[38] Zhou, S. and F. Teng (2013). Estimation of urban residential electricity demand in China using household survey data. *Energy Policy*, 61, pages 394-402.

[39] The Nordic Council of Ministers. The population Nordic cooperation. http://www.norden.org.

[40] O'neill, B. and S. Chen (2002). Demographic determinants of household energy use in the United States. *Population and Development Review*, 28, pages 53-88.

[41] Liddle, B. (2004). Demographic dynamics and per capita environmental impact: using panel regressions and household decompositions to examine population and transport. *Population and Environment*, 26(1), pages 23-39.

[42] Prskawetz, A., Leiwen, J. and B. O'Neill (2004). Demographic composition and projections of car use in Austria. *Vienna Yearbook of Population Research*, pages 175-201.

[43] Liddle, B. (2004). Impact of population, age structure, and urbanization on carbon emissions/energy consumption: evidence from macro-level, cross-country analyses. *Population and Environment*, 35(3), pages 286-304.

[44] Jorgenson, A., Rice, J. and B. Clark (2010). Cities, slums, and energy consumption in less developed countries, 1990 to 2005. *Organization and Environment*, 23(2), pages 189-204.

[45] York, R. (2007). Demographic trends and energy consumption in European Union Na-

tions, 1960–2025. *Social Science Research*, 36(3), pages 855-72.

[46] York, R. editor (2007). *Structural influences on energy production in south and east Asia, 1971–20021*. Sociological Forum: Wiley Online Library.

[47] Okada, A. (2012). Is an increased elderly population related to decreased CO_2 emissions from road transportation? *Energy Policy*, 45, pages 286-92.

[48] Menz, T. and H. Welsch (2012). Population aging and carbon emissions in OECD countries: accounting for life-cycle and cohort effects. *Energy Economics*, 34(3), pages 842-9.

[49] Martínez-Zarzoso, I. and A. Maruotti (2011). The impact of urbanization on CO_2 emissions: evidence from developing countries. *Ecological Economics*, 70(7), pages 1,344-53.

[50] Liddle, B. and S. Lung (2010). Age-structure, urbanization, and climate change in developed countries: revisiting STIRPAT for disaggregated population and consumption-related environmental impacts. *Population and Environment*, 31(5), pages 317-43.

[51] York, R. (2008). De-carbonization in former Soviet republics, 1992–2000: the ecological consequences of de-modernization. *Social Problems*, 55(3), pages 70-90.

[52] Blázquez, L., Boogen, N. and M. Filippini (2013). Residential electricity demand in Spain: new empirical evidence using aggregate data. *Energy Economics*, 36, pages 648-57.

[53] Halvorsen, B. and B. Larsen (2001). Norwegian residential electricity demand—a microeconomic assessment of the growth from 1976 to 1993. *Energy Policy*, 29(3), pages 227-36.

[54] Fan, J., Zhang, Y. and B. Wang (2016). The impact of urbanization on residential energy consumption in China: an aggregated and disaggregated analysis. *Renewable and Sustainable Energy Reviews*.

[55] Sun, C., Ouyang, X., Cai, H., Luo, Z. and A. Li (2014). Household pathway selection of energy consumption during urbanization process in China. *Energy Conversion and Management*, 84, pages 295-304.

[56] Ali, H., Law, S. and T. Zannah (2016). Dynamic impact of urbanization, economic growth, energy consumption, and trade openness on CO_2 emissions in Nigeria. *Environmental Science and Pollution Research*, 23(12), pages 12,435-43.

[57] Yuan, B., Ren, S. and X. Chen (2015). The effects of urbanization, consumption ratio and consumption structure on residential indirect CO_2 emissions in China: a regional comparative analysis. *Applied Energy*, 140, pages 94-106.

[58] Wang, Q., Zeng, Y. and B. Wu (2016). Exploring the relationship between urbanization, energy consumption, and CO_2 emissions in different provinces of China. *Renewable and Sustainable Energy Reviews*, 54, pages 1,563-79.

[59] Ru, M., Tao, S., Smith, K., Shen, G., Shen, H. and Y. Huang (2015). Direct energy consumption associated emissions by rural-to-urban migrants in Beijing. *Environmental Science and Technology*, 49(22), pages 13,708-15.

[60] Aixiang, T. (2011). Research on relationship between energy consumption quality and education, science and technology based on grey relation theory. *Energy Procedia*, pages 1,718-21.

[61] The Nordic Council of Ministers. Total research and development expenditure Nordic cooperation. http://www.norden.org.

[62] The Nordic Council of Ministers. Educational attainment at upper- and post-secondary level Nordic cooperation. http://www.norden.org.

[63] Yohanis, Y., Mondol, J., Wright, A. and B. Norton (2008). Real-life energy use in the UK: how occupancy and dwelling characteristics affect domestic electricity use. *Energy and Buildings*, 40(6), pages 1,053-9.

[64] Cramer, J., Miller, N., Craig, P., Hacket, B., Dietz, T. and E. Vine (1985). Social and engineering determinant and their equity implications in residential electricity use. *Energy*, 10(12), pages 1,283-91.

[65] Frederiks, E., Stenner, K. and Hobman, E. (2015). The socio-demographic and psychological predictors of residential energy consumption: a comprehensive review. *Energies*, 8(1), pages 573-609.
[66] The Nordic Council of Ministers. Denmark—per cent employed of the population between 15-64 Nordic cooperation. http://www.norden.org.
[67] Feng, Y., Chen, S. and L. Zhang (2013). System dynamics modeling for urban energy consumption and CO_2 emissions: a case study of Beijing, China. *Ecological Modelling*, 252, pages 44-52.
[68] Hossain, M., Li, B., Chakraborty, S., Hossain, M. and M. Rahman (2015). A comparative analysis on China's energy issues and CO_2 emissions in global perspectives. *Sustainable Energy*, 3(1), pages 1-8.
[69] Fumo, N. and M. Biswas (2015). Regression analysis for prediction of residential energy consumption. *Renewable and Sustainable Energy Reviews*, 47, pages 332-43.
[70] Aiken, L., West, S. and S. Pitts (2003). Multiple linear regression. Handbook of psychology.
[71] Wang, S. (2003). Artificial neural network. *Interdisciplinary computing in java programming*. Springer, pages 81-100.
[72] Bertolini, M., Bevilacqua, M. and F. Ciarapica, editors (2010). Re-engineering the forecasting phase using traditional and soft computing methods. Industrial Engineering and Engineering Management. 2010 IEEE International Conference. IEEE.
[73] Holt-Winters seasonal method. https://www.otexts.org.
[74] Williams, K. and J. Gomez (2016). Predicting future monthly residential energy consumption using building characteristics and climate data: a statistical learning approach. *Energy and Buildings*, 128, pages 1-11.
[75] Fridedman, J. (1991). Multivariate adaptive regression splines (with discussion). *Annual Statistics*, 19(1), pages 79-141.
[76] Jung, S. and S. Lee (2006). In situ monitoring of cell concentration in a photobioreactor using image analysis: comparison of uniform light distribution model and artificial neural networks. *Biotechnology Progress*, 22(5), pages 1,443-50.
[77] Garson, D. (1991). Interpreting neural network connection weights.
[78] Montano, J., and A. Palmer (2003). Numeric sensitivity analysis applied to feedforward neural networks. *Neural Computing and Applications*, 12(2), pages 119-25.
[79] Bartiaux, F. and K. Gram-Hanssen editors (2005). Socio-political factors influencing household electricity consumption: a comparison between Denmark and Belgium. ECEEE summer study proceedings.
[80] Wyatt, P. (2013). A dwelling-level investigation into the physical and socio-economic drivers of domestic energy consumption in England. *Energy Policy*, 60, pages 540-9.

Chapter 11
Saving Energy—Politics or Business?

Volodymyr (Vladimir) Mamalyga, Ph.D., CEM

The recent history of nations shows that the greatest successes in the field of energy have been achieved by the wealthy countries in Western Europe, North America plus Japan. Recently, greater attention to energy saving and environmental protection is occurring in wealthy countries of the Middle East and the developing countries of China and India. Less wealthy countries lack the financial resources to implement widespread energy conservation and environmental protection improvements. Many leading international companies have transferred production facilities to these countries, reducing their production costs and taxes while avoiding the costs of energy savings improvements and stiffer environmental regulations elsewhere.

Many countries can be justifiably proud of their evolution toward energy efficiency. However, are optimal decisions regarding the use of energy efficient technologies based solely on economics? The answer to this question is certainly not and the reasons are often related to political circumstances rather than economic. This chapter presents a pragmatic approach to feasibility assessments to achieve reductions in energy usage and generate energy cost savings. Case assessments including lamp replacements, wind turbine generators, frequency converters, throttling, and electric motors show that using energy efficient equipment is not always feasible. Their implementation is largely explained by political influences in the project implementation decision-making process.

HISTORICAL INFLUENCES

The first energy crisis in October 1973 was spawned by an oil embargo by the Organization of Arab Petroleum Exporting Countries (OAPEC) in response to the Yom Kippur War. This forced the countries of Western Europe and North America to tackle in earnest the problems of energy saving as one way to ensure the security of energy supplies. Concerns regarding energy sources and costs have since acquired a permanent character in international politics.

Many positive steps towards energy conservation and efficiency have since occurred. Countries have succeeded in growing their economies with simultaneous reductions (rather than growth) in consumption of energy. The assumed correlation between energy use and gross domestic product (GNP) was decoupled. Newer types of energy-efficient equipment were successfully designed and placed in operation on a massive scale. Renewable power generation has become a solution to the environmental issues caused by fossil fuel use.

Moreover, energy saving was transformed from a political choice for the countries suffering from high energy costs and power shortages into a business for manufacturers of equipment, contractors, consulting organizations, banks and energy service companies (ESCOs). People in developed nations began understand the importance of deploying energy saving measures, while their governments encouraged the implementation of new technologies with supportive regulations.

CASE EXAMPLES

Efficient Lamp Replacement

I once witnessed a conversation between an executive of a large Ukrainian enterprise and a salesman of a western company that produced energy-efficient lamps.

Salesman: "Our lamps are the most efficient, reliable, beautiful..."

Executive: "Well, do you know that our enterprise uses still more reliable and efficient lamps?" After the surprised salesman attempted but failed to object, the director added: "*I simply do not turn them on!*" This statement was made without substantiation and calculations, but for the enterprise energy expenses were insignificant. There were other more important projects for the enterprise and decisions to invest in those projects had already occurred.

It is not reasonable to reject proposals that under specific circumstances may prove to be more advantageous. However, to make a purchasing decision simply because it is common practice in western countries would be a mistake. In cases like this, it is expedient to calculate the comparative costs of imported compact fluorescent (CFL) or light emitting diode (LED) lamps with respect to incandescent lamps, determine whether their actual service life in the Ukraine and other post-Soviet countries can be validated, then verify lighting fixture efficiency and daily hours of operation. Perhaps it is better to introduce more daylighting by regularly cleaning the windows.

Economic assessments, such as net present value (NPV), simple payback period (SPP), or internal rate of return (IRR), should determine the expediency of implementing energy saving projects. Unfortunately, dealers of leading manufacturers of energy-efficient equipment seldom have command of the methodology of such calculations.

Example*: To illuminate work areas, one can use either one CFL with the service life of 10,000 hours or ten conventional incandescent lamps with the service life of T = 1,000 hours each. The price of conventional incandescent lamp (PR_{il}) equals \$0.34 USD, while the price of a CFL (PR_{eel}) was \$10.65 (U.S.). The power of one incandescent lamp amounts to P_{il} = 100 W, while that of a CFL is P_{eel} = 21 W. Calculations were conducted for the following values of electricity tariff t_{el}: \$0.02/kWh, \$0.04/kWh, \$0.06/kWh, \$0.08/kWh and \$0.1/kWh. The task is to compare the feasibility of implementing the system of lighting on the basis of one CFL (first alternative) or on the basis of 10 conventional incandescent lamps taking into account the rated service life (10,000 hours/1,000 hours = 10 lamps, second alternative).

The calculations were performed for different values of the rate for bank credits (from 10% to 12% annually) for the case when the system of lighting operates 24 hours per day.

The results of calculating purchase and operation costs for the CFL (*Co1*) and the ten incandescent lamps (*Co2*) for the case of the electricity tariff t_{el} = \$0.02/(kWh) are presented in Figure 11-1. The intersection point of plots for *Co1* and *Co2* has the coordinates interest rate, i = 0.137 and lamp cost *Co* = \$12.95. The analysis shows that the use of the CFL at t_{el} = \$0.02/kWh will be economically viable only in the case of available credit at an annual interest rate less than 13.7%. Difficulties in attracting such investments in Ukraine at the electricity tariffs at the time of the assessment made the use of CFLs uneconomical. Nevertheless, there was a robust market for installing energy-efficient lamps and fixtures in offices of

*This example was made using prices and recommendations from the National Standard of Ukraine DSTU 4065-2001 [2].

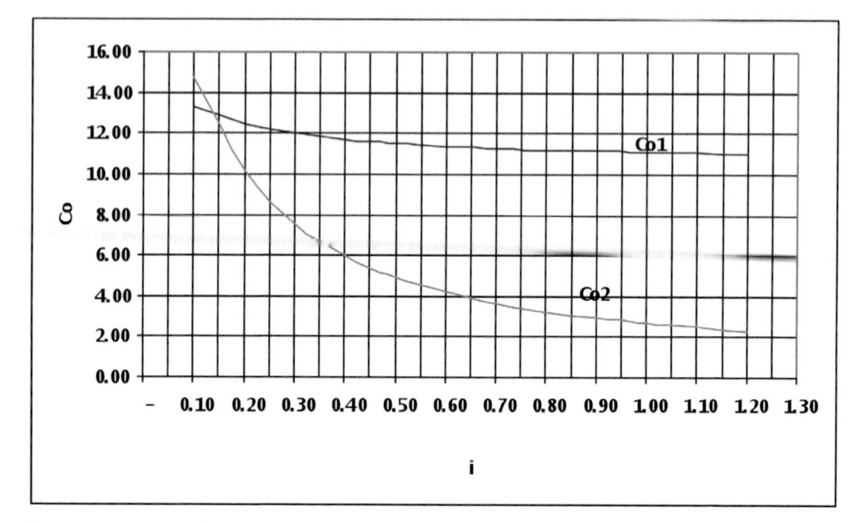

Figure 11-1. Purchase and operation costs of one energy-efficient lamp (*Co1*) and ten incandescent lamps (*Co2*) for case t_{el} = $0.02/kWh at interest rate *i*.

commercial buildings.

The results of similar calculations performed for incremental electricity tariffs are presented in Table 11-1.

Table 11-1. Electric prices, interest rates and lamp costs.

T_{el}, $/kWh	i, rel. units	Co, $
0.02	0.137	12.95
0.04	0.28	13.15
0.06	0.44	13.27
0.08	0.62	13.3
0.1	0.79	13.33

The analysis indicates that the use of CFLs instead of incandescent lamps (given the price for these lamps at the time of assessment and the cost of credit and electricity tariffs) was economically viable only when electricity prices rose above $0.06/kWh.

Similar calculations performed by a company called Ekektromekhanika during an energy audit of the Kiev Metro, where lighting installations operate an average of 20 hours per day, demonstrated that fluorescent lamps were the most efficient even when disposal costs were included, next followed by CFLs, while incandescent lamps were the most expensive. Hence, fluctuations in lamp prices and electricity tariffs change the economics associated with using different kinds lighting equipment which influences the priority of installing new lighting equipment.

Wind Turbine Generators

Almost 10 years ago a Danish commune installed a wind turbine generator (WTG) having the power of 200 kW (at the cost of ≈ €200,000 EUR) using loan proceeds at an annual interest rate of 8%. This WTG produced and average of 5,000,000 kWh annually. With an electricity cost of 0.43 Danish Krone (€0.0581) for one kWh, this production yields an economic value of 215,000 DKK annually. The following expenses are mandatory during each year:

Insurance	10,000 DKK
Service and maintenance	12,000 DKK
Repairs	8,000 DKK

Calculations showed that the payback period of this project was 13.3 years, only slightly less than the service life of such installations which is 15 to 20 years. Using a WTG service life of 15 years, the NPV related to implementation was €13,985, while the IRR was about 9.1%. The internal rate of return is only slightly higher than the interest rate of 8% per annum, which seems insufficient from a business investment viewpoint.

Implementation of such projects from an economic perspective is rather risky, yet may be justified by the minimization of economic and political risks. Such risks include the high probability of energy curtailments, rising electricity prices, and the use of higher cost fossil fuels.

Electric Drive Systems with Frequency Converters

Promotional materials of manufacturers of electric drive systems frequently include a statement similar to this: *"The use of frequency converters in variable drives of pumping units makes it possible to save 20%-50% of the electric power consumed."*

Figure 11-2 shows a comparison of the efficiency of different techniques designed to control the capacity of equipment with fan-type load characteristics [1]. The following notations were used in this figure:

a Throttling
b Blade rotation of the distributor
c Variable hydraulic couplings or electromagnetic slip couplings
d Variation of the resistance in the rotor circuit of the induction motor
e Control of the turbomachine rotation speed free of losses (ideal curve)

The analysis indicates that the most rational technique for controlling the capacity of blade type machines is the efficient variable speed control of drive motors. In order to achieve a 50% saving, the machine in question should operate continuously with the capacity reduced by 40% to 50%. This stipulates the need of pump operation with the efficiency reduced by more 3% to 5%. In this case the installation does not operate under the nominal conditions (i.e., outside the region of maximum efficiency of the pump, motor and frequency converter). At such variable speeds, control ranges of the motor speed ω ($0.5 \cdot \omega_r \leq \omega \leq \omega_r$, where ω_r is rated angular frequency of the drive motor of a blade type machine) reveal that the use of a wound-rotor slip recovery system is more expedient when compared with the use of frequency converters.

At rotational speeds 15% to 20% below their rated speeds, the expediency of using frequency converters on electric motors becomes still more ambiguous. Advocates of frequency converters may argue that this technology has no alternative when there is a need to vary the capacity from zero to the rated value. However, when a motor operates at lower speeds, drive system efficiency can decrease incrementally by 5%, 10% or more. In such cases it is necessary to analyze the possibility of using multiple unit pumping (compressor) stations allowing for step control of the capacity. For example, pumping stations at ore mining and processing mills and water treatment plants commonly use five to ten pumps each. In the power industry, the parallel operation of pumping units on a common pipeline is also used. Several options of controlling the capacity of such pumping stations are possible [2,3]. These include:

- Using a frequency converter for both the capacity control of one of the pumps and the sequential start of all pumping units.
- Application of additional pumps of fractional power with operation controlled by an appropriate converter.

Other options are also possible involving the use of cyclo-converters for starting the main pumping sets. In this case, the capacity control and matching of operation modes of individual pumps can be performed by an economically substantiated combination of different techniques: beginning from throttling in the case of a small speed control range and up to the application of frequency converters.

The use of frequency converters implies the need of applying motors of special design (more expensive than those of the general-purpose design for industry-wide applications and often of higher power for scattering additional losses related to the operation of the converter). As regards electric drive systems the Ukraine, their selection should be performed in accordance with local standards [4,5].

So why use frequency controllers in such cases? Is it energy savings or regulation that drive the decision-making process? Afterwards I began to understand the meaning of the question. I realized that the use of energy efficient equipment can often be a political rather than economic decision.

Throttling

It is generally believed that throttling is always untenable as a cost-effective means of achieving energy conservation. In order to understand the fallacy of this statement, it is sufficient to analyze the position of curves *a* and *e* in Figure 11-2. Accounting for the efficiency of a frequency converter under the rated conditions of no more than 0.95–0.97 for very large (high power) units can be more expedient than using the most advanced frequency converter with capacity control of blade type machines in the range from 95% to 97% of the rated value. Considering the relationship between the costs of a throttle (valve) and frequency converter, it appears that even at the highest prices of electric power, throttling proves to be cost effective as compared with the "frequency converter—motor" system in the variable speed range from about 0.9 to 0.92 to the rated value. From a maintenance and reliability perspective, valves are less troublesome than any converter!

Energy-efficient Motors

Many engineers in industrialized countries believe that energy saving electric motors must replace the motors of general-purpose design. For example, the policy of the European Union (EU) with respect to Poland is to stimulate production of energy-efficient motors. Plants producing energy-efficient motors are entitled to a 20% compensation of their production cost. This program offers an incentive for plants to manufacture and sell more efficient motors rather than motors of conventional design.

Both efficiency and power factor of energy-efficient motors are typically 3% to 5% higher than general purpose motors. Energy saving motors require 30% to 35% more iron, 20% to 25% more copper, and 10% to 50% more aluminum. Using more advanced and expensive technologies to produce energy saving motors, their cost should be 20% to 40% more than that of conventional electric machines. To obtain benefits from the use of energy-efficient motors, we must ensure their operation with maximum possible loads, which implies the use of relatively expensive protection means. In such cases, the cost of protection devices should not exceed roughly 10% of the cost of the electric motor. However, for small motors with rated power of several kW, the protection devices may cost more than the motors themselves. For larger motors with rated from 50 to 100 kW, the cost of protection devices is within 10% of the cost of the motor. However, the efficiency and power factor of these larger motors differs from motors of

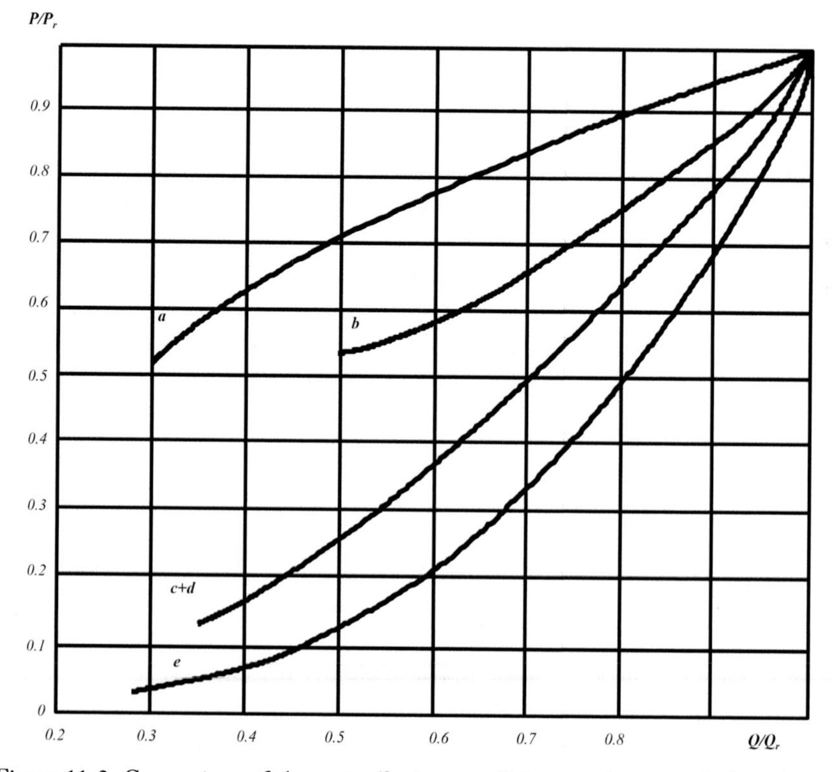

Figure 11-2. Comparison of the cost-effectiveness of the capacity control of mechanisms with "fan" type characteristic of load [1].

general-purpose design by no more than 0.5-1.0%. Therefore, savings are less significant.

Simple calculations indicate that in the Ukraine using current electricity tariffs and the estimated service life of electric motors with rated power of several kW, the use of energy saving electric motors will not be recouped during their entire period of their operation. So, why are energy efficient electric motors so widely used in some western countries? The answer lies in legislation which stimulates the use of energy saving equipment. Countries in the west have higher quality protection devices that allow electric motors to operate at higher load factors and for longer periods. At present, the Ukraine lacks these protection devices and regulated incentives to stimulate the use of energy saving motors. These are reasons why energy saving motors will not be cost-effective there in the near future.

This assessment may be useful for the companies producing energy-effective equipment and those striving to expand their markets in the countries of the former Soviet Union. It is also useful for politicians in developed countries who make decisions on stimulating production and sale of certain types of energy-efficient equipment and technologies. Regardless, it is the project's economic characteristics and corresponding analysis that should determine the expediency of implementing energy saving projects.

SUMMARY

This chapter shows that using energy efficient equipment and technologies is not always economically feasible. Unfortunately, the material resources of many countries are limited and energy efficient equipment is often given a low priority due to implementation costs. The decision to pursue uneconomical technologies is largely explained by the political decision-making process or a regulatory environment favorable to the project. When developing energy saving measures it is advisable to consider the broader economic circumstances that would make energy efficient technologies more justifiable. These might include:

— Term of the project (operating life of the equipment being considered)
— Cost of energy (tariffs)
— Change in tariffs over time
— Cost of money (interest rates on loans and other credit terms)

- The internal rate of return for the enterprise
- Equipment cost for each alternative
- Cost of installation and adjustment of equipment for the alternative
- Duration of equipment operation
- Equipment service life
- Financial and technical risks

References

[1] Bogopolskii, B., Berlovskii, M., Kovalev, S., et al. (1963). Automation of mining fan installations. Moscow: Gosgortekhizdat, page 134.

[2] Mamalyga, V. (1998). Kiev: Enterprise "Elektromekhanika," page 118.

[3] Mamalyga V. (1996). Technical and economic analysis of the results of investigating electric motors of auxiliary mechanisms of power-generating units at the thermoelectric power station and development of the reference requirements for a pilot project. Final report, Supervisor of the work. Kiev: Enterprise "Elektromekhanika," page 29.

[4] Mamalyga, V. (2001, 2004). National Standard of Ukraine DSTU 3886-99, Energy saving and energy audit in electric drive systems, method of analyses and selection. See http://www.epe-association.org/epe/index.php?main=/epe/documents.php%3Fcurrent=847.

[5] Mamalyga, V. (1998). Power saving in electric drivers: Rational, modes of operations and principle of sufficiency in design. Prague, Czech Republic. PEMC 7, pages 186-191.

Chapter 12

Heading Towards Energy Efficiency in the UAE

Maen Al-Nemrawi, CEM, KEO International Consultants

The world's demand for energy is expected to double by 2050. A 50% reduction in atmospheric carbon emissions is needed by 2050 to mitigate the potential effects of climate change. Buildings contribute to our energy dilemma both globally and locally in the United Arab Emirates (UAE). Buildings lead electric consumption in the UAE emirates of Abu Dhabi and Dubai accounting for over 80% of the total annual electricity consumption.

The market analysis shows that 1 billion AED (United Arab Emirates dirham) in savings could be achieved with a mere 10% energy consumption reduction in the buildings sector; this would generate a market potential of over 2 billion AED for all the stakeholders involved in the energy efficiency services and solutions industry by targeting quick return energy efficiency measures. Energy efficiency is a strategy with three essential elements that are needed to be successful: focus, measurements and accountability.

This purpose of this chapter is to discuss how monitoring based commissioning is a viable methodology for achieving quick results by helping to verify energy savings. With commissioning, risks from other energy efficiency initiatives can be reduced.

There are considerable challenges when using energy savings performance contracting (ESPC). These take the form of performance guarantees, financing schemes, legal frameworks, documenting savings and verifying performance. Financing options can be a bottleneck limiting the success of the energy services company (ESCO) business model. Other expenses include high transaction costs, insurance, guarantee and performance bonds which typically amount to 12% to 16% of a project's revenue.

INTRODUCTION

Today, the world faces a global energy dilemma with two major challenges: 1) satisfying the ever-surging global demand for energy; and 2) the increased global pressure to lower carbon emissions while adopting more efficient practices.

The world's energy demand is expected to double by 2050. Both the Intergovernmental Panel on Climate Change (IPCC) and International Energy Agency (IEA) have called for carbon emissions to be reduced by 50% to mitigate the effects of drastic climate changes.

Energy efficiency is often referred to as *the fifth fuel*, the others being coal, petroleum, nuclear power and renewable energy. Energy efficiency is defined as consuming less energy while performing the same function or delivering the same output; it is an important way to address the increasing demand for energy. In practice, energy efficiency can be achieved by various means. All yield reduced energy consumption without compromising operational needs, quality of life and comfort levels. The world's evolution towards greater energy efficiency is an important goal.

RESOLUTION IN FIGURES

A report released in 2015 by the Abu Dhabi Statistics Center indicates that domestic, commercial and governmental buildings in Abu Dhabi account for about 84% of the total electricity consumption in the emirate, the equivalent of 44,386,471 MWh in 2014. The percentage allocation of this electrical usage by sector is shown in Figure 12-1.

Similarly, the Dubai Electricity and Water Authority (DEWA) in Dubai reports the same trend with an overall consumption of 33,659,150 MWh for 2015 is indicated in Figure 12-2.

Comparing the trend of building consumption to 2010, we observe an increase of 35% in consumption in 2014 compared to 2010 levels in Abu Dhabi alone, while the differences were less pronounced in Dubai, with only a 17% increase.

POTENTIAL IN THE UAE

Considering the aforementioned electricity consumption analysis, energy efficiency experts believe that 1 billion AED in savings could be

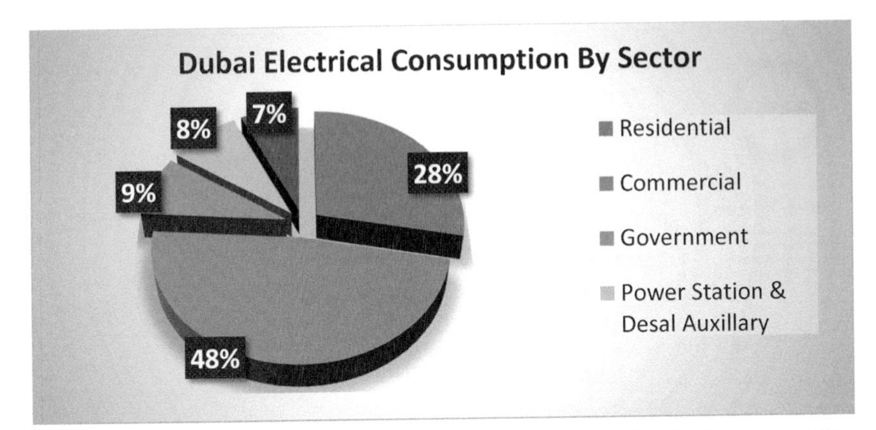

Figure 12-1. Electricity consumption per building category for the Emirate of Abu Dhabi.

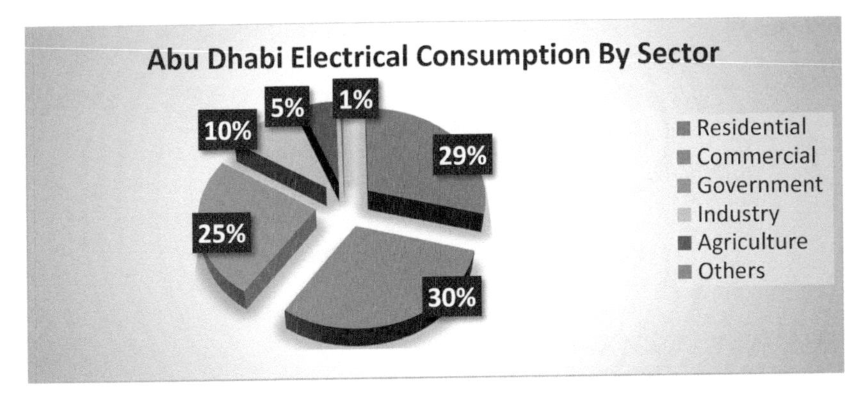

Figure 12-2. Electricity consumption broken down per building category in the Emirate of Dubai.

achieved with a mere 10% energy consumption reduction in the buildings sector (residential, commercial, governmental and private buildings). This is based on the current average electricity tariff in Abu Dhabi and Dubai. Applying a 10% annual saving target for 40% of the existing buildings in these two emirates and assuming a 70% retrofit success rate with a three year simple payback period, this will generate a market potential of over 2 billion AEDs. Figures 12-3 and 12-4 represent the market potential analysis for Abu Dhabi and Dubai.

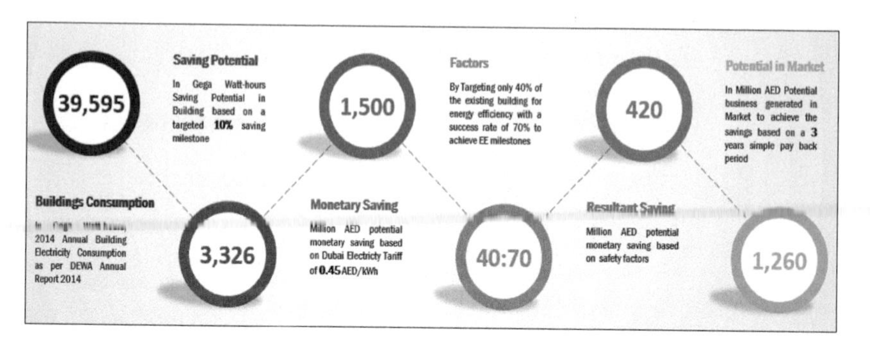

Figure 12-3. Potential in Dubai market.

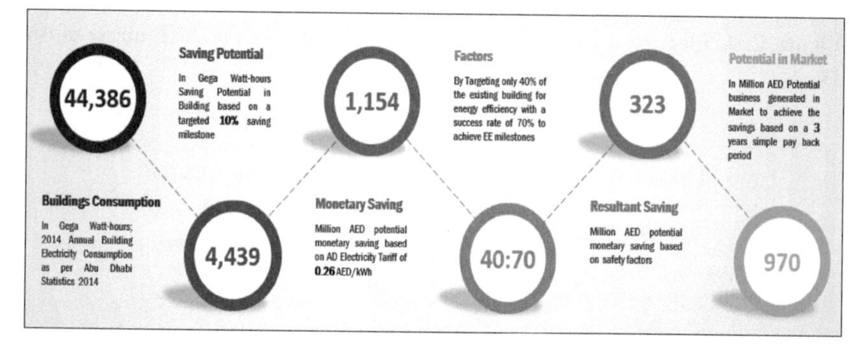

Figure 12-4. Potential in Abu Dhabi market.

THE WAY FORWARD

Energy efficiency is a strategy. Three essential elements need to be considered for it to succeed: focus, measurements and accountability.

Focus—by identifying how the building performs or where the facility lies in the scale of consumption. Priorities should be tied to specific, measurable, achievable, realistic and time (SMART) bounded goals. The priorities and goals could be represented in energy efficiency or energy management policies supported by management. They should be shared with the facility's occupants, users and employees as part of the corporate policies to be implemented.

Accountability—by assigning a team that will take responsibility and ownership for executing the strategy. This team should have a direct reporting channel to management.

Measurements—measuring progress and outcomes is challenging to assess. The ability to achieve the goals and to measure project outcomes represented in the savings is key to success. The unmeasurable is unmanageable.

Energy efficiency can be approached using different practices, schemes and nominations. These approaches differ in execution methods, contractual obligations, term and generated savings. A brief overview of some of these strategies is described below.

PLAN-DO-CHECK-ACT

The elements for an energy efficiency or energy management program can be described as the Plan-Do-Check-Act approach. This is a tiered process initiated by establishing the objectives and processes necessary to achieve the energy efficiency goals and milestones. This is documented in a published energy efficiency policy available to all building occupants.

Success depends on a well-defined energy consumption baseline with energy performance indicators which is supported by a feasibility study for the targeted measures along with a detailed action plan for implementation, measurement and verification. This *Plan* can be a detailed facility investment grade audit conducted by a dedicated energy manager or qualified project manager.

The *Do* phase includes the implementation of the planned energy conservation measures (ECMs). Implementation of these measures is supported by sufficient documentation, training and operational considerations.

A savings *Check* of the ECMs is performed when the implementation is completed. Detailed measurement and verification (M&V) of the energy savings is conducted with an analysis showing the savings based on the energy performance indicators. M&V along with energy monitoring and reporting helps identify deviations from the targeted performance indicators. It enables opportunities to further enhance the building operation and reduce energy consumption. These activities are communicated to management enabling them to *Act* appropriately. It enables building occupants to highlight progress toward energy goals, discuss achievements and implement preventive or corrective actions to further augment monetary savings. This ongoing process helps assure that the building is systematically operated in a manner aligned with the organizational polices.

ISO 50001:2011

ISO 50001:2011—Energy Management System (EnMS) was officially launched in 2011. The standard states that its main objective is "to enable organizations to establish the systems and processes necessary to improve energy performance, including energy efficiency, use, consumption and intensity. Implementation of this standard should lead to reductions in energy cost, greenhouse gas emissions and other environmental impacts, through systematic management of energy." This standard relies on the Plan-Do-Check-Act approach. It details the process to achieve energy efficiency and management goals within an organization.

Such an approach is applicable for organizations with corporate structures. It is suitable for industries in which processes need specific in-house expertise to assure all major energy uses are addressed. This requires in house engagement (top-down) along with a dedicated team led by an energy management champion or the energy manager. This approach needs an internal financing scheme which often initially focuses on no/low cost measures or medium cost measures with simple payback periods of 2-4 years.

MONITORING BASED COMMISSIONING (MBCx)

Another proven approach toward energy efficiency for existing buildings focuses on systematically correcting system inadequacies. Monitoring based commissioning (MBCx) is a tiered process with of goal of providing sustained savings. MBCx involves three primary elements: retro/ongoing commissioning, measurement-based savings, and accounting using metered and monitored data. The process is similar to Plan-Do-Check-ACT with different implementation techniques and reporting structures. MBCx captures savings from three primary sources of costs:

- Savings from persistence and optimization of savings from ongoing commissioning based on early identification of deficiencies through metering and trending;
- Savings from measures identified through metering and trending; and
- Savings from newly identified measures using the energy monitoring and management systems.

MBCx further focuses on low cost measures that do not require high capital investment. The goal is to optimize the capabilities of the control and monitoring systems that control energy consuming electrical and mechanical systems such as lighting, heating, cooling and ventilating equipment. MBCx methodologies include:

- Repairing the basic functions using the existing controls and monitoring system;
- Enhancing the existing controls and operational systems; and
- Upgrading the existing equipment to improve operational efficiency (low hanging fruit).

MBCx has proven to be a viable model. A Lawrence Berkeley National Laboratory study [1] examined the findings for the commissioning of 643 existing and new construction buildings scattered in 26 U.S. states representing 37 different commissioning providers. The findings covered 99 million ft^2 (9.2 million m^2) of building area and proved that a median energy saving of 16% is achievable in existing buildings with median benefit-cost ratios of 1.1 and 4.2 years.

This model is viable in the UAE due to the poor utilization of the potentials in the building energy management systems. Many are not well maintained, commissioned or are operated by non-skilled operators. The operation and maintenance practices in the region are not a priority for most building owners. Maintenance approaches tend to be reactive rather than preventive and predictive. Few facility management staff or companies incorporate energy management and efficiency services as key components of their contractual obligations and work procedures.

MBCx will be more important in the future. Previously subsidized energy tariffs remain low compared to the region's high per capita energy consumption. Enforcement of energy efficiency building codes and regulations has been lax. Utility tariffs have increased as governments have gradually released subsidies for energy and water bills. Such actions have increased utility tariffs in Dubai since 2009 and Abu Dhabi in 2015. Future actions are envisaged to support the movement towards greater efficiency and sustainability. These trends will increase the savings that will result from MBCx.

ENERGY SAVING PERFORMANCE CONTRACTING

Energy savings performance contracting is available for organizations to implement energy efficiency projects. The responsibilities for planning, implementation, savings verification and other operational obligations are delegated to specialized services providers called energy services companies (ESCO). In the UAE, Etihad (Arabic for "Union") ESCO was formed as Dubai's official 'Super-ESCO' by DEWA and the Dubai Supreme Council of Energy in 2013. With a vision of making Dubai one of the world's most sustainable cities, one of their goals is to improve energy and water efficiency in existing buildings using energy savings performance contracting [2].

Performance contracting is a challenging business model with many associated risks including extended contract terms, variable project volumes and guaranteed returns on investment. It is initiated by a management commitment for energy efficiency. Detailed planning and a framework for periodic checks of milestones and key performance indicators are needed for existing buildings. Project implementation, execution and handling require specialized and high caliber expertise. A contractual framework is required for corrective action when deviations from projected savings occur. Periodic assessments are necessary for monitoring, verification and reporting to assure sustained long-term benefits. (Reference 3 provides a detailed discussion concerning performance contracting).

Financing is a key driver toward ESCO business viability. In Dubai, project financing alternatives for ESPC projects are being tested. These take the forms of financing schemes, legal frameworks, and M&V approaches. Financing options are a bottleneck for the success of the ESCO business model due to high transaction costs, high interest rates, insurance, guarantees and performance bonds. These usually amount to 12%-16% of an ESPC project's revenue. The associated financial risks of this model could be mitigated through financing schemes based on successful business case studies such as those funded by the government. Without the mitigation of financial risks, the success of the ESCO industry will be limited and restrained. As ESCO contracts are linked with energy savings and CO_2 emission reductions, there will be more challenges to be resolved in the future.

References

[1] Lawrence Berkeley National Laboratory (2009, July 21). A golden opportunity for reducing energy costs and greenhouse gas emissions. Prepared for the California Energy Commission.

[2] Etihad Energy Services (2017). About Etihad ESCO. http://www.etihadesco.ae/about-etihad-esco, accessed March 14, 2017.

[3] Turner, W. and Doty, S. editors (2013). *Energy Management Handbook*. The Fairmont Press, Inc.: Lilburn, Georgia. Pages 625-644.

Chapter 13

Developing Energy from Sustainable Resources in Jordan

Samer Zawaydeh, Msc, PMP, CRM, REP

Sustainable development is an evolving policy taking different forms in countries throughout the world. The development of renewable energies from solar, wind, geothermal, hydro and tides will improve energy security for oil importing countries and promote sustainability. Renewables also create employment and a supply chain of industries that educate, design, finance, procure, manufacture, manage, construct, commission, and maintain the new clean energy technologies.

This chapter details the status of the development of renewable energy sources in the Middle East and North Africa (MENA) concentrating on the successful efforts in Jordan. The development of Jordan's regulatory system began after the oil crisis in 2008 with renewable energy (RE) and energy efficiency legislation in 2012.

The economic, environmental and social impacts of renewable energy will be discussed as will the lessons learned from adopting strategies to achieve the renewable energy targets in the energy mix. These lessons can be implemented locally by other oil importing countries to improve energy resilience, energy security, and contribute to global climate change adaptation efforts. This chapter concentrates on the development of sustainable energies which are naturally available, socially acceptable and economically feasible in the Middle Eastern country of Jordan.

INTRODUCTION

Renewable energies do not consume fossil fuels and they are available free in nature. Since the 1970s, many countries have directed investments towards the development of renewables.

Solar is Earth's most readily available energy and the ultimate source of most energy resources. Solar photovoltaic (PV) was first developed by Bell Labs in 1954 as a means of producing electricity. Since then, the efficiencies of solar PV cells have improved and costs have declined while the number of the world's electric consumers has increased.

The latest energy crisis in 2008, which increased oil prices to $147 (U.S.) per barrel, caused economic problems in the national economies of oil importing countries while oil exporting countries gained huge monetary reserves. The most rapidly developing renewable energy sources have been wind power and solar PV. The Global Status Report 2015 shows installed PV capacity increasing since 2010.

Hydro power provides the largest amount of renewable electrical energy in the world. New hydro-electric capacity is being added in developing countries.

Generating electricity from wind power is a type of renewable technology that was commercially developed in the 1970s. Since then, the generating capacity of wind turbine generators has continued to increase with longer blade lengths, higher hubs and larger rotor diameters to access greater wind speeds at higher elevations above grade.

Worldwide renewable energy generation capacity is increasing at the rate of approximately 100 GW annually. Renewable energy development in MENA began in the 1990s and within 10 years, Egypt successfully installed over 545 MW of wind powered electrical generation. Currently, Morocco, Tunis, Jordan, the UAE and Lebanon are leading the installation of renewable energy systems.

Energy Status in Jordan

The Ministry of Energy and Mineral Resources (MEMR) Energy Facts and Figures Report 2014 [1] indicates that Jordan imports 97% of its energy requirements with the balance from local production. Jordon's primary energy needs are roughly 8.15 million tons of oil equivalent (Mtoe). 82% of Jordan's primary energy consumption is from crude oil and refined products, 11% from natural gas, 3% is from coal, 2% from renewable energy, 1% from coke, and 1% is imported electricity.

These energy resources are consumed by the transportation (51%), household (21%), industrial (17%) and other (11%) sectors.

About half of the oil consumed is used to generate electricity. Electricity is used by various sectors: household (43%), industrial (24%), commercial (17%), water pumping (14%) and street lighting (2%). The electricity generated in Jordan equals an average annual electricity intensity of 959 kWh per person annually or 5,180 kWh per family, based on an average family size of 5.4 persons.

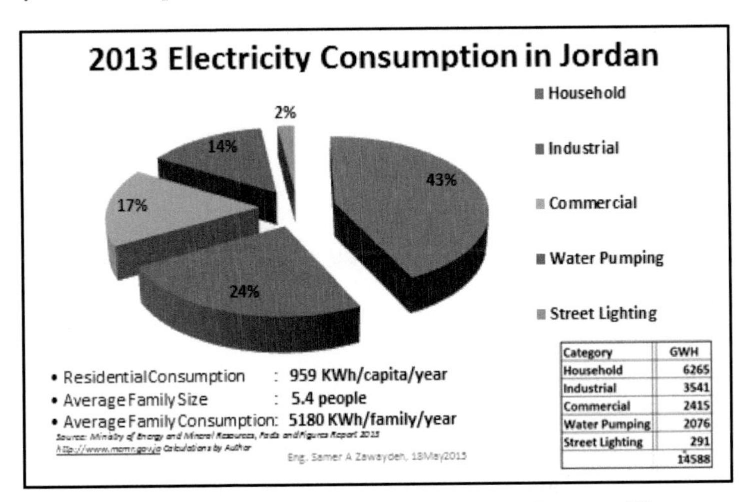

Figure 13-1. Jordan's electricity consumption by sector [1].

Since 2013, the installation of renewable energy resources was permitted after the renewable energy regulatory framework was enacted and it was agreed that electricity from utility scale projects would be sent to the National Electrical Power Company (NEPCO). Subsequently, these intermittent resources are providing a portion of the needed electricity. Consumers are allowed to install renewable energy systems that provide all their annual energy consumption by using on-grid PV and wind systems.

While energy prices were subsidized in the past, the majority of the subsidies were removed after 2008. Currently, fuel prices follow the international market prices for crude oil and national prices are adjusted monthly. The price of crude oil during the past ten years has varied from a high of roughly $147 per barrel to a low of $28 per barrel. Commercial, industrial, hospitality, and other consumers are unable to frequently adjust their commodity prices. This leads to substantial economic losses particularly when oil prices are high.

Jordan's 2020 Primary Energy and Cost Targets

The national strategy is to reduce imported crude oil and products from 82% to the range of 52% to 55% by 2020 [2]. Coal usage is to be maintained in the range of 3% to 5% while natural gas will increase to 20% and oil shale to 10%. The renewable energy share in the energy mix will increase to 10%.

The cost for Jordan's consumed primary energy is found in NEPCO's annual reports. In 2014, this totaled 4,480 million Jordanian Dinars (JDs), an increase from 2010 when the total equaled 2,603 million JDs. Energy costs were 17.1% of Jordan's total GDP in 2013, reaching a high of 21.1% in 2012 as shown in Figure 13-2 [3].

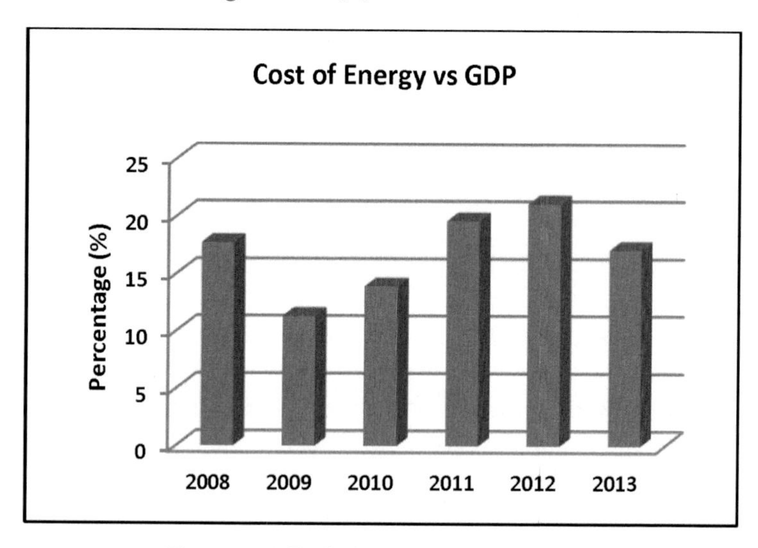

Figure 13-2. Jordan's cost of energy vs. GDP.

The cost per kWh sold in 2013 was 0.145 JD ($0.205 U.S./kWh) and in 2015 it reached 0.156 JD ($0.22 U.S./kWh).

Electrical Energy Consumption and Network Losses

The household sector in Jordan consumes 43% of total generated electricity, 24% is consumed by the industrial sector, 17% by the commercial sector, 14% by the water pumping and 2% for street lighting.

The efficiency of the generating stations in Jordan is improving. Currently, the overall efficiency is 41%, which means that the remaining 59% is lost primarily in the form of heat. The national electricity grid has 17% losses. Technical losses are 2.3% in transmission and 6% in distribution, and the remaining losses are unauthorized uses within the network.

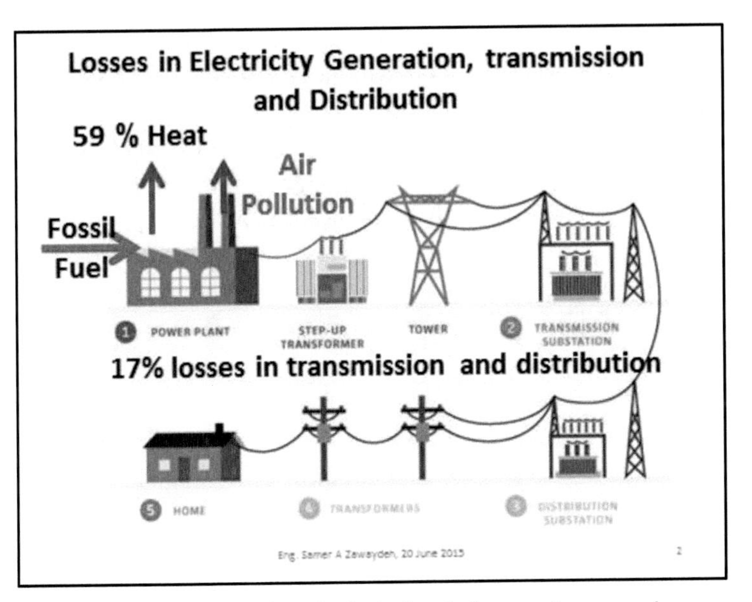

Figure 13-3. Losses from Jordan's electrical generation network.

Jordan's 2020 Renewable Energy Targets

The national plan that was developed in 2007 called for 7% of the primary energy mix to be RE by 2015 and 10% by 2020. The recent RE targets reported by the MEMR are 1,000 MW of wind (65% of total), 600 MW of solar (32%) and 50 MW using biomass (3%) by 2020.

STAKEHOLDERS

The energy sector directly contributes to all the economic sectors in the country. It is very important to have positive interactions among the various stakeholders to achieve their requirements and expectations. The major stakeholders include representatives from both the public and private sectors.

In the public sector, energy policies are established by the Ministry of Energy and Mineral Resources. The energy sector is regulated by the Energy and Minerals Regulatory Commission (EMRC). Electricity is generated by the different generation stations, transmitted by the National Electrical Power Company (NEPCO), and distributed by the electricity distribution companies. The Jordan National Building Council (JNBC) under the Ministry of Public Works and Housing issues the national technical codes and guides those governing building design and construction. The media updates the various stakeholders. Universities are also key stakeholders as they develop the capabilities of students in the energy sector.

The private sector leads the economy and engages the tools for future development in the energy efficiency (EE) and RE sectors. Private sector initiatives are varied. Private universities provide energy-related educational programs with at least ten training centers offering RE and electrical energy courses.

Developers began planning for RE projects after the expression of interest (EOI)#1 was issued in July 2011. A total of 68 companies responded to EOI#1, expressing interest in implementing wind and solar projects. Twelve companies signed power purchase agreements (PPA) with NEPCO in March 2014. Equipment suppliers are offering new products and taking advantage of incentives that include the 0% customs and 0% sales tax on EE and RE systems. Financial institutions are offering green energy loans for EE and RE projects to a maximum 100% of the project value.

Media reports have generated interest in solar and wind projects in Jordan. Energy exhibitions and conferences occur monthly. Such activities facilitate networking and communication among stakeholders and increase the likelihood of progress toward a clean energy economy.

REGULATORY SECTOR

The Renewable Energy and Energy Efficiency Law (REEEL) (13) 2012 was first developed in 2008/2009, issued in preliminary form in 2010, enacted in 2012, revised as REEEL (33) in 2014, and reenacted in 2015. Bylaw (10) providing the customs and tax exemption was also revised and became Bylaw (13) in 2015. Bylaw (73) for energy efficiency was issued to regulate the sector. Figure 13-4 summarizes the development of Jordan's recent energy laws and regulations.

Two bylaws help regulate the development of the RE sector. In 2015, Bylaw (49) supporting the Jordan Renewable Energy and Energy Efficiency Fund was issued along with Bylaw (50) enabling direct proposals.

Wheeling regulations were issued to regulate the generation of RE at locations distant from the points of consumption. The RE Guide was

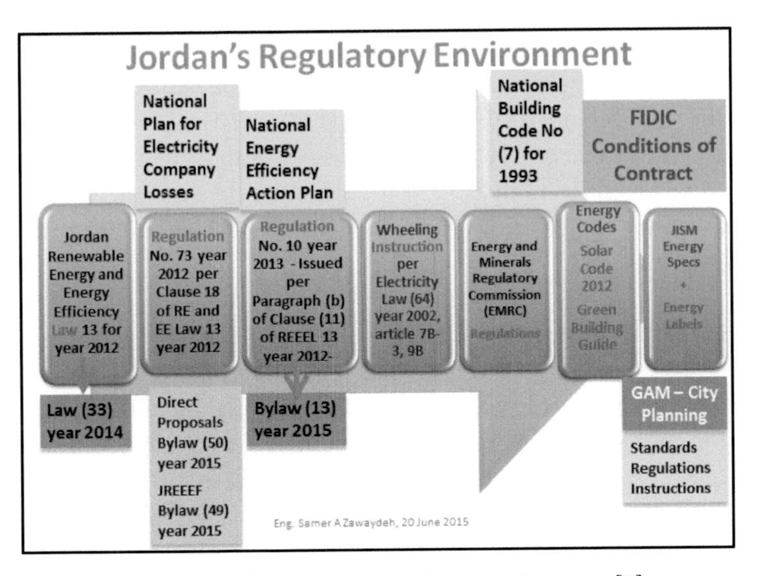

Figure 13-4. Energy sector regulatory environment [4].

issued by the Energy and Minerals Regulatory Commission (EMRC) when the REEEL was issued. This guide assisted consumers with the process of requesting the installation of RE systems, including review, changes, installation and commissioning.

The Jordan National Building Council (JNBC), part of the Ministry of Public Works and Housing (MoPWH), issued the Solar Code in 2012 which outlines the mandatory requirements for the installation of solar water heaters and solar PV. The Jordan Green Building Guide has a section devoted to the energy and renewable energy point structure providing increases in points available for increased percentages of renewable energy supplied. The Jordan Standards and Metrology Organization (JSMO) prepared energy label regulations for electrical equipment.

The National Energy Efficiency Action Plan (NEEAP) was approved in June 2013 and it calls for several projects to be implemented to achieve the targeted requirements for renewables.

COST OF ELECTRICITY

The cost of electricity began increasing in 2008 after fuel prices subsidies were removed. As fuel prices rose, electric companies began to face losses. The reported losses are roughly one billion JD ($1.40 billion U.S.) in 2013 and 1.18 billion JD ($1.65 billion U.S.) in 2014. A national strategy was developed to increase electricity prices through 2017, implement measures to raise money from large consumers and develop energy efficiency measures offset the losses. This plan calls specifically for the implementation of RE projects.

Residential and Industrial Sectors

Jordan's residential and industrial sectors have specific electricity tariffs. The residential sector has a stepped tariff with prices increasing since 2008 for consumers with monthly consumption over 600 kWh monthly. There are seven tariff categories for the residential sector. The first four are subsidized and the tariff is fixed for small consumers. To regulate the market and reduce electric company losses, the tariff was fixed for 2014 through 2017 for each category. Tariffs are shown in Table 13-1.

Electricity prices in Jordan were low until 2008. Afterwards, the number of tariff categories increased along with the unit rates for high use consumers [5]. Electricity tariffs for consumers using less than 600 kWh monthly will increase until after 2017. Residential consumers using more than 600 kWh monthly will experience annual increases in tariffs from 5% to 15% through 2017.

Table 13-1. Residential electricity tariff.

Year – Unit Price (1/1,000 USD/kWh)					
2017	2016	2015	2014	QTY (kWh/ month)	Tariff Category (kWh/month)
46.2	46.2	46.2	46.2	160	1 - 160
100.8	10.8	10.,8	100.8	140	161 - 300
120.4	120.4	120.4	120.4	200	301 - 500
159.6	159.6	159.6	159.6	100	501 - 600
263.2	245.0	228.2	212.8	150	601 - 750
313.6	292.6	271.6	253.4	250	751 – 1,000
414.4	399.0	379.4	362.6		above 1,000

The industrial sector has five different tariff categories:

1) Small—under 10,000 kWh per month.
2) Small—over 10,000 kWh per month.
3) Medium.
4) Large mining.
5) Large other.

Each of these sectors will experience tariff increases over a four-year period. Figure 13-5 shows the current costs of electricity and their expected price increases. Large consumers will experience greater rate increases for electricity than small consumers.

The day rate for large industrial users is 0.27 JD/kWh ($0.38 U.S./kWh) in 2015 and is planned to increase to 0.32 JD/kWh ($0.45 U.S./kWh) in 2017.

Adopting renewable energy projects is a good investment strategy for large consumers because they can reduce their levelized cost of energy (LCOE), avoid the impact of price variability and maintain a competitive advantage.

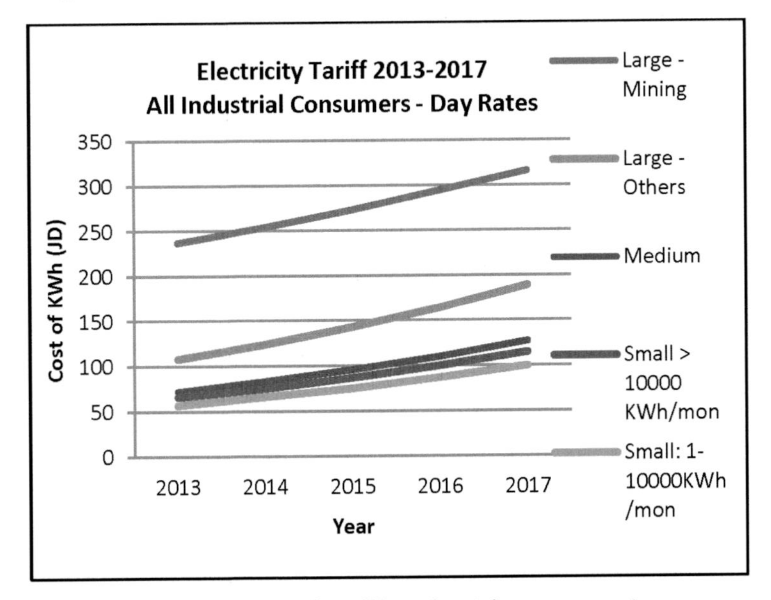

Figure 13-5. Planned tariff for industrial sector categories.

RENEWABLE ENERGY—PHOTOVOLTAIC

The availability of RE resources depends on geographic location and local meteorological conditions. In order to explore the economic feasibility of RE resources, field measurements must be taken for a minimum of twelve months at the locations where utility scale projects are planned. Engineering firms often require that measurements be recorded for three years prior to using data for the design of utility scale projects.

Jordan has over 300 sunny days annually. The German Aerospace Agency (DLR) supported the development of the solar sector in Jordan by installing a solar scientific data measurement station at the Ma'an Development Area. Solar data measured at the industrial park in Ma'an for three years showed an average direct normal irradiance (DNI) of 2,700 kWh/m^2 annually. This is among the highest readings in the world.

Solar Photovoltaic (PV) Installations

As of 2015, there are approximately 747 PV decentralized systems installed in Jordan with a total capacity of 23.6 MW. It is expected that the installations by consumers to completely or partially provide for their annual energy requirements will be over 60 MW by the end of 2015. Utility scale projects being developed with power purchase agreements are expected to provide an additional 200 MW within the next 18 months.

Table 13-2 shows the solar PV projects currently being developed. This is more than 70% of the 2020 Solar RE targets.

Major Investors—Expression of Interest II

Expression of interest (EOI) requests encourage companies to develop utility scale projects. EOI#2 is underway for selection of four companies to each develop 50 MW solar-electric plants. In 2015, 34 companies submitted their responses to the EOI to develop solar plants.

The financial results of the EOI#2 were announced by MEMR in May 2015 as shown in the Table 13-3. The average of the lowest four prices was 0.0482 JD/KW ($0.068 U.S./kWh). This second group of utility scale renewable energy solar PV projects will sell roughly 400 GWH per year at the average price of $0.068 U.S./kWh for the next 20 years.

Table 13-2. Jordan's solar PV projects.

1	*Kawar Consortium*	*53*
2	*Catalyst Private Equity*	*20*
3	*EJRE Projects*	*20*
4	*SunEdison Hellas SA*	*20*
5	*Evolution Solar*	*20*
6	*Shamsuna Power Company*	*10*
7	*Clean Energy Concepts*	*10*
8	*Ennera*	*10*
9	*Bright Power Group*	*10*
10	*Martifer Solar, S.A*	*10*
11	*Scatec Solar*	*10*
12	*Greenland Alternative Energy*	*10*
13	*Philadelphia Solar*	*10*
14	*Spanish Grant*	*3*
15	*Queira*	*65*
16	*Expression of Interest #2*	*200*
17	*Installed Decentralized*	*24*
18	*In progress*	*40*
	Total Solar	***468***

SOLAR PV ECONOMIC DEVELOPMENT

Investment

One of the economic goals for developing countries is to increase foreign direct investment (FDI). The RE sector in Jordan, like those of other countries, has attracted major companies. Over 100 different international companies have competed for RE projects in Jordan since 2012.

Utility scale RE projects initially require large capital investments. The Central Bank of Jordan (CBJ) started EE and RE low interest loans (2.3%) up to 10 million JD ($14 million U.S.) for use of clean energy technologies.

International banks, the World Bank and other major banks in developed countries are eager to provide long term loans at low interest rates because they consider these investments to be secure and use proven technologies. Local banks now offer low fixed interest rates for renewable energy projects for a period of 10 years.

Table 13-3. Results of Expression of Interest #2.

Partnership/bidder	Bid in JOD cents/kWh	Bid in US$/kWh
GI Karnomourakis SunRise PV Systems (Greece)	4.34	0.0613
Saudi Oger (Saudi Arabia)	4.60	0.0649
Fotowatio/ALJ (Spain)	4.89	0.0691
Hareon Swiss Holding (Switzerland)	5.43	0.0767
Evolution Solar/AMP	5.59	0.0789
Philadelphia Solar (Jordan)	5.60	0.0791
SolaireDirect (France)	5.68	0.0802
Neoen (France)	5.74	0.0810
Linuo Group (China)	5.86	0.0827
Activ Solar (Austria)	6.00	0.0847
Mainstream (Ireland)	6.15	0.0868
Kawar-First Solar (Jordan)	6.23	0.0880
Alten Renewable Energy Developments (Netherlands)	6.28	0.0886
Al Salad for RE (JoSolors)	6.36	0.0889
SkyPower-FAS (US)	6.36	0.0898
SunEdison (US)	6.40	0.0904
Elecnor (Spain)	6.41	0.0905
ACWA (Saudi Arabia)	6.58	0.0929
SECI (Italy)	6.60	0.0932
ELF-ENEL (Jordan)	6.6	0.0932
Scatec (Norway)	6.70	0.0946
Hanergy (China)	7.00	0.0988
Spectrum International for Investment (Jordan)	7.30	0.1031
Suncore PV Technology (France)	9.40	0.1327

Bylaw (50) under the REEEL (13) year 2012 was issued in 2015 to regulate investments in RE projects. These bylaws support the RE sector and enhance the transition to clean energy.

Manufacturing and Commercial Development

Renewable energy manufacturing opportunities exist for those with manufacturing experience. The first solar module assembly plant in the Middle East was opened in Jordan in 2008.

Philadelphia Solar began production prior to the issuance of the REEEL. Their Jordanian solar modules were exported to Europe and North Africa. The company expanded production to include mounting structures in 2015. Today, they have a fully automated structure fabrication line to meet the anticipated demand for utility-scale PV projects.

Starting in 2014, electric cable manufacturers developed capabilities to produce PV cables and direct current (DC) cables.

Commercial sector PV developed gradually. It was linked to the issuance of the REEEL in 2012 when there were 40 commercial PV companies operating in Jordan. Today, there are over 500 such companies that directly and indirectly employ over 10,000 people.

University Initiatives

The RE sector needs qualified workers to lead and manage the sector. Job opportunities across the supply chain for the PV sector include sales, estimation, education, design, training, manufacturing, procurement, construction, commissioning, quality control and maintenance.

To supply such needs, several major universities incorporated the required undergraduate and graduate RE university degrees. These universities include:

University of Jordan
Jordan University of Science and Technology
German Jordanian University
Hashemite University
Hussein Bin Talal University
Zaytouneh University
Middle East University

The universities are cooperating with international universities in Europe and the U.S. to develop their training capabilities and their curriculums. The education systems will continuously produce qualified and trained workers for the clean technology sectors in Jordan.

Universities are also investing in utility scale projects. These will support education, help cover their annual energy demand and avoid paying higher electricity costs, presently about $0.38 U.S./kWh. The projects that are being developed include:

Hashemite University (5 MW)
Jordan University of Science and Technology (5 MW)
Hussein Bin Talal University (4 MW)
Yarmouk University (3 MW)
Taffila University (1 MW)
Petra University (1.5 MW
Zaytouneh University (1.5 MW)
Applied Science University (0.5 MW)
Hussein Bin Talal University (50 MW)
Al Beit University (85 MW)
University of Jordan (40 MW)

RE Development Initiatives

National electricity grids are often not designed to accommodate renewable energy. The flow of electricity during the last 60 years was from the generation station to the demand center. Now Jordan's national grid must carry electricity bi-directionally since its consumers are also generating entities. The decentralized RE project owners (commercial and residential) are required to assess the electricity grid near the project prior to site development. To accommodate the new RE projects, this often requires upgrades to the grid infrastructure.

Utility scale RE projects are planned for development in locations with the best resources. The Ma'an Development Area (MDA) was dedicated to the purpose of accommodating developers selected for EOI#1. The MDA is the targeted location for 190 MW of solar PV between 2015 and 2017. The MDA is developing another RE area (5,000,000 m^2) for a future Phase III project.

The REEEL (13) year 2012 established an upper limit to RE development of 500 MW. This limit can be increased depending on the future capabilities of the national grid. This important subject was addressed by international consulting companies from the U.S. and Japan.

The Green Corridor concept will reinforce Jordan's high voltage electricity network, enabling the integration of more renewable generation capacity and improving supply reliability. It consists of two new transmission lines (400 kV/150 km and 132 kV/51 km), upgrading three existing lines (132 kV/100 km) and construction of one new 400/132 kV, 1,200 MVA substation. Construction is anticipated by 2017. A third stage is being considered in southern Jordan between Qatraneh and Aqaba and will be constructed when projects in Aqaba are operational.

Jordan Renewable Energy and Energy Efficiency Fund (JREEEF)

This fund is managed by the MEMR. Bylaw (49) under the REEEL (13) year 2012 was enacted to regulate the development and implementation of monies to support EE and RE projects. The projects implemented by the fund use a revolving loan program. The first project purchased roughly 5,000 solar water heaters (SWHs) and distributed them to 70 community based organizations (CBOs) [6]. These organizations will distribute the SWHs to families that will reimburse installation costs from savings on their electricity bills. Once the consumer repays the CBO the full cost of their SWH system, then another consumer will use the revolving funds to purchase a new SWH.

BENEFITS OF RENEWABLE

The renewable energy resources developed within Jordan are determined by the project owners. Imported fossil fuel energy resources are less attractive than RE systems that provide locally available energy. Energy from RE systems reduces both imported energy expenditures from purchasing fossil fuels and atmospheric carbon emissions.

It is expected that the effects of climate change will be a driver to adopt more of these projects. Several projects in Jordan, including three solar energy projects, are registered under the Carbon Development Mechanism (CDM) and are benefiting economically from carbon emission reductions.

Attractive Financial Incentives

The return on investment for RE projects depends on the cost of electricity for the consumer or the investor. Large consumers will have higher electricity tariffs and the payback period is estimated to be 30 to 36 months. Consumers with lower electricity tariffs will have much longer payback periods on their investments. To assist consumers, the electric distribution companies have a program that allows customers to pay off their investments in new solar PV systems by deducting costs from their monthly electricity bills. This provides a solution for owners, enabling them to own the PV systems a few years after installation.

The Central Bank of Jordan, through the local commercial banks is providing low interest loans for RE and EE projects. This makes the investment in such projects more attractive. In addition, international banks are funding these projects at lower interest rates than other types of investments.

The Ministry of Energy and Mineral Resources has programs that support RE and EE. The MEMR recently announced a plan to support solar PV systems for 6,000 mosques covering 20% of the cost.

Investors in EOI#1 had a fixed feed-in tariff of 0.12 JD/kWh ($0.17 U.S./kWh) for 20 years. This is a very attractive price compared to the current prices 0.048 JD/kWh ($0.068 U.S./kWh).

RE systems are also exempt from taxes according to bylaw (10) year 2012 and bylaw (13) year 2015. This provides a competitive advantage to suppliers and owners who purchase RE systems.

Project Life

RE systems are designed operate for 20 to 30 years, providing predicable energy for an extended period of time. Using PV module manufacturers' equipment degradation information, predicted PV performance can be integrated into the electric grid requirements.

In Jordan, these systems must be cleaned every 2 to 3 weeks depending on environmental and seasonal conditions. The operations and maintenance (O&M) requirements are minimal but provide opportunities for service providers.

Reduction in Fuel Losses in Generation, Transmission and Distribution

Presently, the overall efficiency of all generating stations in Jordan is 41%. This means that 59% of the imported fuel purchased is lost due to system inefficiencies. RE systems eliminate this wasted fuel. This saves money throughout the expected life of the project.

The aggregated loss in the national electricity grid's transmission and distribution networks is approximately 17%. The advantages of onsite RE generation include eliminating these losses and their associated costs.

Fixed Fuel Costs

Jordan's government has issued requests for 400 MW of solar energy. This will allow the prices of electricity to be fixed for the expected life of the projects in accordance with prices in the PPAs. The feed-in tariff for the first 200 MW is 0.12 JD/kWh ($0.17 U.S./kWh) with the second 200 kWh having a lower rate.

Large consumers in Jordan (e.g., banking, telecommunication, universities, large industries, etc.) will have the opportunity to install RE systems to handle 100% of their electricity needs. They will be able to avoid fluctuations in electrical energy prices that result from fossil fuel price variability, improving their competitiveness internationally.

SOLAR ENERGY PROJECT DEVELOPMENT

Residential, Industrial and Commercial PV Systems

Residential systems are relatively small, rated typically less than 200 kWp. The systems can be installed within 2 to 3 months after the application to the EMRC has been approved.

Jordanian industries vary in size. Large industries consume about 10% of total electricity. These industries need to adopt renewable energy resources to help offset increasing fossil fuel costs.

The largest grid-connected commercial solar PV installation is the 1 MW ground mount system installed in May 2014 at the Ma'an Development Area.

Refugee Camps

Several hundreds of the housing units in a refugee camp that were supplied at the beginning of the war in Syria in 2011 were equipped with off-grid PV systems. These consisted of 0.5 kWp modules with battery storage.

The camp grew afterwards and there is now a project to install 6 MW of PV in phases to provide for the complete needs of the camp (about 12 GWh annually).

Two Axis Tracking Pilot Projects

Tracking technologies increase the electricity produced by PV systems. The Hashemite University was one of the first in Jordan to install pilot projects using different types of solar PV modules and designs. The expected increase in the annual energy yield for the two axis systems ranges from 15% to 45%.

Figure 13-6. Two axis tracking system (Hashemite University).

Water Pumping

PV off-grid and on-grid systems are also used for water pumping applications. Consumers with agricultural tariffs pay the subsidized rate of 0.06JD/kWh ($0.084 U.S./kWh) and often do not view solar PV systems as an attractive investment. However, water pumping applications are often feasible alternatives for other consumers.

Transportation

In Jordan, 51% of the final energy consumption is consumed by the transportation sector [1]. Pilot projects have been initiated to investigate the use of electric vehicles in Jordan.

A solar PV electric charging station was installed at the Royal Scientific Society in 2009. The station generates electricity that is stored in batteries which supply the energy required for electric automobiles.

In 2015, the Greater Amman Municipality announced that they will install ten charging stations. Large suppliers also made electric cars available in Jordan for high end consumers such as owners of BMWs.

Solar Water Heaters

The use of solar water heaters started in Jordan in the late 1970s. Over 1.12 million m^2 of collector area have been installed [8]. This installed capacity reduced the country's primary energy demand by 145,000 tons of oil equivalent (TOE) annually [1]. The SWHs are directly contributing to 1.8% of the 2020 RE target. Most of these systems are single family domestic hot water heating systems.

Jordan's engineering, manufacturing, operation and maintenance capabilities for solar hot water heating systems have not advanced rapidly enough to meet the growth of consumer demand. The solar hot water technologies for multi-family, commercial, industrial, and for combined cooling and heating systems need to be further developed to match the progress of these systems in countries like Spain and Greece.

While there are roughly 50 small equipment suppliers, there are just a few large manufacturers to cover market demand. Engineering of larger systems will continue to be important for progress, enabling SWHs to make a greater contribution to Jordan's 2020 RE targets.

WIND PROJECT DEVELOPMENT

Wind energy projects have been located at Hofa and Ibrahimieh since 1998. Utility scale projects started in 2010 and the first completed project was at the Fujeij research station (1.65 MW). The results recorded at the research station indicated that the area has substantial potential for wind power.

These projects were followed by the completion of the Taffila wind farm (117 MW) in May 2015 with 38 turbines. These are Vestas 112x3.0

Figure 13-7. Taffila Wind Farm.

MW turbines with a hub height of 94 meters and blade length of 56 meters. The balance of the project was constructed by a local contractor with 250 workers who completed the excavation, foundation, and road work in 18 months. This required special expertise and equipment from outside Jordan, especially for the heavy lift transporting and cranes. The wind farm will be operated by 40 people for the next 20 years. The expected annual energy yield is 400 MW.

The Ma'an wind energy project (80 MW) and several other projects are pending and are expected to start construction in 2016 (see Table 13-4).

Further development of 700 MW of wind energy projects by 2020 is needed in order to achieve Jordan's RE targets.

Table 13-4. Wind energy projects.

#	Project Name	Size	Status
	Hofa	1.125	Old Project
	Ibrahimieh	0.32	Old Project
1	Fujeij Research Station	1.65	On Line
2	Taffila	117	Commissioned
3	Ma'an	80	Under Construction
4	Fujeij	99	Awarded
5	Rajef	83	Study stage
6	Ibrahimiah	6	Study stage
7	Lamsa	100	Study stage

GEOTHERMAL ENERGY PROJECTS

One of the largest geothermal projects in the Middle East is located at the American University of Madaba. It was designed and constructed by a company that specializes in this field. The details of the projects are shown in Figure 13-8.

While geothermal energy is available, the technology is not yet cost effective, and there are few interested consumers. There are only about ten geothermal residential and commercial projects installed by building owners in Jordan.

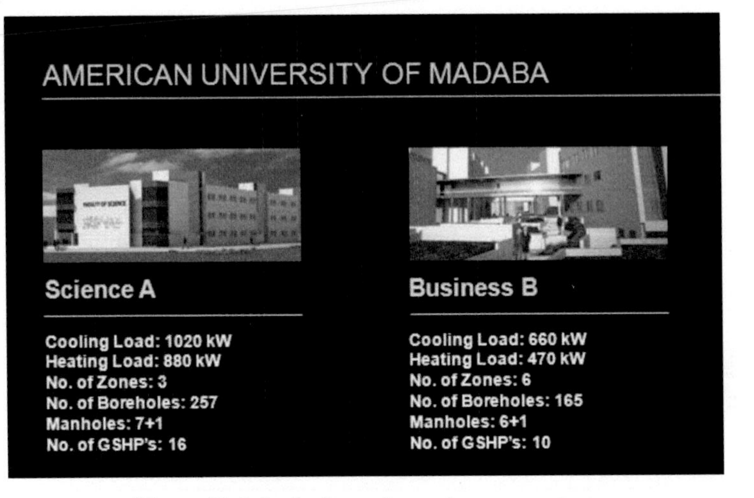

Figure 13-8. Jordan's geothermal energy project.

REDUCING CARBON EMISSIONS

Combating climate change is a key driver of the world movement to adopt renewable energy. In Jordan, the energy production sector contributes 74% of total greenhouse gas emissions with the industry share being just 8% [9]. The RE systems planned for development in Jordan will reduce future carbon emissions.

The U.S., China and India have agreements to reduce atmospheric CO_2 levels. Jordan has a relatively small population (6.6 million) and average energy intensity. While not a major carbon emitter, the country can have an important leadership role in the Middle East. Jordan's CO_2 emissions are reported annually by the International Energy Agency (IEA). The reported 2011 CO_2 emissions for Jordan total 19.8 million metric tons.

Average CO_2 emissions per kWh from electricity generation as reported by Jordan to the IEA from 1990 to 2010 [7] peaked in 1992. By 2010, each kWh of electricity generated on average 581 grams of CO_2.

The installed SWH capacities in Jordan reduce the emissions of CO_2 by 370,000 tons annually. This is one of Jordan's most important contributions to global sustainability.

Wind energy in Jordan will provide the largest share of renewables with roughly 1,000 MW of electrical capacity installed by 2020. To achieve the

1,000 MW target, the installation rates in the Table 13-5 were assumed for this assessment. The estimated energy generated includes the Taffila Wind project's design annual energy yield. Table 13-5 estimates the CO_2 emissions avoided based on the average CO_2 generated from each kWh in Jordan.

Table 13-5. CO_2 emission reductions from wind energy.

Year	MW	GWH/ year	GWH/ year accumulated	CO_2 tons avoided
2015	117	400	400	232,400
2016	80	274	674	391,306
2017	203	694	1,368	794,530
2018	200	684	2,051	1,191,795
2019	200	684	2,735	1,589,060
2020	200	684	3,419	1,986,325
Total	**1,000**			**6,185,415**

Jordan is planning to install a 50 MW plant to continuously generate electrical energy from biomass using supplies of waste materials. The estimated reductions of CO_2 from the operation of this plant are shown in Table 13-6.

Table 13-6. CO_2 emission reductions from biomass energy.

Year	MW	GWH/ year	GWH/year accumulated	CO_2 tons avoided
2018	50	400	400	232,400
2019	50	400	800	464,800
2020	50	400	1,200	697,200

The large-scale projects (200 MW) will start construction by May 2015. The estimated CO_2 reductions due to renewable energy projects are shown in Table 13-7.

Using a computer simulation for different types of solar PV modules installed in Amman with identical sizes, tilt angles and orientation, the expected average 25-year life cycle electricity yield is 41,000 kWh per KWp after accounting for any degradation as specified in manufacturer data sheets. For one sample project, the annual energy yield was calculated for several solar PV suppliers, which translated into the 25-year expected total

Table 13-7. 2020 CO_2 emission reductions from renewables.

Project Name	Size	Operation Date	CO_2 Emission Reduction (Ton)	No. of Years	Total Reductions (Ton CO_2)
EOI#1	200	2016	200,000	5	1,000,000
EOI#2	200	2018	200,000	3	600,000
Misc.	100	Varies	100,000	2	200,000
SWHs	1.1M M^2	2012	350,000	9	3,150,000
Wind	1,000	2020			6,200,000
Biomass	50				700,000
Total					11,850,000

energy yields as shown Figure 13-9.

Using the average CO_2 emissions in Jordan (580g of CO_2/kWh), the approximate reduction in CO_2 per kWp is estimated to be 24 metric tons during the expected project life of 25 years. Based on the national plan to install 600 MW by 2020, and calculating the CO_2 emissions saved during the following 20 years, the expected emission reductions would equal 14 million metric tons of CO_2.

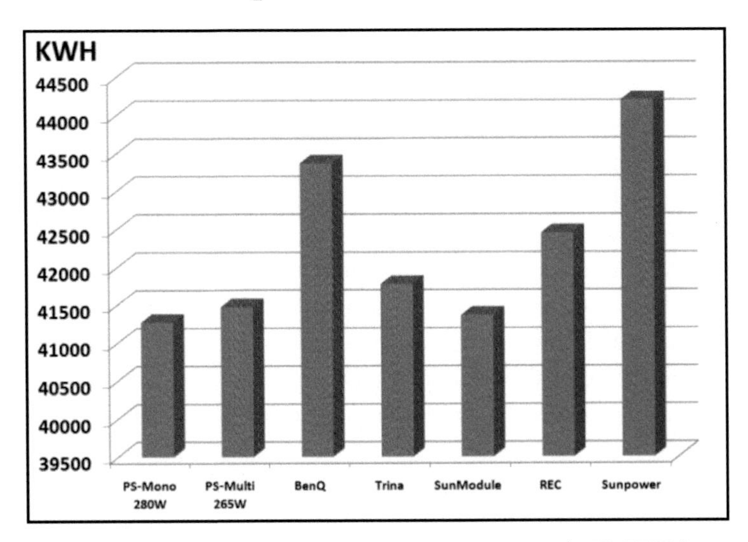

Figure 13-9. 25-year PV modules total energy yield (kWh/KWp).

Table 13-8.
2020 CO_2 emission reductions due to 600 MW of solar installed for 20 years.

Criteria	Amount	Unit
1kWp	41,000	kWh/25 years
	24,108,000	g CO_2
	24,108	kg CO_2
	24.108	ton CO_2
1MW	24,108	ton CO_2
600MW	14,464,800	ton CO_2
	14.4648	Million ton CO_2

SOCIAL IMPACTS

Energy directly impacts the lives of people. Consumers are more severely affected by increases in energy prices, directly impacting all other economic sectors. Some sectors are more resilient than others. For example, the commercial sector increases the price of their products as energy prices increase. The manufacturing sector is unable to readily increase the prices of their goods since products from exporting countries such as China and Saudi Arabia may be more competitive. This is why many industries that heavily depend on electricity have closed their businesses in Jordan.

The many stakeholders in the renewable energy sector must be fully integrated within the economy to successfully implement their projects. This process begins with the education of engineers, technicians and labor.

The stakeholders working in the energy sector need to upgrade their skills to work with renewables. A vocational education center in Ma'an started a year-long training program for 30 technicians in 2014. The program will help them gain skills for employment on utility scale solar PV projects. Several Jordanian universities are developing curriculums to accommodate RE education. As the market in solar and wind energy systems matures, more specialization will eventually develop into a multi-tier career map [10].

The RE technologies available will impact the projects being planned and developed to achieve the 2020 targets and beyond. Knowledge of RE technologies and grid capacities will enable stakeholders to develop an accurate means of developing system capacities. This will lead to social and environmental benefits including new employment, education, regional competitiveness, and greater availability of financing to develop such projects.

SUMMARY AND CONCLUSIONS

Sustainable development suggests that our proposed solutions be economically feasible, environmental friendly, socially acceptable, and have long term impacts. The RE sector is achieving these requirements. RE is a sustainable energy solution which improves a country's mix of energy resources.

RE technologies are commercially available and can be competitive with fossil fuel systems. Clean renewable energy technologies will continue to improve and further reduce the cost of energy.

The regulatory framework is the key in any country for the development of RE. Continuous improvement based on favorable public and private stakeholder experiences will continue to evolve. As stakeholders become more aware, they will be able to make more informed decisions.

Technical knowledge and workforce development are the most important factors in the development of solar PV.

There are many economic and environmental advantages for using RE systems. These include job creation, energy security, reduction in losses, access to electricity, and reductions in greenhouse gas emissions. RE systems will be among the alternative solutions adopted by countries rather than continuing to use fossil fuels and contributing to atmospheric pollution and climate change. RE systems must be adopted to avoid the instability associated with international fuel price variability.

Developing frameworks and incentives for consumers, industries and governments to use renewable energy is necessary to maintain steady and predictable economic growth. Since energy is integrated within the different economic sectors and some businesses are unable to quickly respond to fluctuations in fossil fuel costs, they are forced out of business. The regional development of the RE sector will help ensure the availability of skilled employees.

References

[1] Ministry of Energy and Mineral Resources (2014). Energy facts and figures report.

[2] Nasdaq online database (2015). Crude Brent Oil 10 year prices, http://www.nasdaq.com/markets/crude-oil-brent.aspx?timeframe=10y.

[3] National Electrical Power Company (2014). Annual Report. www.nepco.com.jo.

[4] Zawaydeh, S. (2013, December 9). Energy efficiency is the challenge for sustainable development. Third International Conference of Renewable Energy Engineers.

[5] Zawaydeh, S. (2013, December 10). Status of green finance in Jordan. USAID Jordan Water Reuse and Environmental Conservation Project Green Finance Conference at the Royal Scientific Society. Amman, Jordan.

[6] Ghazal, M. (2014, Apr 13). Jordan Times. http://www.jordantimes.com/news/local/project-provide-affordable-solar-water-heaters-families.

[7] U.S. Energy Information Administration (2014, November 19). PV energy yield dependency factors http://www.eia.gov/todayinenergy/detail.cfm?id=18871.

[8] Mathner, F. and W. Weiss (2014). Solar heating and cooling programme, IEA Solar Water Heaters Worldwide. http://www.aee-intec.at/0uploads/dateien1016.pdf.

[9] UNFCCC (2009). Jordan's Second National Communication to the United Nations Framework Convention on Climate Change http://unfccc.int/resource/docs/natc/jornc2.pdf.

[10] U.S. Department of Energy (2015). Solar career map. http://energy.gov/eere/sunshot/solar-career-map.

Chapter 14

Innovative Solution for Petroleum Refining in Nigeria

Nathaniel Umukoro, Ph.D.

Nigeria, a major petroleum producer and exporter, suffers from lack of refining capabilities, sometimes creating supply shortages. To satisfy demand, petroleum products must be imported from other countries. The situation is dire in areas of the Niger Delta. This has contributed to the development of artisanal refining of petroleum—a homegrown solution. Several studies have considered various aspects of this locally innovative strategy, often focusing on its environmental problems. Little attention has been given to how local refining can be harnessed to ameliorate the problems associated with petroleum shortages in Nigeria. This chapter examines the benefits of local, innovative refineries and argues that they be legitimized and regulated.

INTRODUCTION

The production of crude oil is Nigeria's most important source of foreign exchange and non-renewable energy. It contributes over 90% of the foreign exchange earnings of Nigeria and about 80% of capital and recurrent expenditure. The Nigerian oil industry has predominantly served the interests of both domestic and international elites. This usually results in the export of crude oil and shortages of refined petroleum products in Nigeria. Despite having four petroleum refineries, Nigeria exports crude oil but must import refined petroleum products for domestic consumption. The importation of refined petroleum products has been associated with problems such as fuel subsidy scams. Fuel subsidies cost Nigeria large sums of money that could otherwise be used for education, health, agriculture, rural development, transport, land and housing. Given Nigeria's consumption of about 45 million liters of premium motor fuel daily, the existing refineries produce only 12 million liters daily and are unable to meet local requirements [1].

The situation is more challenging in riverine areas of the Niger Delta. This has contributed to the innovation of a locally contextualized strategy for petroleum refining, usually called artisanal refineries. Several studies on artisanal refineries in the Niger Delta have focused on their negative aspects. For example, a study conducted by United Nations Environment Program in 2008 identified artisanal bunkering and artisanal or illegal refining of petroleum as a source of environmental pollution similar to oil spills and gas flaring by oil companies. Naanen and Tolani examined the social context of illegal oil bunkering and artisanal refining in the Niger Delta and viewed the situation as a private gain and public disaster because of the environmental pollution associated with these activities [2]. Asimiea and Omokhua also examined the environmental impact of illegal refineries on the vegetation of the Niger Delta [3].

As most studies have focused on the negative aspects of artisanal refineries, the positive aspects of artisanal refining are yet to be effectively researched. This chapter considers the positive aspects of this innovative strategy and how it can be harnessed to ameliorate the problems of petroleum refining in Nigeria, seeking to answer the following questions:

- In what ways have locally contextualized innovative solutions to the problem of petroleum refining benefited the riverine communities of the Niger Delta?
- How can this innovative strategy be scaled up to ameliorate the problem of petroleum refining in Nigeria?
- What benefits will Nigeria derive from legitimizing and institutionalizing an innovative petroleum refining strategy in the Niger Delta?

These questions were answered using data from both primary and secondary sources. Primary data for answering these questions were derived through observation, interviews and focus group discussions with selected individuals engaged in petroleum refining using local approaches. Key informant interviews were also conducted with selected staff of non-governmental organizations, universities and other research institutions in Nigeria. Secondary data were obtained from books, journal articles and reports. This discussion begins with an overview of petroleum production in Nigeria.

PETROLEUM PRODUCTION AND THE NIGERIAN ECONOMY

Prior to Nigeria's independence in 1960, about 80% of its labor force was engaged in agricultural activities. Agricultural products accounted for 85% of Nigeria's foreign exchange earnings. While Nigeria continued to be primarily an agricultural country after its independence, the increases in oil production and international crude oil prices during the 1970s led to the neglect of the agricultural sector which provided employment for majority of the population.

Petroleum Development in Nigeria

The early development of the petroleum industry in Nigeria can be traced to 1908, when the German Bitumen Corporation, started petroleum exploration activities in the Araromi area of western Nigeria. These pioneering efforts were affected by the outbreak of World War I in 1914. After the war, oil exploration resumed in 1937, when Shell D'Arcy (the forerunner of Shell Petroleum Development Company of Nigeria) was awarded sole concessionary rights for the territory of Nigeria. After years of investment and search, crude oil was discovered in commercial quantities in 1956 at Oloibiri in the Niger Delta area of Nigeria.

Production of crude oil in commercial quantities began in 1958 from the Oloibiri field, about 72 km (45 miles) east of Port Harcourt in the Niger Delta. The initial crude oil production rate was 5,100 barrels per day (bbl/d). Nigeria became politically independent in 1960 and afterwards oil production became characterized by agitations from host communities and human rights violations by the government. Scholars have documented and analyzed the confrontations that have characterized oil production which involved the host communities, multinational corporations and the government. For example, Obi studied how oil extraction and the dispossession of the people of the Niger Delta resulted in incidents of violence—clashes between rival armed groups, militias and government troops—characterized by killings, sabotage of oil pipelines and installations, and a thriving transnational trade in stolen oil (or illegal oil bunkering) [4]. Frynas examined the consequences of foreign oil production activities in rural communities of the Niger Delta [5]. The study gives a comprehensive overview of the environmental and social impact of oil operations which were little understood. Frynas further argued that the country's environmental and land

this supply situation is compounded by poverty and lack of access to petroleum products in the challenging and difficult terrain of the Niger Delta.

Despite the uncertainties associated with identifying the origin of the artisanal refining process, there is a consensus from those interviewed that it was likely the technology originally used for refining a local gin called ogogoro was adapted for crude oil refining. The adaptation became widespread because of the inability of people to earn adequate revenue from agricultural activities and the production of the local dry gin. This was due to oil pollution that destroyed farm land, aquatic life, and the raffia palm used for producing ogogoro. Some of the respondents asserted that most people engaged in artisanal refining of crude oil in the Niger Delta did not set out originally to refine crude oil. They started with bursting pipelines to siphon petrol and diesel for sales. When these activities were becoming really dangerous—many people were being arrested or incinerated at the point of siphoning—they began artisanal refining activities as an alternative. According to the respondents, the government security agents in the Niger Delta did not realize that artisanal refining of crude oil was occurring. The military joint task force (JTF) initially concentrated on fighting militants. Members of the JTF were surprised to learn that militants in the mangrove forests were obtaining fuel to operate their boats and generators. It was later discovered that the militants were getting supplies of refined products from artisanal refiners. This marked the beginning of efforts to locate and destroy artisanal refineries in the Niger Delta.

These theories of the origins of artisanal refineries notwithstanding, more salience is often placed on the notion of existential exigency and the pressures placed on artisanal refiners to earn a living. Artisanal refineries respond to the perennial scarcity of petrol, diesel and kerosene in Nigeria which are used for vehicles, producing electricity and domestic cooking [11]. The need for these products is imperative due to inadequate electricity supplied from the national grid and sub-national systems. In parts of the Niger Delta where regular consumer petroleum products cannot be easily obtained, products from illegal refineries have become indispensable. Established marketers of petroleum products are also known to patronize artisanal refined pans which they mix with products from other sources in underground tanks for filling (gas) stations, dispensing product to unsuspecting members of the public. Products from artisanal refineries are also used by industrialists who depend on electric generating plants to keep their businesses operating in the absence of a constant electricity supply. With the end of the insurgency and the commencement of the amnesty program, artisanal refining continues as a source of income for demobilized insurgents and idle Niger Delta youths. The raw materials needed are accessible and plentiful.

HARNESSING THE HOMEGROWN SOLUTION

Local refining of petroleum using artisanal refining methods has increased in the past few years. Riverine communities, especially those in remote areas of the Niger Delta, have obtained steady supplies of refined petroleum products such as kerosene for the past decade from local refineries. The refined products are sold in commercial quantities to other parts of Nigeria, thus providing sources of income and employment for youths in rural areas of the Niger Delta. Participants in focus group discussions generally espoused the view that artisanal refining of petroleum fills an economic vacuum—local communities suffer from the impacts of oil extraction but see none of the economic benefits. The failure of the Nigerian government to provide basic public services and security in the Niger Delta has resulted in a breakdown of the social contract. In the face of corruption by political elites, communities view artisanal refining as means of surviving economically in the absence of mainstream livelihoods. They assert that the local refineries help balance the activities of the region's militants and pirates since thousands of jobless people participate. Key informants and participants in focus group discussions acknowledged the environmental consequences of these activities. Regardless, they view their activities as the only available option to survive the economic hardships they confront. They also feel a strong sense of ownership toward using their artisanal refining technologies and do not want to abandon their activities.

The Role of Research and Development

Empirical research and surveys of business activities show that innovation spawns new and improved products and services, higher productivity and lower product prices. Economies with consistently high levels of innovation have high levels of economic growth [12]. The experiences of economically advanced countries indicate that research and development (R&D) is crucial for harnessing homegrown innovation for sustainable development. In February 2012 Barack Obama, a former president of the

U.S., stated, "We need to build a future in which our factories and workers are busy manufacturing the high-tech products that will define the century... Doing that starts with continuing investment in the basic science and engineering research and technology development from which new products, new businesses, and even new industries are formed." He emphasized that investment in technology and future capabilities is transformed into new products, processes and services. The economic growth of any nation depends on the capacity to educate, innovate, and build long-term national investments in basic and applied R&D. Mutually reinforcing and complementary investments in R&D by both the private and public sectors work in concert to support the development, production, and commercialization of new products and processes.

The benefits associated with promoting innovative activities are captured in the following statement of former U.S. president George W. Bush in 2004, "America leads the world because of our system of private enterprise and a system that encourages innovation. And it's important that we keep it that way. See, I think the proper role for government is ... to create an environment in which the entrepreneurial spirit flourishes... the government can be a vital part of providing the research that will allow for America to stay on the leading edge of technology... I think we ought to encourage private sector companies to do the same, invest in research." Since innovation has long been recognized as an important driver of economic growth, it is pertinent that the Nigerian government learn from the experiences of countries such as the U.S. by promoting innovative R&D. National investment in R&D includes investments by governmental entities, colleges, universities, businesses and non-profit organizations.

Various institutions in the Niger Delta specializing in petroleum related activities have important roles in future R&D. These institutions include the Federal University of Petroleum Resources (FUPRE), the Petroleum Training Institute (PTI), the Delta State University Department of Petroleum Engineering and the University of Benin. Key informants indicated that the artisanal refinery operators have begun supporting R&D. They have discovered that pollution from their refining activities can be minimized by using gas cookers to heat crude oil during the refining process instead of using wood or wasted crude which often produces pollution in the form of a thick smoke.

Participants in focus group discussions and key informants from the institutions indicated that it is possible for the locally developed processes for petroleum refining to be enhanced in ways that will improve the quantity and quality of the refined products. They identified challenges for R&D associated with artisanal refining. One is inadequate research funding for educational institutions which creates difficulties for those interested in collaborating with local refiners to improve the processes. It is imperative that organizations both within and outside Nigeria assist researchers who are interested in finding ways to improve the local technologies for petroleum refining. Developed countries have demonstrated that funding innovative projects solves problems. American research universities have been a model of innovation throughout the world, addressing complex economic, social, scientific and technological problems [13]. Universities contribute to the quality of a region's economic infrastructure by developing knowledge-linking activities to enhance the commercialization of new technologies, support organizational and community change, and assure the education and competency of workers and professionals [14].

In the 1970s, China and India funded research for innovative solutions and became economic super powers. They looked inwards, finding solutions to their challenges, and implemented those that they believed would work best given their circumstances. They developed their own technologies and constructed production facilities that improved over time. Today, these countries not only provide their indigenous technological needs but also export technologies and industrial products.

The improvement of the indigenous technology for petroleum refining through R&D facilitates air and water pollution reduction. If the quality of these refineries is enhanced, the government of Nigeria can offer licenses to the entrepreneurs enabling them to engage in their activities legally. The legalization of the artisanal refineries will make it possible for the entrepreneurs to purchase crude oil at a price stipulated by the government, reducing crude oil theft by the refiners.

Economic Impact of Failing to Harness Homegrown Petroleum Refining

The inability of the Nigerian government to improve the performance of state-owned refineries has made Nigeria a major importer of petroleum products. The failure of the government to harness the ingenuity of artisanal refinery operators in the Niger Delta has adverse effects on the Nigerian state. It increases expenditures for the importation of petroleum products. Ploch observed that "Nigeria imports an estimated $10 billion of fuel annually for domestic consumption" [15]. In 2012, "Nigeria consumed 270,000 bbl/d and in 2013, she imported slightly more than 84,000 bbl/d of petroleum products" [16]. Nigeria purchases fuel from distant countries including Venezuela, the U.S., Canada, Brazil, the Netherlands and the United Kingdom. Nigeria also imports premium motor spirit (PMS) from non-oil producing countries including the Niger Republic, Cote d'Ivoire, the Netherlands, India, Korea, Finland, Singapore, France, Israel, Portugal, Italy, Sweden, Tunisia and others [17].

Closely related to large government expenditures on imported refined petroleum products is the issue of subsidies. Studies indicate that subsidies associated with importing such products enable the embezzlement of public funds [18]. For example, a parliamentary probe in 2012, determined "that graft in the fuel subsidy scheme cost Nigeria $6.8 billion between 2009 and 2011" [19]. If artisanal refineries are legalized, monies expended on subsidies can be used to provide health services, public works, youth employment, urban mass transit, vocational training, and infrastructure improvements, such as roads, rail, water resources and electricity.

Failing to harness the homegrown solution for petroleum refining also encourages environmental pollution in the Niger Delta that is caused by government security agents when they destroy artisanal refineries. Currently, the federal government of Nigeria uses its Joint Task Force (JTF) to fight against oil theft and the proliferation of artisanal refineries. This is not an effective long-term strategy. The JTF activities temporarily interrupt some of the refining operations, yet camp owners and workers interviewed did not view JTF's activities as a major threat to their business. After the destruction of artisanal refineries and their products by the JTF, camp owners quickly rebuild their operations in new locations. All the respondents during focus group discussions indicated that the major concerns about the activities of the JTF are the effects of the destruction of artisanal refineries and crude oil on the environment.

When the security forces seize crude oil and equipment used for artisanal refining there does not seem to be a safe and environmentally friendly method for disposal. Rather, they are indiscriminately set ablaze in the open, often in close proximity to homes and businesses. Burning of artisanal refineries and associated equipment and supplies is a JTF practice openly acknowledged in their press releases. When the JTF decides to terminate an artisanal refinery operation, all confiscated crude oil containers and tankers found at the site are destroyed by combustion. Boats laden with crude oil or refined products are burnt or emptied into the waterways and wetlands causing adverse environmental impacts. One respondent noted that "when the armed forces destroy artisanal refineries and the refined products thick smoke covers the skies, dangerous gases are emitted into the atmosphere thereby causing environmental pollution."

Benefits of Harnessing the Homegrown Solution for Petroleum Refining

Improving the capacities of artisanal refineries in the Niger Delta through R&D will benefit the Nigerian government by creating additional revenue. Studies have shown that countries that increased refining capacities seem to have gained from their linkages with oil production [20]. Some may argue that since production from artisanal refineries is low, they may not notably impact in the Nigeria economy. This is not necessarily true. An analysis of Chinese national oil companies operating in Africa and central Asia suggests that new technologies enable some oil projects to operate profitably [21]. In countries such as Chad, where the French and U.S. oil companies felt that refinery projects would not be viable, oil projects have instead proved to be beneficial to the local economy [22]. A recent analysis by the International Monetary Fund found that a new refinery built by the Chinese National Petroleum Corporation (CNPC) in Chad is economically viable, providing adequate margin between the costs of production and the income from refined products [23,24].

Benefits that the Nigerian state can gain from supporting larger production volumes from artisanal refiners include increased economic diversification resulting from linkages with other firms, a more reliable supply of petroleum products, creation of new employment, reduction of poverty and reduced income inequality. More refineries are needed in Nigeria for strategic reasons. They will help improve the nation's gross domestic product.

Since the Nigerian government is advocating increased participation in agricultural activities, harnessing the decentralized homegrown solution for petroleum refining complements increased local agricultural productivity. Farm machinery requires petroleum. The byproducts of petroleum refining are useful for agricultural purposes. Most pesticides and many fertilizers are made from petroleum. The use of chemical pesticides and fertilizers increases agricultural productivity.

Legalizing and increasing production from artisanal refiners will foster improved security in the Niger Delta. Unemployed youths are influenced to engage in militant activities and the destruction of oil pipelines, forcing the federal government to depend mainly on the importation of refined products from other countries. Artisanal refineries offer alternative employment opportunities. The frequent crises in the Niger Delta region between 2002 and 2006 paralyzed the oil sector, making Nigeria further dependent on imported petroleum products. Funso Kupolokun, the group managing director of the Nigerian National Petroleum Corporation, once asserted that Nigeria depended 100% on imported petroleum products when the nation's four refineries were closed due to a damaged supply line while militants fought for local control of the Niger Delta's oil [25].

CONCLUSION

The homegrown solution for petroleum refining in the Niger Delta can be harnessed for petroleum refining in Nigeria. This requires the efforts of the Nigerian government and research institutions in Nigeria. Federal government actions to destroy the indigenous refineries should be reconsidered. Hindering development prevents operators from improving artisanal refineries and supporting more focused research and development.

It is also important to review extant laws on refining of petroleum products. Such laws prohibit the use of indigenous technology for petroleum refining in the Niger Delta since they are not recognized by the state as credible for licensing purposes. For example, Section 3(1) of the Petroleum Act states that no refinery shall be constructed or operated in Nigeria without a license granted by the minister. The government through a well-articulated policy framework should formalize the activities of the indigenous refineries by licensing their operations in Nigeria.

Rather than signing agreements with foreign countries to establish modular refineries (as Nigeria did with the U.S. in 2012), the government should harness homegrown solutions to oil refining, creating more petroleum products in the Niger Delta. If the local refiners are encouraged, the government could work with them and sell crude to them for refining. Legitimizing this type of strategic engagement could end petroleum theft, reduce environmental pollution and create more employment opportunities.

References

[1] Ibe-Kachukwu, E. (2016). Refineries back in production but need $700 million dollars.
[2] Naanen, B. and P. Tolani (2014). Private gain, public disaster: social context of illegal oil bunkering and artisanal refining in the Niger Delta. Port Harcourt: Niger Delta and Environmental Relief Foundation.
[3] Asimiea, A. and G. Omokhua (2013). Environmental impact of illegal refineries on the vegetation of the Niger Delta, Nigeria. *Journal of Agriculture and Social Research*, 13(2), pages 121-126.
[4] Obi, C. (2010). Oil extraction, dispossession, resistance, and conflict in Nigeria's oil-rich Niger Delta. *Canadian Journal of Development Studies*, 30(1,2), pages 219-236.
[5] Frynas, J. (2000). *Oil in Nigeria: conflict and litigation between oil companies and village communities*. LIT/Transaction Press. Hamburg/New Brunswick, New Jersey.
[6] Ako, R. and P. Okonmah (2009). Minority rights issues in Nigeria: theoretical analysis of historical and contemporary conflicts in the oil-rich Niger Delta region. *International Journal on Minority and Group Rights*, 16(1), pages 53-65.
[7] Klieman, K. (2012). U.S. oil companies, the Nigerian civil war, and the origins of opacity in the Nigerian oil industry, 1964-1971. *Journal of American History*, 99(1), pages 155-165.
[8] Nigerian National Petroleum Corporation (2017). History of the petroleum industry in Nigeria. http://www.nnpcgroup.com/NNPCBusiness/Businessinformation/OilGasinNigeria/IndustryHistory.aspx, accessed 10 February 2017.
[9] Asuru, C. and S. Amadi (2016, July). Technological capability as a critical factor in Nigerian development: the case of indigenous refineries in the Niger Delta region. International *Journal of Innovative Research in Education, Technology and Social Strategies*, 3(1), pages 2,467-8,163.
[10] Murdock (2012, July 12). Nigerians are stealing corporate oil, refining it themselves, and selling it on the black market. Business Insider. http://www.businessinsider.com/nigeria-illegal-oil-refining-2012-7.
[11] Omoweh (1995). Shell, environmental pollution, culture and health in Nigeria: the sad plight of Ughelli oil communities. *African Spectrum*, 30(2), pages 115-120.
[12] Atkinson, D. and A. McKay (2007). Digital prosperity: understanding the economic benefits of the information technology revolution. The ITI Foundation. Washington, District of Columbia.
[13] Cole, J. (2010). *The great American university: its rise to preeminence, its indispensable national role, why it must be protected*. Public Affairs: New York, New York.
[14] Walshok, M. (1997). Expanding roles for research universities in economic development. *New Directions for Higher Education*, 97, pages 17-26.
[15] Ploch, L. (2013). Congressional research service, Nigeria: current issues and U.S. policy. U.S. Congressional research publication. http://www. gamji.com/article6000/NEWS6719.htm, page 9, accessed 5 November 2016.
[16] U.S. Energy Information Administration (2013). Nigeria country study. Washington, District of Columbia. U.S. Government Press, page 13.
[17] Chimezie (2009). Importation pricing of petroleum products in Nigeria: the shame of a nation. http://www.theeconomistng.blogspot.com/.../importation-pricing-ofpetroleum.htm., page 7, accessed 30 January 2017.
[18] Okonjo-Iweala, N. (2011, December 5). Fuel subsidy removal. The Nation.
[19] Agande, B. (2012). Oil subsidy must go—Jonathan. The Vanguard, Friday, 16.
[20] Clark, J. (1990). The political economy of world energy: a twentieth century perspective. Harvester Wheatsheaf.
[21] Jiang, W. (2009, September 10). Fueling the dragon: China's rise and its energy and resources extraction in Africa. *The China Quarterly*, 199, pages 585-602.

[22] Dittgen, R. and D. Large (2012, May). China's growing involvement in Chad. escaping enclosure? SAIIA occasional paper 116. China in Africa project. SAIIA. http://www.saiia.org.za/occasional-papers/chinas-growing-involvement-in- chad-escaping-enclosure, accessed 10 August 2013.

[23] International Monetary Fund (2013). Chad. 2012 Article IV Consultation. IMF country report no. 13/87 Chad. http://www.imf.org/external/pubs/ft/scr/2013/cr1387.pdf, page 12, accessed 10 February 2017.

[24] Baur, S. (2014). Refining oil—A way out of the resource curse? Working paper series, Department of International Development, London school of economics and political science.

[25] Izere, I. (2006). Fuel scarcity/price increase: lawmakers failed Nigeria to produce at maximum capacity. http://www.legaloil.com/News.asp?Month=4&Year=0&TextSearch, page 1, accessed 15 February 2017.

Chapter 15

Potential of Photovoltaics for Water Heating in South Africa

Ognyan Dintchev, G.S. Donev, J.L. Munda, O.M. Popoola, M.M. Wesigye, and S. Worthmann

Electric water heating (EWH) is the most common technology used to heat water in residential and commercial buildings in the country of South Africa. Water heating typically accounts for roughly 40% of the energy consumption of a typical South African home. The high capital costs of solar thermal water heating (SWH) systems in South Africa have made implementation of this technology very expensive. Recent reductions in the costs of solar photovoltaic (PV) systems have made them economically competitive with conventional electrical systems. The relatively steady annual electrical generation output of solar PV systems when compared with the seasonal performance of the conventional SWH has created opportunities for solar PV to capture a larger share of the local water heating market. This chapter reviews and assesses the three primary water heating technologies currently used in South Africa and provides an energy and financial analysis of each technology.

SOLAR WATER HEATING IN SOUTH AFRICA

South Africa has been experiencing serious electricity generating capacity shortages for some time. Load shedding has become a constant threat in South Africa and new ways of providing energy storage and alternative electricity supplies are necessary to resolve this problem. The residential and commercial sectors use a substantial amount of electricity for water heating. The primary technologies for water heating available are electric water heating and thermal applications of solar water heating. South Africa has one of the world's highest solar thermal potentials. However, SWH has not evolved to be the country's preferred water heating option. The reasons for this are financial, technological, legislative, economic, cultural and social.

The recent decline in the costs for solar photovoltaic modules and improvements in module efficiencies create opportunities for PV to be used for water heating. However, additional research and publicity are needed to develop, adapt, and apply this technology in South Africa [1]. The National Solar Water Heating Program (NSWHP) was launched in 2009 by the South African government with a goal of encouraging the installation of one million SWH systems by 2015 [2,3]. Its primary objectives were:

- Reduce electricity demand and greenhouse gas emissions.
- Protect poor people from electricity tariff increases.
- Facilitate a local manufacturing industry.
- Create employment.

Over 400,000 systems were installed in residential dwellings by the conclusion of the NSWHP. Some of the problems that developed included:

- The intended reduction in electricity usage was not achieved since it focused on locations with comparatively low electricity consumption.
- Low quality imported products have dominated installations.
- Poor quality installations due to lack of training.
- Unreliable verification of the number and location of installed systems due to lack of monitoring.
- Lack of maintenance support which resulted in users reverting to electric systems.

Two new goals have now been adopted. The Department of Energy (DoE) has set an intermediate aggregate target of 1.75 million SWH installations by 2019. The National Development Plan established a cumulative target of five million installed SWHs by 2030 [2]. These goals were based on a revised framework that emphasized local content requirements for new SWH installations and aimed to achieve more sustainable SWH systems [2].

Despite these goals, solar water heating in South Africa has received inconsistent support from the government in restructuring its SWH policies. The technology failed to become a preferred solution for consumers considering alternative technologies and renewable energy solutions. The main cause of the low market penetration of SWH technologies is their extremely high price. This is linked with the higher cost of capital in South Africa when compared to countries with similar solar energy potential. Affordable SWH systems are under-designed and incapable of providing adequate energy and monetary savings to offer an acceptable return on investment.

Another major obstacle to SWH development in South Africa is the absence of a developed market for quality (preferably locally made) solar collectors, storage tanks and other components that enable customers to assemble a system that matches their requirements. Repair and replacement of faulty system components is difficult. As a result, the market offers predominantly compact solar water heating units at inflated prices [4].

Compared to solar alternatives, electric water heating (EWH) is an established and regulated business in South Africa. It is supported by favorable legislation and the involvement of insurance companies. Appliance prices are very low [4].

CONVERTED OR MODIFIED WATER HEATERS

Conventional EWHs are easily replaceable and usually for a low cost. EWH manufacturers do not provide options for easy conversion of EWH hot water tanks into similar SWH storage tanks which increases the costs of solar conversions. This often justifies the relatively high electricity costs for consumers reluctant to scrap their still operational EWHs.

Regardless, there are commercial methods to convert conventional EWHs into solar but the conversion costs are higher than comparable new SWHs [6]. There are also commercial solutions to convert an existing EWH into a hybrid one using solar photovoltaics [3]. The technology replaces the flanges of an EWH with a dual heating element: one 900W for the direct current (DC) output from a PV array and a titanium electric water heater (3kW) for 230 volt alternating current (VAC). The unit uses a maximum power point tracker and a controller that allows system performance to be optimized in the absence of solar DC input from the PV array. However, all these additions increase the unit's costs comparable to the cost of a SWH [4,7].

Typically, the price of converted solar water heaters is comparable to conventional SWHs. The cost comparisons among EWHs, SWHs, converting EWHs into SWHs and converting EWHs into PV water heaters for 100 liter to 300 liter units are provided in Figure 15-1. The prices shown are averages at the middle range of the market. The extreme high and extreme low prices were not considered in this comparison [4,6].

Options for Residential Water Heating

Our study was conducted on behalf of the secretariat of the South African-German Energy Partnership managed by Deutsche Gesellschaft für Internationale Zusammenarbeit (GIZ) GmbH and funded by the German Federal Ministry for Economic Affairs and Energy. Its principle objective was to compare the three options for residential water heating (electric water heater/geyser, solar water heater, solar PV system powering existing electric geyser) in terms of thermal performance, efficiency and cost [1].

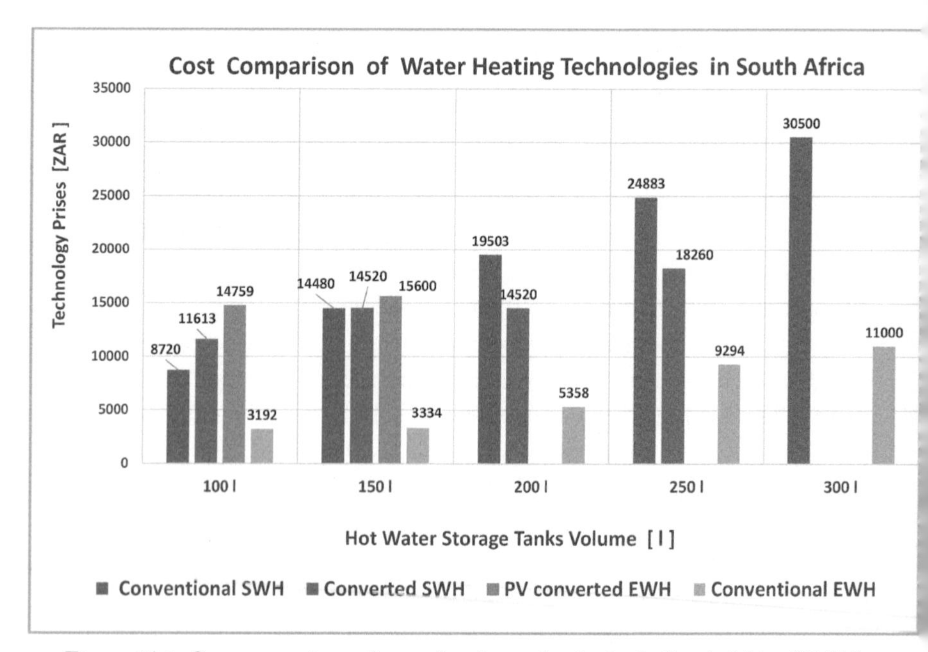

Figure 15-1. Cost comparison of water heating technologies in South Africa ($1 U.S. = 14 ZAR in January 2017).

The selected systems investigated used specifications typical for the medium cost market in South Africa.

HOT WATER HEATER STUDY RESULTS

The methodology of the study was based on validating the simulation models by physical back-to back testing of the three technologies. This testing was intended to show how the three technologies perform given similar conditions. The test results were valid for the days of testing during the summer month of December 2015. Considering the time of the year, temperature and weather conditions, the following parameters were measured, recorded and tabulated in Table 15-1.

Table 15-1. Test performance results.

Parameter	**Unit**	**Technology**		
		Electric Water Heater	Solar Water Heater	PV Water Heater
Thermal Energy Output per day	kWh	6.76	6.36	4.91
Performance (EWH as 100%)	%	100.00	94.08	72.65

Simulation Results

The three water heating systems were individually simulated. To validate the simulation model, the system specifications were comparatively similar to the systems physically tested. The simulations were performed using the 2015/16 version of Polysun 8.1 simulation software [5]. The simulated energies were: 1) the grid input fraction for the EWH and as back up energy for SWH and PV water heating; 2) the solar fraction yield for heating water for the SWH; and 3) the solar yield of the solar PV array at the AC inverter output terminals. The simulations were limited to the specific designs and typical selections of the three technologies at a specific location in Pretoria, South Africa. The results and findings should not be considered to be representative for EWH, SWH and PV water heating technologies. The simulated system specifications are presented in Table 15-2.

The annual simulated energy performances of the EWH, SWH and PV water heating are illustrated in Figures 15-2, 15-3 and 15-4. The simulations were performed with the arrangement that the three technologies deliver the same amount of hot water.

Table 15-2.
Specification of simulated water heating systems.

System Geographic Location: Pretoria, South Africa Latitude: -25.75 °, Longitude:28.2 °, Elevation:1,402 m				
No	Item	Electric Water Heater	Solar Water Heater	PV Water Heater
		Specifications		
1	Hot water storage tank Volume	150 l	150 l	150 l
2	Electric Heating element	230 V, 50 Hz, 3 kW	Back up: 230 V, 50 Hz, 3 kW	230 V, 50 Hz, 3 kW
3	Standing losses over 24 h	2.59 kWh	2.59 kWh	2.59 kWh
4	Thermostat settings	50 °C	50 °C	50 °C
5	Solar Collector		1x flat plate, gross area:2m ^{2}	
6	Circulation pump		Eco, small 72 l/h	
7	PV Modules			3 x 300 W PV modules
8	Grid-tied inverter			2.2. kW
	Purchase Price	R 3,000.00	R 20,515.00	R 23,412.00

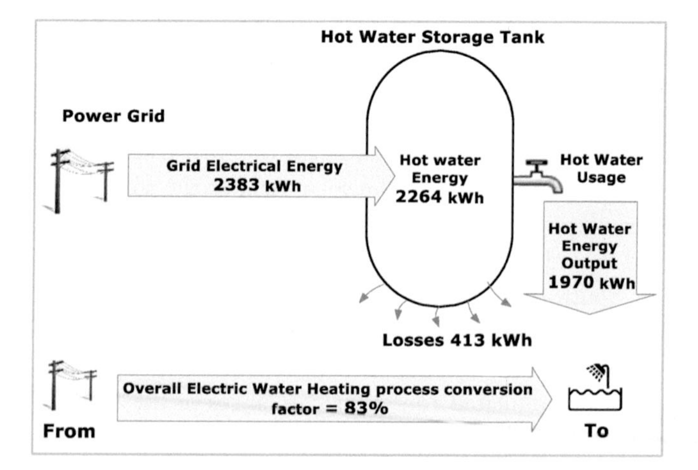

Figure 15-2. Annual simulated thermal performance of the electric water heater.

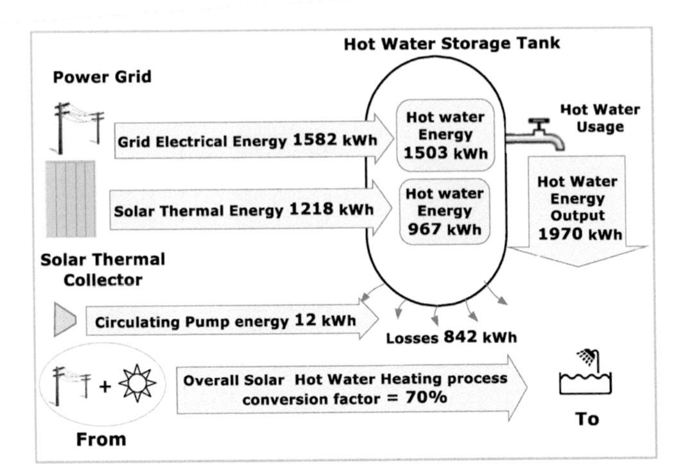

Figure 15-3. Annual simulated thermal performance of the solar thermal water heater.

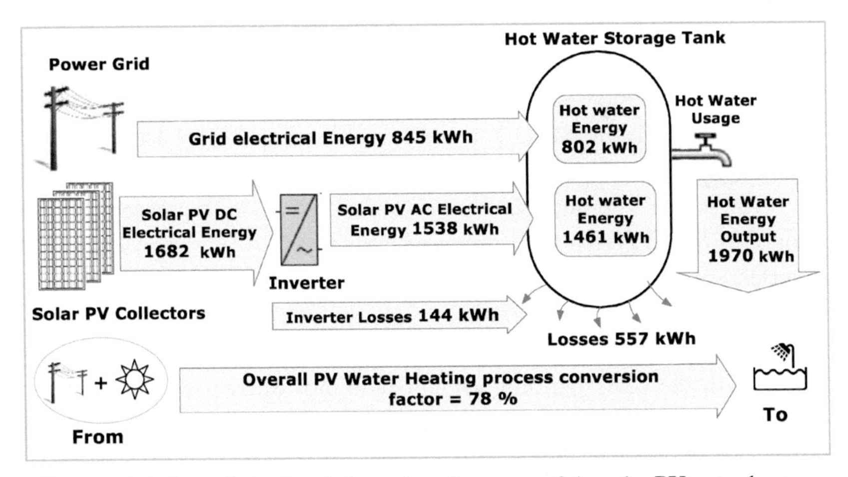

Figure 15-4. Annual simulated thermal performance of the solar PV water heater.

WATER HEATER COMPARISONS

Performance comparisons are shown in Figure 15-5. It is evident that the PV solar yield of 1,538 kWh/a is higher than the corresponding SWH yield of 1,218 kWh/a. The percentage of annual grid electricity consumption of the solar PV water heating system (35%) is less than the corresponding fraction of the SWH (56%). The solar component of the solar PV water

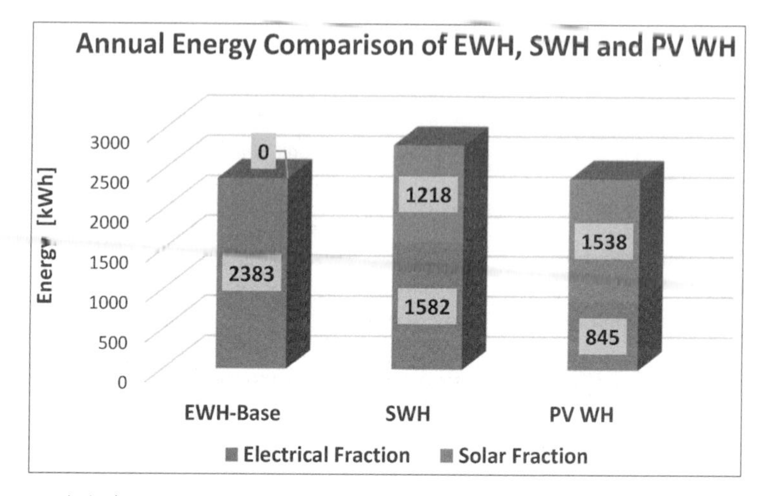

Figure 15-5. Annual energy balance comparisons for water heating using the three technologies.

heater is superior to that of the SWH.

The grid energy input saved (not used) for water heating is shown in Figure 15-6. Both the SWH and the solar PV water heater use less energy from the grid than the EWH.

The solar PV annual yield is not as dependent on the seasonal conditions as is the SHW. When compared annually, the PV solar yield is 10% superior, with a yield of 55% compared to the SWH's corresponding yield of 45%.

The annual average daily PV AC yield is shown in Figure 15-7 together with the corresponding average daily hot water usage. It is evident that only 60% of the PV AC yield is used directly for heating water while the balance is exported to the grid.

ENERGY AND FINANCIAL ASSESSMENTS

Comparing the SWH to the PV Water Heater

The financial analysis of the SWH and PV water heating technologies was based on identical assumptions [7]. The same parameters obtained from the simulations are used as inputs/drivers to the financial analysis (see Table 15-3). The financial drivers per system are the grid input per system saved and related monetary costs [7].

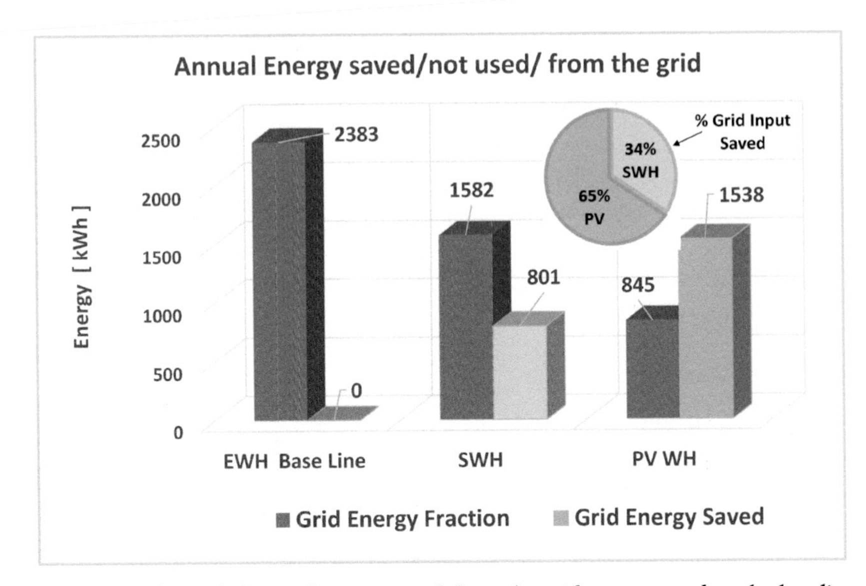

Figure 15-6. Annual electrical energy saved from the grid as compared to the baseline of the electric water heater.

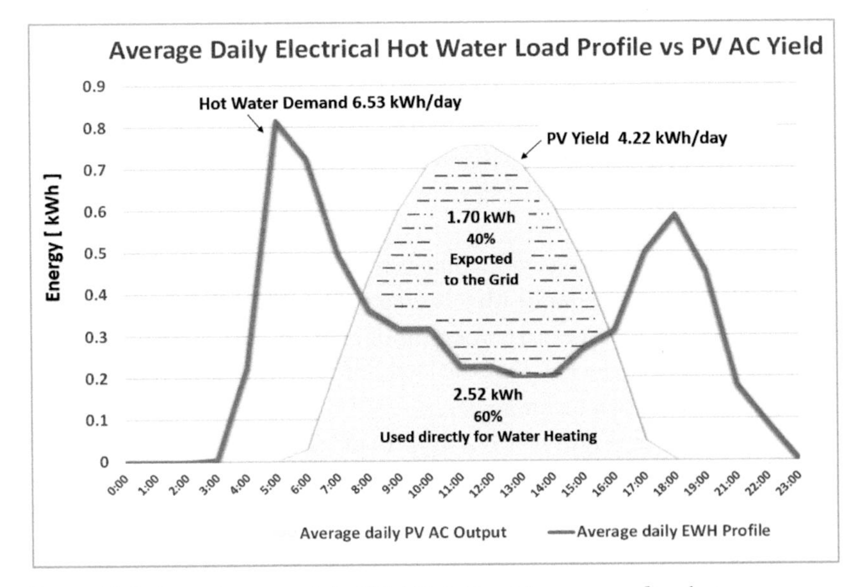

Figure 15-7. Annual average daily PV AC yields with corresponding hot water usage.

Table 15-3. Macro and per system drivers.

Drivers - Macro	Unit	Macro Driver	
Household HW consumption	liters / annum	54470	
Grid input for WH - without any system	kWh / annum	2383	
Grid electricity unit cost - current	ZAR / kWh	1.50	
Current Interest rate	% / annum	9.50%	
Current Inflation rate	% / annum	5.00%	
Current Grid electricity inflation rate	% / annum	15.00%	
Drivers - Per System	**Unit**	**SWH**	**PV**
Purchasing cost - once off	ZAR	20,515	23,412
Installation cost - once off	ZAR	1,000	1,000
Maintenance cost coefficient	% / annum	2%	1%
Expected useful life	years	15	20
Grid input for WH - with system	kWh / annum	1582	845
Grid input saved	kWh / annum	801	1538
% of Grid input saved	%	34%	65%

Energy Performance

From the simulations and comparisons it is evident that the SWH's total energy used for water heating is greater than the EWH's and the solar PV water heater's counterpart when producing the same quantities of hot water. This is due to the fact that the losses of the SWH are higher than the losses of the other two technologies.

As shown in Figures 15-2 and 15-3, the annual grid electricity consumption of the solar PV water heater is less than the corresponding fractions of the SWH and the solar fraction per yield of the solar PV water heater is greater than those of the SWH. Figure 15-6 shows that grid input energy saved by the solar PV water heater is greater than the thermal SWH.

The annual electricity saved (not used) from the grid of the two technologies for water heating demonstrates the superiority of the solar PV water heater—1,538 kWh compared to 801 kWh for the SWH. The percentage of energy savings of the two technologies compared to the EWH are 65% and 34% respectively (see Figure 15-6).

The solar energy performance comparison indicates that the thermal SWH output is affected by the climate whereas the solar PV WH performs steadily throughout the year. Also, the PV annual solar yield (55%) is higher than that of the SWH (45%).

In summary, for the specific selected South African conditions, the solar PV water heating technology offers better energy performance than the SWH. This results from its greater solar energy harvesting potential and its relatively constant energy output during the year. It is also easier to install than residential electric water heating systems.

Lifecycle Costs and Savings Analysis

Next considered are lifecycle analysis methodologies that include the simple payback period, net present value, return on investment and internal rate of return. Each offers a methodology to perform lifecycle financial analysis. A summary of the financial outputs is shown in Table 15-4. An interactive decision tree was developed to compare the financial performance of the three technologies during their useful life. The cost per liter of heated water is calculated by considering the solar and grid fractions for water heating as shown in Table 15-5.

Simple payback period (SPP)—The SPP is the total installation costs divided by the difference of annual savings less annual operating costs. The result is positive when annual savings exceed annual costs. The SPPs for the SWH and the solar PV water heater are 10.5 and 6.9 years respectively. The solar PV water heater is superior by 3.6 years (see Table 15-4),

Net present value (NPV)—The NPV is the difference between the present value of cash inflows and the present value of cash outflows. The discontinued NPV payback periods for the SWH and the solar PV water heater are 16.0 years and 9.6 years respectively, making the solar PV water heater superior by 6.4 years. The SWH fails to achieve a payback of its investment within its useful lifecycle. Over the lifecycle of the SWH, the NPV is negative value (R–2,056) while the solar PV water heater is a positive value (R43,466). The solar PV water heating system with its higher NPV of R43,446 provides better financial performance as it returns a greater monetary sum. The solar PV water heater is superior by a difference of R45,522 or 2,214% (see Table 15-4).

Table 15-4. Financial analysis outputs.

Analytic Criteria	Unit	SWH	PV	Amount	%	Preferable
Efficiency						
Grid Input saved	%	34	**65**	31		PV
Payback						
Payback - simple, theoretical	years	10.54	**6.96**	-3.59	-34	PV
Payback - discounted, theoretical	years	15.98	**9.57**	-6.41	-40	PV
Achieves NPV payback over useful life	Y/N	**NO**	**YES**			PV
Life Cycle						
Implementation cost (initial capital outlay)	ZAR	**21,515**	23,412	1,897	9	SWH
Useful life	years	15	**20**	5.00	33	PV
Cost per heated liter - system only	ZAR	0.09	**0.04**	0.06	-60	PV
System NPV (ROI amount)	ZAR	2,056	**43,466**	45,522	2214	PV
ROI coefficient	%	-7.98	**164.85**	172.83	2166	PV
IRR	% / annum	8.33	**20.86**	12.53	150	PV

Performance omparison

		Margin
Higher proportion of grid input saved	**PV**	92%
Lower initial capital outlay	**SWH**	-9%
Shorter simple payback period	**PV**	-34%
Shorter discounted payback period	**PV**	-40%
Longer useful life	**PV**	33%
Lower cost per heated litre	**PV**	-60%
Higher NPV	**PV**	2214%
Higher ROI	**PV**	2166%
Higher IRR	**PV**	150%

The cost per heated liter of water is calculated on a NPV basis as system total costs (including purchase, installation, plus the NPV of the annual system maintenance costs adjusted for inflation) divided by total system output (liters of hot water supplied during the system's useful life). The SWH and solar PV water heater have values of R0.09/l and R0.04/l respectively, making the solar PV water heater superior with R0.06/l (see Table 15-5).

Return on Investment (ROI)—The ROI can be used to calculate the proceeds received from an income source as a proportion of all costs attributable to the same source of income. The ROIs for the SWH and solar PV

Table 15-5. Interactive decision tree.

1 Item / Option	Unit	Grid (EWH)	SWH	PV
a. option useful life-years	Years	n/a	15	20
b. adopted useful life-years	Years	15		
c. grid unique startup costs	ZAR	3,000		
d. grid unique recurring costs (annual)	ZAR	200		
2 Life-Cycle		**Grid**	**SWH**	**PV**
a. startup costs	ZAR	3,000	21,515	23,412
b. life cycle costs NPV(total costs of system)	ZAR	72,231	4,259	2,955
c. kWh grid supplement(cost paid to utility)	ZAR	0	49,002	41,038
d. option total cost NPV	ZAR	75,231	74,776	67,405
e. option per annum cost NPV (15,15,20ys)	ZAR	**5,015**	**4,985**	**3,370**
f. liters heated	kl	817	817	1,089
g. cost per liter (solar+grid)	ZAR/l	**0.092**	**0.092**	**0.062**

water heaters have values of -8.0% and 164.9% respectively. The ROI of 164.9% means that for the useful lifecycle of the solar PV water heating system, the income generated exceeded all system costs by a factor of roughly 1.7. The solar PV water heating system is the better performing system as it returns an amount which is greater in proportion to all costs divided by the funds invested.

Internal Rate of Return (IRR)—The IRR is the annual return received on the investment during the product lifecycle (also known as the discount rate) expressed as a percentage. The IRR for the SWH is 8.3%. The IRR for the solar PV water heater is 20.9%. The SWH's yield is less than an interest-bearing investment which is risk neutral. The IRR of the solar PV water heater is greater than the SWH by 12.5% and is higher than the macro interest rate of 9.5%.

CONCLUSIONS

Data from Table 15-4 indicate which water heating system meets each comparative criterion and by what percentage margin it outperforms the other system. The solar PV water heating system demonstrates better performance in most criteria:

- The solar PV water heater supplies more total output. Based on the higher proportion of grid input saved, the solar PV water heater outperforms the SWH by a margin of 90% (i.e., the proportion of grid kWh saved exceeds that of the SWH by a factor of 0.9).
- The solar PV water heater returns its investment quicker in both SPP and NPV terms. The SWH fails to achieve a discounted payback over its useful life.
- The solar PV water heater offers a higher IRR and ROI. It demonstrates a lower per heated liter of water cost over a longer lifecycle.
- The solar PV water heater's initial capital outlay is 13% higher than that of SWH. This is the single criterion for which PV can be deemed as underperforming. If the cost of the storage tank (electric water heater) for a solar PV water heater is not included in the purchase price (the household uses an existing electric water heater) then solar PV water heating is superior in all criteria.

Based on the above comprehensive energy and financial analysis including parameter constraints and calculations, when compared to SWH, the solar PV water heating system constitutes the best capital investment for the specific South African conditions that were considered.

References

[1] GIZ, South Africa (2015). Long term performance monitoring of a randomly selected group of residential SWH systems under the one million SWH program in South Africa. Version 1, revision 1. Study supported by DoE GIZ. https://www.giz.de/en/worldwide/312.html.

[2] Department of Energy, Republic of South Africa (2015). http://www.energy.gov.za/files/swh_overview2.html.

[3] Eskom (2007). Solar water heating rebate program (2007). www.eskom.co.za.

[4] Kwikot (2015). About Kwikot. http://www.kwikot.co.za/About.php.

[5] Polysun (2016). Polysun® simulation software. http://www.velasolaris.com/english/home.html.

[6] Gerserwise (2015). Intelligent energy control. http://www.geyserwise.com.

[7] Galena Hill Systems.

Chapter 16

SWOT Analysis of Iran's Energy System

Alireza Heidari, Alireza Aslan
Ahmad Hajinezhad and Seyed Hassan Tayyar

Due to the importance of energy to societies, analysis of local, regional and national energy systems at different levels is salient for researchers and policy makers. This chapter considers Iran's extraction, processing, conversion, transmission, distribution, and consumption of energy. Iran is one of the world's fastest growing energy consumers and its economy is highly dependent on energy exports. The patterns of Iran's present energy system impact its economy and development. Our research assesses the energy system in Iran by identifying and analyzing its strengths, weaknesses, opportunities and threats (SWOT). The chapter considers these from internal and external perspectives and concludes that energy fulfills an important role in policy and development. This assessment can help policy makers and researchers identify challenges and create suitable and robust strategies.

INTRODUCTION

What roles does energy have in policymaking and the development of today's modern societies? The economies of most oil exporters, including Iran's, are highly dependent on oil and gas revenues. The impact of energy on economic growth, public policy, national security, and government revenues seems undeniable. To assess the influence and role of energy in Iran, analysis of its energy system is necessary [1]. Energy system analysis considers the types of energy carriers from energy resources to end use energy systems. It includes the processes of extraction, conversion, transmission, distribution, and consumption of energy. Energy system analysis can be divided based on sectorial, regional, national and global attributes [2]. The energy system in Iran is complex, requiring careful and systematic analysis [3]. It faces many challenges, including the following:

- Management of the country's vast oil and gas resources.
- The government's budgetary dependence on oil and gas revenues.
- Misappropriated funding and investment in the oil and gas sector.
- The country's low productivity and high energy intensity (especially in non-industrial sectors).
- High household sector energy consumption.
- Foreign policy issues such as sanctions on oil and gas exports.
- Low energy prices combined with high internal energy subsidies.
- The sharing of oil wells with neighboring countries.
- Complexity of energy production.
- Large geographic area and physical energy security issues.
- Inaccurate statistics and estimates regarding proven resources.
- Political behavior about resource extraction.
- Unpredictable behavior of consumers to variable energy prices (price elasticity).

Given these challenges, an analysis of Iran's energy system requires a strategic and systematic approach. Such an approach must consider various aspects of the problem and the impact of any effects on the whole system and its environment. SWOT analysis is one methodology that meets these criteria. The origin of this analysis methodology are largely credited to Albert Humphrey, a U.S. management consultant who led research projects at Stanford University beginning in the 1960s to evaluate strategic plans. SWOT analysis is a strategic planning tool that provides a detailed analysis of the state of a system. It assesses the internal factors (strengths and weaknesses) and external factors (opportunities and threats) of a system from different levels (e.g., organizational, industry or regional). The extracted factors help decision and policy makers develop appropriate strategies for different situations or conditions [4]. Identifying the internal aspects of the system as strengths and weaknesses and the external ones as opportunities and threats is an appropriate methodology to analyze Iran's energy system.

This chapter reviews Iran's present energy system and analyzes the interactions between the energy system and public policy in Iran. We consider the strengths and weaknesses of the system, the external opportunities, and the threats facing Iran's energy sector by using a SWOT analysis. Our analysis is applied for each segment of the energy system including management and energy economics, energy supply, transmission, conversion and energy demand.

IRAN'S ENERGY SYSTEM

A county's energy flow provides information on the primary parameters of energy production, conversion and final consumption. An energy flow diagram is a beneficial way to begin a status assessment and analysis of Iran's energy system, a country with large oil and gas reservoirs. As shown in Figure 16-1, much of Iran's energy demand in 2012 is supplied by domestic oil and gas resources.

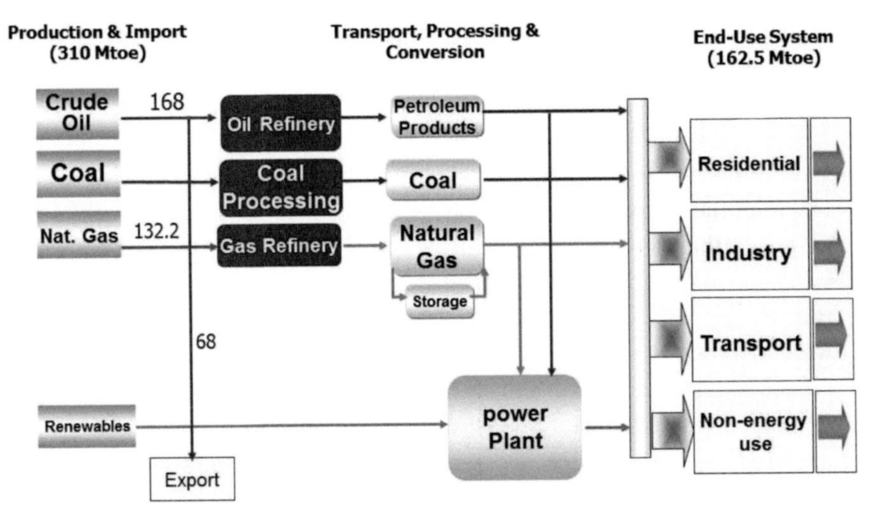

Figure 16-1. Energy flow diagram for Iran in 2012.

Iran is ranked among the world's top five countries with the largest deposits of proven oil and natural gas reserves. It also ranks among the world's top ten oil producers and top five natural gas producers [5]. Iran produced 3.2 million barrels per day (bbl/d) of petroleum and related liquids in 2013 and more than 5.6 trillion cubic feet (Tcf) of dry natural gas in 2012. Iran's oil is consumed primarily by its power plants. The excess is exported. The majority of the country's foreign exchange earnings and government budget are dependent on oil exports [6,7].

According to the International Energy Agency (IEA), from 310 million tons of oil equivalent oil (Mtoe) production in Iran in 2012, 168 Mtoe was consumed by domestic oil production (54% of total) and 132 Mtoe was consumed by domestic gas production (42.5% of total). Exports totaled 68 Mtoe, and 162.5 Mtoe were consumed by end users [8]. Figure 16-1 illustrates that while Iran had a small share of renewable sources in its primary energy supply, the export of crude oil is important to Iran's economy.

Due to sanctions against Iran and declining global oil prices, Iran's oil export revenues decreased over recent years. This directly impacted economic growth and social conditions in the country [9]. To understand the challenges of the energy system in Iran, our analysis takes two approaches. First, we consider the efficiency of Iran's energy system and energy policy challenges. Second, we assess the strengths and weaknesses, threats and opportunities of each sub-sector of Iran's energy system (including the management and economics of energy, energy supply, transmission and conversion and final demand).

Efficiency of Iran's Energy System

Efforts toward energy efficiency in oil exporting countries are often minimal. After the second oil crisis in 1978, oil importing countries planned projects to increase the efficiency of their energy systems. However, oil exporting countries often failed to develop robust programs to increase their energy efficiency. More than 600 million barrels of crude oil per year were lost by Iran's energy sector during 2012. The losses were mainly from power plants that converted fossil fuels to electricity [10]. While many developed countries use systems such as combined heat and power (CHP), combined cycle processes, and renewable energy to minimize their losses, these technologies are not yet deployed in Iran [11,12].

More than a million barrels of oil equivalent daily is lost in transmission and distribution processes in Iran, which is equal to twice the energy consumption of Greece or the annual total primary energy supply in Sweden [13]. The losses in Iran's other economic sectors are also undesirable.

Since a country's energy consumption is highly dependent on its industrial and commercial sectors, developed countries often consume more energy. An indicator of a country's energy system is its per capita energy index. However, this index may not reflect the efficiency of a country's energy system. While Iran may have comparatively high per capita energy consumption, most energy is used by non-industrial sectors or processes with high rates of losses. Such energy consumption does not result in gross domestic product (GDP) growth but in wasted energy due cultural issues, dissipation and use of inefficient energy technologies.

For our purposes, an energy intensity index is a better indicator than the energy use per capita index [14,15]. This index reflects total energy consumption in tons of oil equivalent per gross domestic product (GDP) in thousands of U.S. dollars (USD). While the energy intensity index in Japan and the U.K. equaled 0.10 in 2012, the global average was 0.16. Using purchasing power parity (PPP) is useful for comparing domestic markets since PPP accounts for a county's relative cost of local goods, services and inflation. Turkey had a GDP of $1,015 billion (U.S.) PPP with a 2012 energy intensity of 0.12. Nearby Iran's was $1,053 billion (U.S.) with a PPP energy intensity of 0.21. This shows that energy efficiency in Iran is very low compared to Turkey. Iran requires about two million barrels of crude oil per day greater energy consumption to produce a similar GDP purchasing power parity then does Turkey [16-18].

For Iran, a 20% energy reduction in consumption from a multi-year national program (using oil prices at $50 per barrel) offers the opportunity for more than $16 billion savings.

Challenges of Policy-making

Policies are important for managing energy consumption. While an analysis of energy systems in terms of efficiency shows that the energy system is inefficient, this does not mean that policies are absent. There are policies in place that make Iran's challenges difficult to resolve.

Many energy exporting countries are challenged by high government subsidies for energy. These subsidies maintain low prices for energy to facilitate development, thus increasing production and employment. However, subsidies can lower energy system efficiency. Consumers often benefit from energy subsidies when energy is subsidized by the government [19]. For example, the price of gasoline in Iran was $0.10 per liter (USD) in 2015. That same year the price of gasoline in the U.S. was $0.76 per liter, $2.52 in Turkey, $1.58 in Brazil, and $1.15 in India [20]. In many of these countries rather than gasoline being subsidized by the government it was being taxed.

Fuel subsidies result in increased demand. By 2010, despite large oil and natural gas reserves, Iran became one of the world's largest importers of gasoline [21]. In response, the Iranian parliament in 2011 enacted regulations to make oil and gasoline prices equal to the marginal cost or the free on board

(FOB) price in the Persian Gulf. The revenues from the liberalization were allocated to industrial and welfare infrastructure development [22].

Iran's targeted subsidy plan failed to achieve its intended goals. Sanctions against Iran led to the devaluation of its national currency while the U.S. dollar had a three-fold increase in comparative value. The difference between energy prices in Iran and its neighbors caused pressure on the government. As energy prices increased, purchasing power declined and inflation ensued. As a result, the liberalization of prices for energy carriers was challenged.

A large part of the Iranian government's budget is derived from oil revenues which represent a large share of its GDP. As economic development in one sector increases other sectors may decline (the Dutch disease) and such is the case with Iran's economy. It is a primary weakness of Iran's management system and its energy-based economy. One of the country's strategic priorities is to resolve this situation [6,23]. Strategies include attracting foreign and domestic investment, increasing non-oil products and exports, taxing production and forming a foreign exchange deposit bank [9,24].

Other challenges Iran faces include fuel smuggling, conflicts with neighboring countries, physical insecurities and market uncertainties.

The Importance of Strategic Approaches to Energy

Energy clearly effects economic development, national security, sustainable development, health and social welfare. Therefore, analysis of the energy system based on the strategic approaches is essential.

There are a variety of approaches to strategies. They include ways to meet required levels of achievement, ways to create opportunities for companies or industry, ways of creating strengths and avoiding pitfalls from weaknesses. For a strategy to be successful and meet targets, it is essential to identify opportunities and avoid threats. System strengths are used to overcome system weaknesses.

SWOT is a strategic planning tool used to analyze and evaluate the internal and external situation of systems. The SWOT analysis methodology is an effective tool for identifying environmental conditions and the ability of the system to adjust [4]. Figure 16-2 provides a SWOT analysis matrix.

A SWOT analysis can be implemented for different fields and levels. Due to the importance of a country's energy system and the variety of opportunities, threats, strengths, and weaknesses of such systems, SWOT analysis can be used to identify better strategies and tactics. After determining and evaluating a system's internal strengths and weaknesses and external opportunities and threats, suitable strategies can be identified and used to define action plans, schedules and budgets.

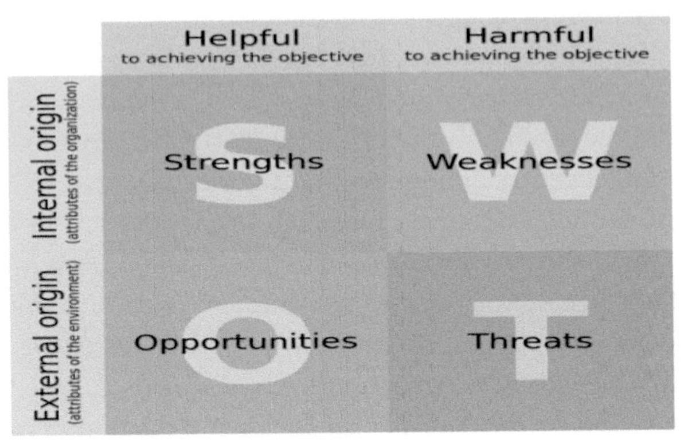

Figure 16-2. SWOT analysis matrix.

Research Using SWOT Analysis for Energy Assessments

Past research has been conducted using SWOT to assess energy systems for renewable energy (RE) and policy development. Wei-Ming Chen et al. used SWOT analysis to compare energy policies that promote RE in Japan, South Korea and Taiwan [25]. All three countries are considering RE as an alternative energy source to improve energy security. It examined these countries' strengths, weaknesses, opportunities, and threats in the context of advancing renewable energy policies and technologies, expanding domestic RE installations, and improving their strategic position in international markets as exporters of clean energy technologies [25]. Terrados et al. developed a regional energy plan using a SWOT analysis. This study explained the energy component of the strategic plan for the province of Jaén in southern Spain, which was squarely focused on the development of renewable energy resources, mainly solar and biomass [26].

Markovska et al. analyzed the strengths, weaknesses, opportunities and threats of the energy sector in Macedonia to identify its status and to plan future actions towards sustainable energy development [27]. Camille Fertel et al. used a SWOT analysis of Canadian energy and climate policies on the themes of energy security, energy efficiency, technology and innovation. Their analysis showed that there is a lack of consistency in the Canadian energy and climate strategies beyond the application of market principles. They suggested increasing cooperation among the various agencies by using a combination of intergovernmental policy tools [28].

SWOT analyses have been performed to assess Iran's RE systems. Mahmood Haji-Rahimi and Hamed Ghaderzadeh introduced The Challenges of Sustainable Management in Renewable Natural Resources in Iran by documenting positive and negative aspects regarding the management of rangelands as a renewable resource using a SWOT framework [29]. Faezeh Moradzadeh used a SWOT analysis to study the RE situation in Iran. He discovered challenges including research and development (R&D), investments, green subsidies, strategic planning and innovation in RE. He also found substantial capacity for renewables and a skilled labor force [30, 31].

There are articles that discuss the different parts of Iran's energy system using SWOT analysis. Ali Faridzad published an article considering the strengths, weaknesses, opportunities and threats for oil and gas upstream sectors and political challenges for the production of oil and gas in Iran [32]. Numerous on-topic conference papers, collegiate research articles and academic reports have been published in Persian [24,33,34].

METHODOLOGIES AND RESULTS

To analyze the current situation of the energy system in Iran using SWOT, the internal strengths and weaknesses of the energy system plus the external opportunities and threats are identified. The study's methodology uses applied research. The research is descriptive since we wanted to explain the energy system in Iran based on strategic factors.

To perform our research, we reviewed the academic and practical papers, reports, and statistics related to Iran's energy system. After that, the opportunities, threats, strengths and weaknesses were identified by using available documentation and consultation with professionals and experts.

Energy systems are often divided into three main technical sectors: 1) the energy supply system; 2) conversion and transport; and 3) final energy consumption based on the reference energy system (RES). In addition to these sectors, study of the energy system's macro-management and macro-economy are essential for a comprehensive analysis. To identify and analyze Iran's energy system and perform our SWOT analysis, we divided it into four parts: 1) management and economy of energy; 2) primary supply; 3) processing, conversion and transport; and 4) end use system (final consumption). Next, the internal strengths and weaknesses, and external threats and opportunities, are identified based on a survey for each of four the parts (see Figure 16-3).

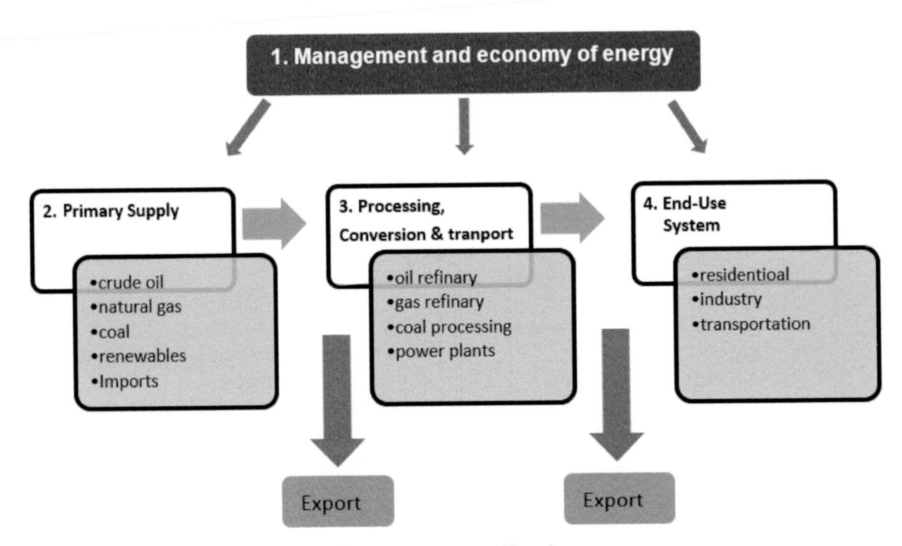

Figure 16-3. Four segments of Iran's energy system.

Factors of Energy Management and Economics

Management and economy of energy refer to the approaches of policy makers to energy problems, regulations, procedures, and the impacts of energy on a country's economy. This segment is important because it affects other sectors of the energy system including, production, exports, revenues and global rankings. A SWOT analysis of the management and economy of energy is found in Table 16-1.

Factors of the Primary Energy Supply

The primary energy supply system is the first part of the reference energy flow diagram. It shows energy resources, fossil energy reservoirs, and the country's renewable energy potential. This system includes the exploration and exploitation of oil and natural gas, sources of RE, imports of energy and use of energy resources. This is known as the upstream energy sector.

Due to the geopolitical position of Iran, as well as rich reserves of oil and gas and RE potential, an analysis of the effective factors on primary energy supply system is necessary. Our SWOT analysis of the primary energy supply in Iran is provided in Table 16-2.

To supply the northern power plants with gas and oil (e.g., Neka power plant) and residences in the northern provinces during cold seasons, Iran imports energy using a swap contract agreement with Turkmenistan.

Table 16-1. SWOT matrix management and economy of energy.

	Strengths	Weaknesses
Internal factors	- Multiple sources of oil and gas - Comprehensive reservoir study in 2008, by the Presidential Technology Cooperation Office - Start of feasibility studies to evaluate the use of intelligent wells and reservoirs in the country - Start of preparing the reservoir database - Access to the high seas - Possession of the Strait of Hormuz - Low value of the domestic currency to foreign currency to boost exports - Develop a comprehensive system of research and technology in the oil ministry - Develop and implement policies, including short-term energy efficiency labeling and energy labeling - Large span geographic extent and borders for transit - Significant reserves of oil and gas fields in the Caspian and Zagros - Proximity of oil resources to the export terminals - Independency and political stability in comparison with neighbors and competitors	- Dependency on oil budget - High share of oil in GDP - Lack of liquefied natural gas (LNG) technology development for export - Lack of cooperation with regional competitors - Lack of international companies in the oil and gas industry - Marketing problems, especially in the field of natural gas - Political and non-technical production of resources - Improper pricing of petroleum products - Subsidies for petroleum products - Lack of attention to foreign investment - Unknown status of foreign investment - Investments in different areas, under the pressures of the members of parliament and officials - Inaccuracies in statistics and estimates about the established and proved resources - Idealism of the objectives and policies of the country's oil and gas and to ignore the facts - High dependence of economy on oil revenues - Lack of a comprehensive model of the energy vision - Lack of adequate training and upbringing of oil executives, and few managers - Rentier government with the Dutch disease - Low GDP (excessive imports) - No good extraction of shared wells with neighbors - Lack of interest in R&D - Lack of technology policy and planning in the country, particularly in the oil and gas sector. - Lack of long-term planning strategies for energy efficiency as well as comprehensive information and statistics - Legal barriers and lack of transparency of business rules in order to attract foreign investors, particularly in the oil and gas industry

Table 16-1 (*continued*). SWOT matrix management and economy of energy.

	Opportunities	**Threats**
External factors	- Oil and gas sector has the largest economic advantage - The impact of Iran on the region's political and economic decisions - Europe's energy supply problems from Russia - High price of petroleum products at the border compared to domestic prices - The possibility of transferring oil to Asia - The potential of Caspian gas pipeline to Europe - Long-term contracts to sell gas to neighbors at guaranteed prices - Geopolitical situation in the Persian Gulf region - Increasing demand for oil and gas in the world - Worldwide oil peak production - Possibility of changing the oil and gas industry of Iran to a technology hub (hardware and software) - Uncertainty about the rate of RE substitution in place of oil and gas in developed countries	- A drop in world oil prices by harvesting unconventional technologies - Smuggling fuel out of the country - Lack of adequate security for foreign investors - Geographic tension from neighbors and lack of physical security of energy - Direct relationship between foreign policy and external sanctions on oil and gas supply - Lack of proper position in the global natural gas market - The lack of a strong position in oil producing and exporting countries (OPEC) - Powerful competitors and producers in the area (neighborhood) - Good relations between competitor countries and buyers - Sanctions about buying energy from Iran - Buyers hostile relations with Iran - The lack of regional cooperation in oil and gas projects - Lack of participation of international organizations in the financing of joint projects such as the natural gas pipeline project - Iran's poor relationship with the world's science and technology - Lack of proper communication with credible oil companies in the world - Absence in the global petrochemical market - The lack of experienced international contractors

Due to increasing energy consumption, and the inefficient technologies used for oil extraction, treatment, transfer and distribution, Iran faces future domestic energy supply challenges and decreasing oil exports. The country also needs to improve oil extraction technologies and renewable energy's share of total primary energy supply. About 25% of Iran consists of deserts which receive daily solar irradiation about 5kWh/m^2. If 1% of these areas were covered by solar collectors, the energy obtained would be five times more than annual gross electricity production in Iran [35].

Table 16-2. SWOT matrix of the primary energy supply system.

	Strengths	Weakness
Internal factors	- Number of engineers trained at different levels of BS, MS and Ph.D. - Significant reserves of oil and gas fields in the Caspian and Zagros - Multiple sources of oil and gas - Oil sector skilled manpower - Upstream rules to support the development of infrastructure - Potential of renewable energy - A vast desert with a high solar radiation, such as Yazd province - The potential of wind across the country - High biomass potential in provinces such as Khuzestan - Technical knowledge of wind turbines in the country - Renewable energy research centers and organizations in the country - Low cost of oil production due to land resources in southern areas of the country - Shallow waters of the Persian Gulf and low cost of production of offshore resources - A sufficient basic infrastructure in the oil sector - Over one hundred years of experience in the oil industries - Oil and gas development potential across the country. - Huge oil and gas and petrochemical projects - Enough gas for injection into oil fields for enhanced oil recovery (EOR) - Technical possibilities to recover the gas injected into oil reservoirs after the removal of oil - Safety of domestic gas production systems - Secure nationwide network of gas transmission and distribution	- Low extraction from the shared fields with neighbors - Low-efficiency technology and old equipment - Old oil fields and need to revive - Oil resource pressure falling - The lack of ownership for monitoring oil and gas reservoirs - Weakness in exploration - Weakness in extraction - Weakness in drilling - Weakness in operation - No possibility of rapid extraction of oil and gas shared fields - Lack of upstream enforcement to support the development of infrastructure - Lack of attention to the protection of reserves - Lack of adequate allocation of gas for injection into oil reservoirs - Lack of the necessary investments for optimized development of oil and gas fields - Failing to distinguish between types and how to invest in oil and gas sector with other sectors of the economy - Failure to attract foreign investment in the oil and gas industry - Weakness in completion of oil and gas and petrochemical projects and no more income - Centralized energy supply sources and lack of distributed generation - Lack of attention to issues of passive defense - Lack of development in the energy storage systems - Lack of the necessary investments for commercializing renewable energies.

Table 16-2 (*continued*). SWOT matrix of the primary energy supply system.

	Opportunities	Threats
External factors	- The most attractive sectors for foreign investment - Ability to swap oil and gas with northern countries like Turkmenistan - Neighboring producer countries in the region to import energy (secure supply) - Possibility of changing to the oil and gas industry technology hub in the area (hardware and software) - Geopolitical situation in the Persian Gulf region - Ability to export technical and engineering services in the upstream field of oil and gas - Conventionalizing the unconventional technologies such as shale gas and sand oil - Ability to import modern technology, mining and oil and gas exploration - Ability to import the renewable energy technologies - Shared interests in the Persian Gulf to cooperate with neighbors - Attracting international oil companies for the supply of energy resources	- Poor quality of imported gas from Turkmenistan - Hostile relations with neighbors - Sanctions and international pressure by the western powers and the lack of modern technologies - Absence of large foreign investors in the oil and gas sector because of international political pressure - Lack of access to technologies in the oil and gas sector due to international sanctions by the western nations - High speed discovery and extraction of neighboring countries, especially in the shared field of oil and gas reservoirs (south Pars) - Weak foreign policy, especially in the region - Physical threats (e.g., foreign military attacks on vessels and oil wells) particularly in the Persian Gulf

Analysis of the Factors of Energy Transfer and Conversion

Transfer and conversion of energy is the central segment of the reference energy system and the interface between the primary energy supply system and the final consumer. The system consists of power plants, refineries and power transmission networks to deliver the energy to final consumption. This is also known as the middle-stream field of energy.

Ninety-nine percent of the energy sector in Iran is public, and privatization efforts are remarkably slow. The infrastructure required for private investment does not guarantee a transparent and reliable situation. For example, for the construction of a private power plant the necessary infrastructure includes natural gas pipelines and connection to the electrical grid. The gas distribution network is exclusive and controlled by the government, the sole purchaser of electricity. This reduces the bargaining power of plant

Table 16-3. SWOT matrix of the energy transfer and conversion system.

Strengths	Weakness
- Developed transportation - Extensive network of gas pipelines - Widespread power grid - Placement of compressed natural gas (CNG) stations across the country to exploit the country's transportation system - High potential of energy and electricity production in the country - Diversification of production of oil products - Safe generation, transmission and consumption systems of gas - Replacement of oil and gas to fuel power plants in different seasons - The combined cycle gas power plants to increase the efficiency of thermal power plants - Power-trained specialists in universities - Infrastructure engineering and supply of raw materials in the interior of the country - High energy saving potential in consumption and waste by power plants and refineries - Numerous domestic cable and metallurgical industries	- Lack of development of renewable energy - Lack of LNG technology development - Low efficiency of power plants and equipment - Lack of safe development of rail transport - Remote areas of energy networks - Lack of development of renewable power to remote areas - Low capacity of oil products - Insufficient number of refineries and investment - Lack of proper transfer of gas and oil in cold seasons in the cold regions cause electricity and gas outages - Old gas pipelines - Inadequate inspection of gas and electricity transmission and distribution lines and products - Lack of construction of new and modern refineries - Usage of old technology, facilities and refineries and exhaustion - Lack of downstream petrochemical units that can bring petrochemical products into the global market. - Problems to convert natural gas into LNG and gas-to-liquid (GTL) - Increase of low efficiency gas turbine power plants - Lack of investment in R&D - Uneconomical petroleum products for domestic needs

Table 16-3 (*continued*).
SWOT matrix of the energy transfer and conversion system.

Opportunities	Threats
- Variety of available fuel for power plants - Growing Iraq and Afghanistan electricity markets to buy from Iran - Good market to offer petroleum products - Increasing demand for energy in developing countries - Ability to export technical and engineering services including construction, especially in the downstream oil and gas pipelines and oil refineries - Ability to export power plant technical knowledge to its neighbors - Opportunity to overhaul outdated power plants and refineries in the presence of renewable alternatives - The possibility of clean development mechanism (CDM) projects to attract financial support to improve efficiency and reduce pollution from power plants and refineries	- Sanctions and the impossibility of importing new technologies - Centralized power plants and refineries plus the risk of military attacks - Lack of development of passive defense - Reducing the amount of fuel going into power in cold seasons and the risk of electricity outages - Not responding to increasing demand from fossil fuel power plants if there are no renewable power plants constructed - No entry of foreign investors - No entry of private investors - The lack of an overhaul because of the full load demand on power plants and refineries

owners. The same is also true for refineries since prices, market structure, and regulations are government controlled. Finally, due to the lack of government funding to support law enforcement, unsuccessful experiences and projects, and high-risk market conditions, there is a lack of investor interest.

Factors in Final Consumption

The final consumption (end use) system is the last segment of the reference energy system. It contains all energy consumers. Domestic energy demand is determined by this sector. The SWOT analysis of this sector is provided in Table 16-4.

One of the problems with the energy demand system is the energy losses in the distribution network from theft. Stolen electricity is considered as illegal distribution network losses since it is not recorded by meters and fails to provide income.

Some locations in Iran have problems with the physical security of energy transfer. Indeed, energy inefficient technology in consumer equipment,

Table 16-4. SWOT matrix of the final consumption (end use) system.

Strengths	Weakness
- Developed transportation - Extensive network of gas pipelines - Widespread power grid - Placement of CNG stations across the country to exploit the country's transportation system - High potential of energy and electricity production in the country - Diversification of production of oil products - Safe gas generation, transmission and consumption systems - Replacement of oil and gas to fuel power plants in different seasons - Combined cycle gas power plants to increase the efficiency of thermal power plants - Power-trained specialists in universities - Infrastructure engineering and supply of raw materials in the interior - High energy saving potential in consumption and waste by power plants and refineries - Numerous domestic cable and metallurgical industries	- Lack of development of renewable energy - Lack of LNG technology development - Low efficiency of power plants and equipment - Lack of safe development of rail transport - Remote areas of energy networks - Lack of development of renewable power to remote areas - Low capacity of oil products - Insufficient number of refineries and investment - Lack of proper transfer of gas and oil in cold seasons in the cold regions causes electricity and gas outages - Old gas pipelines - Inadequate inspection of gas and electricity transmission and distribution lines and products - Lack of construction of new and modern refineries - Using old technology, facilities and refineries and exhaustion - Lack of downstream petrochemical units that can be closer to petrochemical products in the global market. - Problems converting natural gas into LNG and GTL - Increase of low efficiency gas turbine power plants because of the quick launch - Lack of investment in R&D - Uneconomical petroleum products for domestic needs

Table 16-4 (*continued*).
SWOT matrix of the final consumption (end use) system.

Opportunities	Threats
- Variety of available fuel for power plants of the country - Growing Iraq and Afghanistan electricity market to buy from Iran - Good market to offer petroleum products - Increasing demand for energy in developing countries - Ability to export technical and engineering services including construction, especially in the downstream oil and gas pipelines and oil refineries - Ability to export power plant technical knowledge to neighbors - Opportunity to overhaul outdated power plants and refineries in the presence of renewable alternatives - The possibility of CDM projects to attract financial support to improve efficiency and reduce pollution from power plants and refineries	- Sanctions and the impossibility of importing new technologies - Centralized power plants and refineries coupled with the risk of military attacks. - Lack of development of passive defense - Reducing the amount of fuel going into power in cold seasons and the risk of electricity outages - Not responding to the increasing demand for fossil fuel power plants if there will be no renewable power plants - No entry of foreign investors - No entry of private investors - The lack of a full overhaul because of the full load demand on power plants and refineries

lack of attention to the efficiency of energy, and improper use of equipment are the greatest problems facing the energy consumption system.

CONCLUSIONS AND POLICY IMPLICATIONS

Energy has an important role in national policy and development. To assess the influence and role of energy in Iran, the study and assessment of the country's energy system is required. An energy system analysis considers the types of energy carriers from energy resources to end use energy systems. This study uses a SWOT analysis to evaluate the internal strengths and weaknesses of Iran's energy system and the external opportunities and threats it faces. The system is divided into four sectors and the SWOT analysis is applied for each sector of the energy system, including management and energy economics, energy supply, transmission and conversion, and the energy consumption system.

While the analysis shows that Iran's energy system is inefficient, system strengths can overcome weaknesses by exploiting available opportunities. The strengths of the system include available fossil energy sources, high potential for renewable energy and a skilled workforce. Weaknesses of this system consist of misapplied policies, consumer behaviors and energy inefficient equipment. Systemic problems include disproportionate energy intensity, increasing levels of energy consumption in non-industrial sectors (e.g., transportation), high rates of growth in electricity consumption and excessive pressures on the natural environment.

Opportunities in Iran's energy system include following the principles of sustainable development to develop and implement a comprehensive energy policy. Others include a reasonable extension of fossil-fuel energy sources, importing and using high efficiency technologies and equipment, reducing excessive energy consumption, and developing RE resources.

The threats facing Iran's energy system include international sanctions and energy policy challenges with its neighbors and customers. Iran needs to increase its GDP and reduce dependence on oil revenues to overcome its internal energy challenges. The country needs to reform its approach to foreign diplomacy to enable an increase in oil and gas exports while importing technical knowledge and more efficient energy technologies.

SWOT results are a primary input for strategic plans. The results of this assessment can be used to develop strategies that will enable Iran to improve its energy system efficiency and attract foreign investment. To this end, Iran needs to liberalize its energy prices and subsidies.

References

[1] Rostamihozori, M. (2002, February 14). Development of energy and emission control strategies for Iran. Dissertation der Universität Fridericiana zu Karlsruhe.

[2] Amirnekooei, K., Ardehali, M. and A. Sadri (2012, October). Integrated resource planning for Iran: Development of reference energy system, forecast, and long-term energy-environment plan. *Energy,* 46(1), pages 374-385.

[3] Kazemi, A., Reza, M., Mehregan and H. Shakouri (2011, October 23). Energy supply model for Iran with an emphasis on greenhouse gas reduction. The 41st International Conference on Computers and Industrial Engineering. University of Southern California. Pages 491-492.

[4] David, F. (2011). *Strategic management: concepts and cases.* 13th edition: Prentice Hall. Page 178.

[5] International Energy Agency (2014). Total petroleum and other liquids production. http://www.eia.gov/countries/cab.cfm?fips=ir.

[6] International Monetary Fund (2010, January 11). Islamic Republic of Iran staff report for the 2009 Article IV Consultation.

[7] Iranian National Bank (2010/11). Balance of payments. *Economic Trends,* 16, third quarter, page 16.
[8] International Energy Agency (2015). Energy statistics for country, Iran. http://www.iea.org/countries/non-membercountries/iranislamicrepublicof.
[9] Eslamifar, G., Shirazi, A. and A. Mashayekhi (2012, July 22-26). A system dynamics model to achieve sustainable production of oil in Iran. 30th International Conference of the System Dynamics Society. St. Gallen, Switzerland.
[10] International Institute of Energy Studies (2009). Iran hydrocarbon balance. Page 1,387.
[11] International Energy Agency (2009). *Co-generation and district energy: Sustainable energy technologies for today and tomorrow.* OECD/IEA Publications: Paris, France.
[12] Veerapen, J. (2011*). Co-generation and renewables solutions for a low-carbon energy future.* OECD/IEA Publications. Paris, France.
[13] International Energy Agency (2010). Statistics, Sweden, indicators for 2010. http://www.iea.org/statistics/statisticssearch/report/?country=SWEDEN&product=indicators&year=2010.
[14] International Atomic Energy Agency (2005). Energy indicators for sustainable development: guidelines and methodologies. Vienna, Austria.
[15] Yanagisawa, A. (2011, January). Trade-off in energy efficiencies and efficient frontier – Relationship between GDP intensity and energy consumption per capita and what it means. *IEEJ Energy Journal.* Japan. https://eneken.ieej.or.jp/data/3618.pdf.
[16] International Energy Agency (2012). Indicators for Japan. http://www.iea.org/statistics/statisticssearch/report/?country=JAPAN&product=indicators&year=2012.
[17] International Energy Agency (2012). Indicators for Turkey. http://www.iea.org/statistics/statisticssearch/report/?country=TURKEY&product=indicators&year=2012.
[18] International Energy Agency (2012). Indicators for Iran. http://www.iea.org/statistics/statisticssearch/report/?country=IRAN&product=indicators&year=2012.
[19] United States Congress (2006). Energy and the Iranian economy. http://en.wikipedia.org/wiki/Energy_in_Iran, page 50.
[20] International Energy Statistics (2013). *GTZ – International fuel prices.* Eighth edition, GIZ publication.
[21] Energy Information Administration. International energy statistics. http://www.eia.gov/cfapps/ipdbproject/iedindex3.cfm?tid=5&pid=62&aid=3&cid=regions&syid=2008&eyid=2012&unit=TBPD.
[22] Iranian Parliament (2010, January 5). The Iranian targeted subsidy plan.
[23] International Atomic Energy Agency (2014). Country nuclear power profiles 2014. Islamic Republic of Iran, updated 2010. http://www- pub.iaea.org/MTCD/Publications/PDF/CNPP2014CD/countryprofiles/IranIslamicRepublicof/IranIslamicRepublicof.htm.
[24] Bijan, Z. (2012). Strategies for overcoming the oil-dependent economy. Lecture, University of Tehran.
[25] Chen, W., Kim, H. and H. Yamaguchi (2014, November). Renewable energy in eastern Asia: Renewable energy policy review and comparative SWOT analysis for promoting renewable energy in Japan, South Korea and Taiwan. *Energy Policy,* 74, pages 319-329.
[26] Terrados, J., Almonacid, G. and L. Hontoria (2007, August). Regional energy planning through SWOT analysis and strategic planning tools: impact on renewables development. *Renewable and Sustainable Energy Reviews,* 11(6), pages 1,275-1,287.
[27] Markovska, N., Taseska, V. and J. Pop-Jordanov (2009, June). SWOT analyses of the national energy sector for sustainable energy development. *Energy,* 34(6), pages 752-756.
[28] Fertel, C., Bahn, O., Vaillancourt, K. and J. Waaub. (2013. December). Canadian energy and climate policies: a SWOT analysis in search of federal/provincial coherence. *Energy Policy,* 63, pages 1,139-1,150.

[29] Haji-Rahimi, M. and H. Ghaderzadeh (2008, March). The challenges of sustainable management in renewable natural resources in Iran: a SWOT analysis. *American-Eurasian Journal of Agricultural and Environmental Sciences,* 3(2).

[30] Wuppertal Institute for Climate, Environment and Energy (2010, October). German-Iranian co-operation VI: feed-in laws and other support schemes in international perspective.

[31] Faezeh, M. (2009). Strengths, weaknesses, opportunities and threats for renewable energy in Iran. Thesis, Azad University.

[32] Ali, K. (2013, March). Strategic priorities of Iranian oil and gas policies using SWOT analysis. Ninth international energy conference. Tehran, Iran.

[33] Ansari, A. and M. Hadjarian. Strengths and weaknesses of Iran oil and gas industry. Strategic management class lecture. Allame Tabatabaei University. www.hajarian.com/esterategic/tahghigh/ansari3.pdf, in Persian.

[34] Tehran University (2012, November). An analysis of the status of the economy's dependence on oil revenues and practical solutions to overcome it. Congress of Economics, Faculty of Economics.

[35] Dehghani, M. and M. Feylizadeh. (2014, September). An overview of solar energy potential in Iran. *International Journal of Current Life Sciences,* 4(9), pages 7,173-7,180.

Chapter 17

Sustainability and Environmental Management for Ports

Atulya Misra, Karthik Panchabikesan, Elayaperumal Ayyasamy and Velraj Ramalingam

Seaports are global hubs for the transportation of goods. They have an important role in today's global societies and are crucial nodes in transportation networks. Sustainable energy use impacts people, the world's environment, and is relevant to the operation and maintenance of ports.

In this chapter, an inventory of greenhouse gas (GHG) emissions in the Port of Chennai is made by accounting for the various port facilities, the housing areas, and the fishing harbor, all managed by the Port of Chennai. GHG emissions are quantified by following the guidelines of the Intergovernmental Panel on Climate Change (IPCC) and the World Port Climate Initiative (WPCI). Our estimate of GHG emissions for the financial year 2014-15 indicates that 280,558 metric tons of CO_2e/year were generated by the port and port related activities. The detailed estimation of energy consumption and emissions generated by the individual systems are useful for energy engineers when implementing energy conservation measures and renewable energy technologies. Implementation of GHG mitigation strategies for all port-related activities will help achieve substantial GHG reductions, reducing the adverse impacts of global climate change.

INTRODUCTION

The environmental threats posed by global warming in the 21st century diversely impact the health and economies of communities. The Intergovernmental Panel on Climate Change fourth assessment report states that among the GHG emissions from industrial activities, CO_2 emissions are one of the major causes of climate interference [1]. Seaports act as global hubs for the transportation of goods. Hence, they have an important role in today's globalized society and are nodes in global supply networks. Sustainable practices relevant to the operation and maintenance of ports impact the world's people and its environment. Due to growing international trade, maritime emissions are expected to more than double by 2050 if no mitigation actions are initiated.

The activities in and around seaports use vast amounts of energy that contribute to GHG emissions. The development of a structured baseline for greenhouse gases helps identify the areas where improvements could be made to mitigate GHG emissions. GHG inventories can be used to evaluate the effectiveness of the mitigation measures over a period of time. A carbon footprint represents the amount of GHG emissions an organization or event directly or indirectly releases over a measured period. Carbon footprints can be used for tracking emission trends and providing the information required to mitigate seaport GHG emissions. The ultimate challenge for the marine and port related industries is to develop environmentally sound, cost-effective, and practical solutions to achieve near zero carbon emission technologies.

The major sources of air pollution at ports and their impact on health were extensively studied by Bailey and Solomon [2]. They suggested a broad range of mitigation approaches such as switching to cleaner versions of diesel fuel, restricting truck idling hours, transitioning to alternative fuels and replacing older diesel equipment with newer equipment. They also proposed measures that incorporate emerging technologies such as shore side power for docked ships, zero emission technologies (e.g., fuel cells), and automated container handling. Mora et al. [3] performed an extensive environmental analysis of a sample port considering 21 major port activities (sea traffic, land traffic, fishing activity, dredging, waste disposal, etc.) which often have adverse environmental impacts. Sustainable environmental management indicators such as air quality, atmospheric contaminant emissions, gas emissions, noise pollution, and quality of spilled waste water were also indicated.

Schrooten et al. [4] studied the effects of recent international efforts to reduce the emissions from seagoing vessels. Their investigations revealed that CO_2e (carbon dioxide equivalent) emissions between 2004 and 2010 would likely increase by 2% to 9%, while NO_x emissions would increase as much as 8%, and SO_x emissions would decrease by 50%. Wang et al.

[5] compared global ship emission inventories using an automated mutual assistance vessel rescue data set and determined that the world's cargo fleet accounted for about 80% of total commercial fleet emissions. They also generated spatial proxies of global ship traffic with two global ship reporting data sets as proxies to geographically resolve global ship emissions. Based on bottom-up and activity-based methodologies, Meyer et al. [6] estimated the atmospheric emissions by international merchant shipping in Belgian's North Sea. Their estimate included the four primary Belgian seaports and compared their results with international emission estimates.

Han et al. [7] identified the status of pollution mitigation measures implemented in the shipping sector and developed an environmental evaluation scheme by investigating the actual conditions of environmental pollution from ship and port areas. GHG emissions were estimated for the Port of Barcelona by Villalba et al. [8] to be 331,390 metric tons in 2008, half of which were attributed to vessel movement (sea-based emissions) and the other half to port related activities (land-based emissions). They also reported that the biggest polluters were auto carriers with 6 kg of GHG emissions per ton of cargo handled. Fitzgerald et al. [9] used a cargo based methodology and estimated that the international maritime transport of New Zealand's imports and exports consumed 2.5 million metric tons of fuel during the year 2007, generating 7.7 million metric tons (Mt) of CO_2e emissions. Liao et al. [10] investigated the variations in carbon dioxide emissions by moving containers from established ports through the emerging Port of Taipei in northern Taiwan. They suggested the adoption of an analytical approach to understand the prospective CO_2e reduction in the route selection of inland container transportation. Gibbs et al. [11] analyzed secondary data and information on actions taken by ports in the United Kingdom (UK) to reduce their emissions. Their studies focused on operations at five major UK ports and they determined that emissions from shipping at berth were ten times greater than those from the ports own operations. Moreover, it was found that shipping emissions associated with seaborne trade at those ports were more important than those generated by port operations. Chang et al. [12] measured the greenhouse emissions, particularly from ocean going vessels in the Port of Incheon. They calculated the categories of GHG emissions and the movement of vessels from their arrival in port, to docking, cargo handling and departure. Their results revealed that among various types of vessels, international car ferries were the most substantial emitters, followed by fully loaded container vessels.

India is the third largest emitter of greenhouse gases in the world, after China and the U.S. Between 1994 and 2007, India's GHG emissions nearly doubled and continue to increase. Port authorities have a unique responsibility to adopt sustainable practices that preserve natural resources while ensuring economic growth. Prior to this assessment, ports in India have lacked GHG inventories of their activities. This chapter offers a detailed inventory that estimates the carbon footprint for the financial year 2014-15 of activities in the Port of Chennai, along with its housing colony and fishing harbor. Mitigation strategies to reduce GHG emissions are also discussed.

PORT OF CHENNAI

Chennai, the fourth largest Indian port in terms of throughput, is situated on the Coramandel coast and has a handling capacity of 86 million metric tons annually. The port has three docks namely Ambedkar Dock, Bharathi Dock and Jawahar Dock. Ambedkar Dock has facilities to handle steel, packed cargo, granite and other project cargos. Bharathi Dock handles containers, iron ore and petroleum oil. Jawahar dock handles food grains, coal, fertilizer and dolomite. The port has a 7 km entrance channel with 24 berths for cargo handling distributed along three docks. It has six transit sheds and four warehouses with a total covered space of 43,194 m^2. There are 52 tanks for oil storage with a capacity of 166,469 kL which are connected through pipes from the oil jetty. Tank farms are located at various locations, both within and outside of the customs boundary areas. Container terminal 1 has 300 reefer (refrigerated vessels) points and container terminal 2 has 120 reefer points for cold storage. The Chennai port has minimal land area availability when compared to India's other major ports, while the cargo handling capacity is much greater. The port has one of India's longest container handling quays with a length of 1,717 m plus an 885 m long quay at Bharathi Dock.

The port has five tug boats for maneuvering merchant vessels, four mooring launches and two dredgers to maintain the depth of the quay. The photographic views of container ship, dredger, mooring launch and tug boat are shown in Figures 17-1 (a), (b), (c) and (d) respectively. Cargo is moved to the marshalling yard for transportation to various locations using diesel locomotives. Railway locomotives (700 HP and 1,400 HP Locos), though mainly diesel driven, are a hybrid variety that allow the transmission of power to the locomotives using an alternator. The equipment used for material handling includes fork lifts, rail mounted gantry cranes, shore-to-ship cranes, transfer cranes and front-end loaders for loading bulk cargo. Electric cranes on the quays are used for handling dry bulk cargo and dredging along berths to salvage any cargo that has slipped or spilled along the berth while handling. Rail mounted gantry cranes are electrically driven unlike the transfer cranes which are diesel. Forklift trucks, pay loaders, excavators, locomotives, generators, and mobile cranes are all diesel driven. The container terminals use electric forklift trucks, which are battery powered except for the hydraulic operation of cylinders.

The port authorities control activities from vessel berthing to the transportation of cargo from the port. They also employ private companies including Chennai Container Terminal Private Limited (CCTL) and Chennai International Terminals Private Limited (CITPL) for container

(a) Container ship (b) Dredger

(c) Mooring launch (d) Tug boat

Figure 17-1. Images of marine vessels at Port of Chennai.

handling on a build, operate and transfer (BOT) basis. The port is powered by electricity received from the Tamil Nadu electricity board. Diesel fuel is used by trucks, generators, rail locomotives and cranes. The fishing harbor has a handling capacity of 2,000 metric tons of fish annually, of which nearly 200 metric tons are exported. Port of Chennai acts as the largest car terminal in India and has a container terminal with a capacity of 1.6 million 20-foot equivalent units (TEU) per annum. The port has facilities to handle petroleum, oil and lubricants (POL), fertilizers, containers, iron ore, food grains and other types of project cargos. The port primarily handles containerized cargo and petroleum that together account for 90% of the total cargo.

ENERGY CONSUMING PORT ACTIVITIES AND CO_2e EMISSIONS

The carbon footprint of the Port of Chennai was estimated for the year 2014-15 by applying ISO Standard 14064-1 and WPCI guidelines [13,14]. These guidelines provide methodologies to evaluate carbon footprints. The direct emissions (scope 1), indirect emissions (scope 2) and the other indirect emissions by port tenants and users (scope 3) have been considered in the assessment. Other indirect emissions accounted in scope 3 include emissions from the fishing harbor and housing colony, which are managed by the Port of Chennai.

The scope 1 emissions include diesel fuel usage by transportation, the operation of port-owned fleet vehicles (tugs, dredgers, pilot and mooring launches), electricity generation by diesel generators and material handling equipment such as cranes and fork lift trucks. The emissions are estimated using Equation 17-1 and based on port activities per the guidelines of WPCI [14]. The emission factor (EF) for diesel fuel consumption used is 2.68 kg of CO_2e/liter.

$$Emissions = \sum_{i=1}^{n} (Diesel\ consumption)_i \ x\ EF \qquad (17\text{-}1)$$

In Equation 17-1, *n* is the number of diesel consuming equipment. The scope 2 emissions are evaluated based on the purchased electricity for the operation of port-owned equipment such as cranes, pumps, reefer containers, machineries in workshop and for building air conditioning, lighting and other uses. Based on the sources of electricity generation in Tamil Nadu, the generation quantities and the emission factor for end user consumption, the import/export of electricity and aggregate technical and commercial (AT&C) losses can be estimated using 1.33 kg CO_2e/kWh. The scope 2 emissions estimate uses an activity based approach and is given by Equation 17-2.

$$Emissions = Electrical\ Energy\ Consumption\ x\ EF \qquad (17\text{-}2)$$

Scope 3 includes several large categories of emissions that contribute to the greatest portion of emission inventory. These categories include the electricity and diesel consumption for privately operated cranes and other wharf equipment, diesel fuel used in fishing harbor, and electricity consumption in ice factories. The emissions due to merchant vessel operations inside the port, truck movements for intermodal transportation of cargo, the use of liquefied petroleum gas (LPG), plus power and petrol use in the port's housing colony also contribute to scope 3 emissions. The GHG emissions due to various port activities are depicted in the Figure 17-2.

Merchant vessel auxiliary engines are operated in the port to provide electricity to the ships and to power the ship cranes for material handling, contributing to emissions. Most of these ships have one or more boilers that are used for fuel heating and producing hot water or steam. A hybrid approach is used to estimate GHG emissions from merchant vessels, based on WPCI guidelines [14]. In this inventory, the emissions from sea transit are not considered. Only the emissions from maneuvering and berth hoteling within the boundary of the Port of Chennai are considered. Anchorage hoteling is not considered as few merchant vessels are subjected to anchorage hoteling in the Port of Chennai. During the maneuvering phase, the emissions from main vessel engine are estimated based on Equation 17-3.

$$Emissions = \sum_{i-1}^{n} (MCR\ x\ LF\ x\ operating\ duration)_i\ x\ EF \qquad (17\text{-}3)$$

In Equation 17-3, n denotes the number of merchant vessels, *MCR* is the engine's maximum continuous rated power in kW, and *LF* is the load factor. The load factor is estimated using Equation 17-4.

$$LF = (\text{Maneuvering speed/Ship's maximum speed})^3 \qquad (17\text{-}4)$$

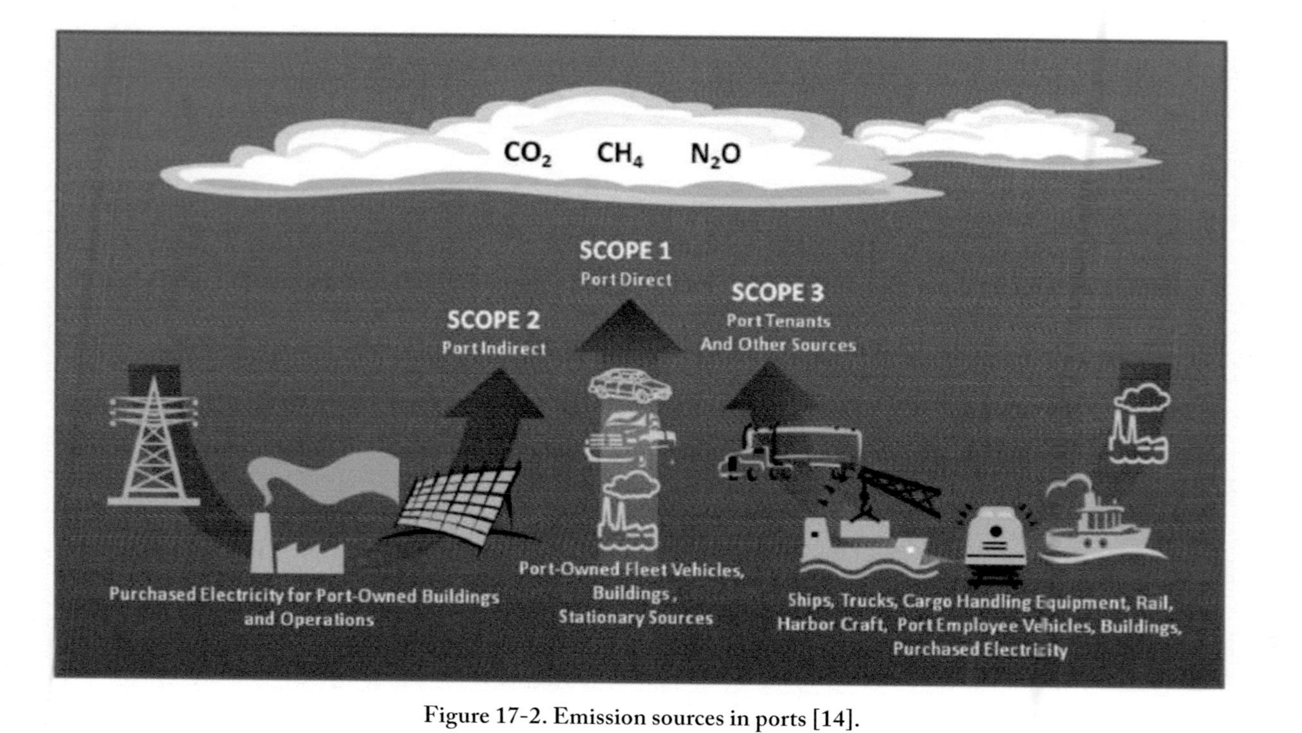

Figure 17-2. Emission sources in ports [14].

The main vessel *MCR* and maximum rated speed are obtained based on world fleet averages, per WPCI guidelines [14]. The maneuvering speed of merchant vessels within the boundary of the Port of Chennai is four knots. The operating time or maneuvering time is taken as the sum of pre-birth time and outward navigation time. These data were provided by the Chennai port authorities. It is assumed that the average year of manufacture of the merchant ships that visited the Port of Chennai in the financial year 2014-15 is 2000 or newer and use medium speed direct drive propulsion. Merchant vessels are assumed to operate their main engines on residual oil (RO), an intermediate fuel oil (IFO 380), or one with similar specifications and an average sulphur content of 2.7%. The assumed GHG emission factor is 0.69 kg CO_2e/kWh, based on the IVL, the 2004 Swedish Environmental Research Institute study [15]. The emissions from vessel auxiliary engine and boiler during the maneuvering phase are estimated using Equation 17-5. The details of auxiliary engine and boiler capacity are obtained from the Port of Los Angeles (POLA) inventory of air emissions [16].

$$Emissions = \sum_{i=1}^{n} (AS \ x \ Act \ x \ EF)_i \tag{17-5}$$

In Equation 17-5, i corresponds to an operating auxiliary engine or boiler, AS is the auxiliary system (engine or boiler capacity) in kW, n denotes the number of auxiliary systems in operation, and Act is the operating/maneuvering time (hours).

The maneuvering time is the same as that used to estimate GHG emissions from the main engine during maneuvering. The emission factor for the auxiliary engine (692.8 g CO_2e/kWh) and boiler (994.8 g CO_2e/kWh) are based on IVL 2004 [15] and the Euro NATO Training Engineer Center (ENTEC) emission factor for steam boilers respectively [17].

During the hoteling phase, the main engine is turned off and only the auxiliary engines and boilers are used. The GHG emissions during the hoteling phase are estimated based on Equation 17-6, with the operating time signifying the time at berth. The berth time for all merchant vessels visiting the Port of Chennai was provided by the port authority and was used to obtain the GHG emissions. Emissions from on-road vehicles include emissions mostly from trucks. Since the details regarding the fuel consumption of non-port-owned trucks used for transporting cargo are not available, a surrogate is used to estimate GHG emissions (shown in Equation 17-6). Per WPCI guidelines for heavy duty vehicles such as trucks, an emission factor of 4.6 kg CO_2e/h is estimated for idling periods while 1.0 kgCO_2e/km is used for on-terminal running activities [14].

$$Emissions = \sum_{i=1}^{N} (Act)_i \; x \; EF \qquad (17\text{-}6)$$

In Equation 17-6, N is the number of vehicles, i is the counter for vehicles and EF is the emission factor (either kgCO_2e/h or kgCO_2e/km). Other scope 3 emissions from sources such as vehicles used for employee transportation and LPG consumption in the housing colony are calculated based on the activity, given by Equation 17-1 with the emission factors for the various fuels consumed obtained from WPCI [14].

RESULTS AND DISCUSSION

The emission reduction approaches for the individual ports varies based on a variety of governance and operational models. We next evaluate the electricity and fuel consumption by various port activities and estimate their annual GHG emissions using various WPCI [14] and IPCC guidelines [1]. The ways to reduce these emissions using novel methodologies are also discussed.

Emissions from Port-owned, Diesel-consuming Vehicles and Equipment

Scope 1 emissions were entirely based on the use of diesel fuel in port activities such as transportation, maneuvering, dredging and generation of electricity. A large quantity of diesel is consumed by the operation of tug boats that berth merchant vessels and by dredging boats that deepen the quay. Mooring launches are used for mooring ships. Pilot launches transport the pilots to merchant vessels so they can navigate them to the berths and maneuver them safely in and out of the port. Nearly a quarter of the total volume of the cargo is handled with port-owned diesel locomotives. Diesel consumption and emissions by port-owned vehicles and equipment are summarized in Table 17-1. Tug boats accounted for 62% of the total diesel consumption for driving their heavier propulsion engines, followed by diesel locomotives and dredgers, each accounting for roughly 10% of total diesel consumption.

Table 17-1.
Diesel consumption and emissions by port vehicles and equipment.

Sources of Emissions	Diesel Consumption (litres/year)	CO_2 Emissions (tons/year)	Total Scope 1 Emissions (%)
Generator	16,000	43	0.6
Vehicles	142,186	381	5.2
Mobile equipment	163,143	437	6.3
Diesel locomotives	261,487	701	10.2
Pilot Launches	125,896	337	4.9
Mooring Launches	12,337	33	0.5
Dredgers	243,199	652	9.5
Tugs	1,606,692	4,306	62.4
Others	2,380	6	0.1
Total	**2,573,320**	**6,896**	**100.0**

Emissions from Port-owned Electricity Consumption

The scope 2 emissions were comprised entirely of the emissions from the electricity purchased from the Tamil Nadu Electricity Board for all port-owned operations. Electricity is used in lighting and air conditioning for buildings and for powering the cargo handling equipment. The connected load of motors that are used by electric cranes and by water and oil distribution pumps account for 37% and 36% of the total power distribution respectively. The lighting and air conditioning systems accounted for the greatest portion of the remaining electricity consumption. The total electricity consumption of the Port of Chennai under scope 2 was 5,654,053 kWh and the annual CO_2e was estimated to be 6,389 metric tons.

Emissions from Port Tenants and Users

The emissions from private operators of port and port users are categorized under scope 3 (indirect). Any emissions not accredited to scope 1 and scope 2 emissions are included in scope 3. These carbon emissions include those from all remaining sources (e.g., merchant vessels, diesel usage in the fishing harbor, LPG usage in housing colony, and electricity consumption by tenants). These activities have minimal potential for emission reduction, since they are not under the operational control of the port. However, enforcing mandatory rules on the tenants may reduce the CO_2e emissions.

Emissions by Port Tenants and Users

Figure 17-3 shows the electricity consumption and its equivalent CO_2e emissions from the various activities of port tenants and users. The total electricity consumption by the port and its tenants is 21,693,589 kWh annually which creates 24,513 metric tons of CO_2e emissions annually. The Port of Chennai is the hub for containers, cars and project cargo on India's east coast. Being the country's second largest container port, it generates business opportunities for handling high capacity containers. Terminals 1 and 2 is have high electricity consumption, accounting for 68.3% of the total. This is due to the electricity used by cranes for container movement and transfer cranes for intermodal transportation of containers. The housing colony is the third largest consumer of electricity with a consumption of 2,795,872 kWh annually. The fishing harbor uses 10.8% of the total electricity, of which the largest portion is used for the ice production. Figure 17-3 shows the GHG emissions of various port areas.

Diesel fuel is the major source of power for merchant vessels during their maneuvering and hoteling operations and also powers the fishing boats and trucks used for intermodal transportation of cargo in the port. Diesel operated cranes in container terminals 1 and 2 also contribute to total emissions. The details of the GHG emissions from the usage of diesel in merchant vessels and trucks are described below.

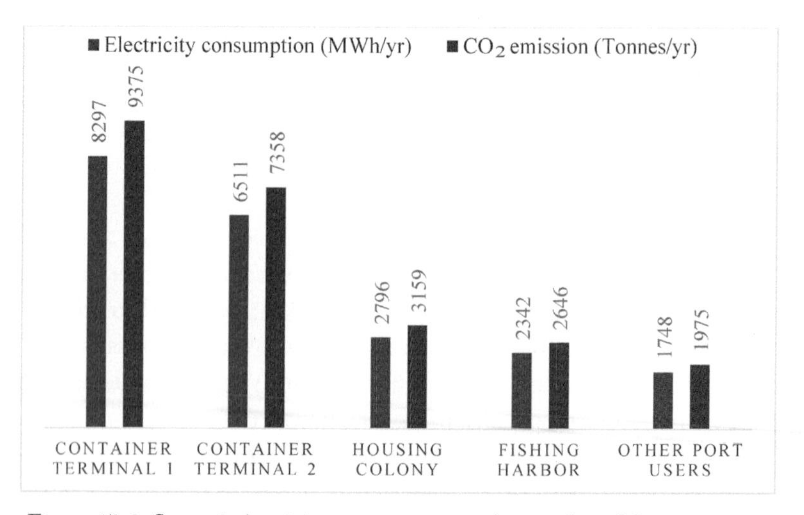

Figure 17-3. Scope 3 electricity consumption and equivalent CO_2e emissions.

Emissions from Merchant Vessels

The Port of Chennai handles all types of cargos ranging from liquid bulk, dry bulk and break bulk (dry bulk carriers), with containers being the major business provider. Nearly 2,000 vessels call at the Port of Chennai, of which approximately 40% are container ships. The emissions from merchant vessel engines during the maneuvering and hoteling periods are a major contributor to emissions. High capacity marine auxiliary engines and boilers are operated in the port to provide electricity to the ships for lighting, air conditioning, other on-board activities, and powering the ship cranes for material handling. Hence a large amount of fuel is consumed by the vessels which are docked at the port for various purposes. GHG emissions by various types of merchant vessels during the maneuvering and hoteling periods while calling at the Port of Chennai are detailed in Table 17-2.

Figure 17-4 shows that the tankers contribute 56% of total emissions followed by container ships (16%). Figure 17-5 shows the percentage of emissions during maneuvering and hoteling periods by the auxiliary engines and boilers. A total of 85.7% of the GHG emissions from merchant vessels occur during hoteling while emissions resulting from maneuvering are just 14.3% of the total. The tankers generate maximum emissions during their hoteling phase as their boilers are of very high capacity in comparison with the boilers in the other vessels.

Table 17-2.
GHG emissions by merchant vessels while maneuvering and hoteling.

Vessel Type	No. of Calls/year	Emissions During Manoeuvring (ton/year)			Emissions During Hoteling (ton/year)		Total (ton/year)
		Main Engine	Aux. Engine	Boiler	Aux. Engine	Boiler	
Dry Bulk	165	200	772	217	1,796	2,269	5,253
Container	780	298	4,282	995	12,121	7,180	24,876
Tanker	429	1,508	5,722	4,092	8,743	67,035	87,100
General Cargo	293	141	1,646	225	9,656	2,631	143,00
RORO	89	102	1,040	250	866	375	2,633
Cruise	48	18	629	154	14,843	5,817	21,461
Total	1,804	2,267	14,091	5,933	48,024	85,307	
Total Emissions During Manoeuvring and Hoteling		22,291			133,332		155,623

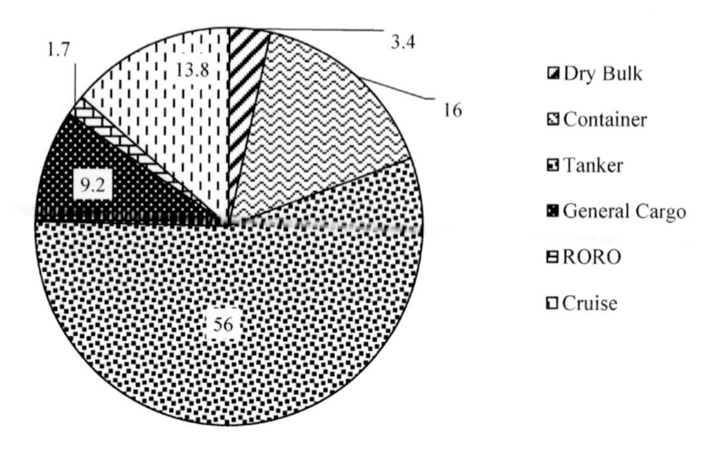

Figure 17-4. Percentage of GHG emissions by types of merchant vessels.

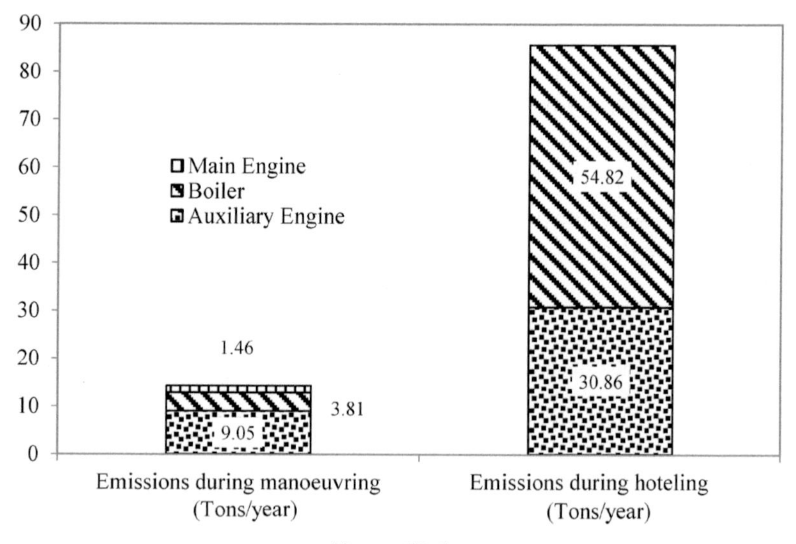

Figure 17-5.
Percentage of merchant vessel GHG emissions during maneuvering and hoteling.

Emissions from Trucks

Heavy vehicles (e.g., trucks, lorries, tankers, tippers and trailers) are used inside the port for the intermodal transportation of cargo. These vehicles move to various locations in the port depending on the type of cargo they are handling. The port has three gates (Gates 0, 2 and 10, referred to as

A, B and C respectively) for the movement of vehicles to facilitate the import and export of goods. Details concerning the number of vehicles entering and leaving the port and their average distance travelled were obtained from the port authorities. The average idle time per trip was estimated to be 30 minutes and the GHG emissions were estimated based on WPCI 2010 guidelines. The details of the estimated GHG emissions from the trucks during both idling periods and during on-terminal activities are provided in Table 17-3. The total CO_2e emissions from trucks used to transport cargo was estimated to be 6,343 metric tons annually, of which nearly one-third of the emissions were from vehicle idling.

Table 17-3. GHG emissions by the trucks operating inside the port.

Gate	No. of Vehicles	Average Distance Travelled (km)	Emissions During Idling Periods (tonnes/year)	Emissions During On-terminal Activities (tonnes/year)
A	487,788	5.0	1,135	2,491
B	238,264	4.4	555	1,071
C	170,193	4.0	396	695
Total			2,086	4,257

Emissions from the Fishing Harbor and Others

The Port of Chennai fishing harbor is the second highest consumer of diesel, consuming nearly 26,145 kL/year. The fishing harbor hosts 900 mechanized boats and roughly 1,200 fishing boats. Apart from the supplies for the domestic market, it also exports 2,000 metric tons of fish annually, contributing appreciably to emissions from diesel usage. The container terminals 1 and 2 are the other major diesel consumers, consuming 1,600 kL/year and 2,150 kL/year, respectively. Other port users consume about 7.5 kL/year.

Emissions from Use of Petrol and LPG

A survey was conducted in the housing colony to estimate the GHG emissions from fuel consumption for transporting employees, the consumption of LPG for residential cooking, the port canteen and the Central Industrial Security Force (CISF) barracks. The housing colony consists of 1,336 residences. In a sample of 10% of these, a questionnaire was used to survey LPG and petrol consumption. For the annual consumption of 7,603 liters of petrol for employee transportation, it is estimated that 17.3

metric tons of CO_2e annually is released into the atmosphere. LPG is the only source of fuel for cooking in the Port of Chennai. The various areas of LPG usage include the CISF barracks, port canteen and the housing colony. Total LPG consumption is 214 metric tons, accounting for 638 metric tons of CO_2e emissions annually with nearly 70% of the total generated by the housing colony.

Summary of Scope 3 Emissions

The percentage of GHG emissions based on scope 3 sources are provided in Table 17-4. Diesel usage by the port tenants contributes 90.6% while electricity consumption contributes 9.2%. Other scope 3 emissions such as petrol usage in the housing colony and LPG consumption were insignificant when compared to other emission sources.

Table 17-4. Total scope 3 GHG emissions.

Source of emissions	GHG emissions (tons/year)	% of total emissions
Diesel consumers		
Merchant vessels	155,623	58.2
Fishing harbour	70,069	26.2
Container terminal 1 & 2	10,050	3.8
Trucks	6,343	2.4
Other port users	20	0.01
Electricity consumption by port tenants	24,513	9.2
Petrol usage	17.3	0.01
LPG consumption	637.7	0.2
Total	267,273	100

CARBON FOOTPRINT IN THE PORT OF CHENNAI

The emission scenario of the Port of Chennai is provided in Figure 17-6. Scope 1 and 2 emission categories represent less than 5% of the port's overall emissions, while scope 3 emissions associated with tenants account for the majority of the port-wide emissions. GHG emission reductions from all port-related sources are necessary to minimize the impact of port-related operations on climate change. Though scope 1 and 2 emissions are of lower magnitudes, they are easier for the port to control and therefore a good place to begin.

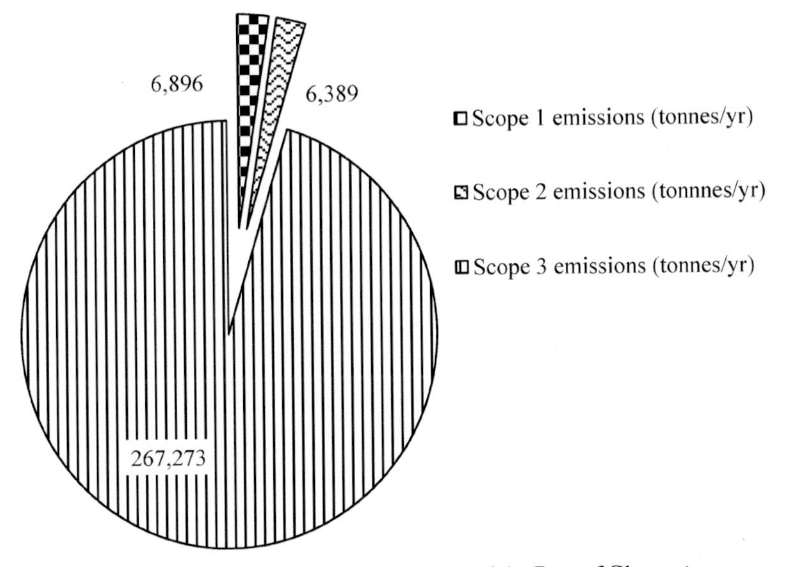

Figure 17-6. GHG emissions scenario of the Port of Chennai.

CONCLUSION

The carbon footprint of the Port of Chennai was estimated for the financial year 2014-15 using ISO Standard 14064-1. In conformance with the standard, the scope 1 (direct) emissions from diesel usage, scope 2 (indirect) emissions due to electricity consumption and scope 3 (indirect) emissions caused by the energy used by port tenants, the fishing harbor and the occupants from the housing colony were considered. The GHG emissions from the Port of Chennai total 280,558 metric tons of CO_2e annually and the scope 3 emissions account for 95.3% of the total. Scope 1 and scope 2 emissions together accounted for 4.7% of total emissions.

The detailed information made available in this chapter will be useful in implementing future energy conservation measures, renewable energy technologies, and integrating them with smart grids. The implementation of GHG mitigation strategies for all port-related activities will achieve significant GHG reductions, thus reducing the adverse impacts of global climate change. The information will also be useful for policy makers in enforcing mandatory measures and adopting commercially available energy and environmentally friendly technologies.

Acknowledgement

The authors acknowledge the administration and the technical staffs of the Port of Chennai for their financial support and assistance rendered during the data collection.

References

[1] Intergovernmental Panel on Climate Change (2006). Guidelines for national greenhouse gas inventories.
[2] Bailey, D. and G. Solomon (2004). Pollution prevention at ports: clearing the air. *Environmental Impact Assessment Review* 24, pages 749-774.
[3] Peris-Mora, E., Orejas, J., Subirats, A., Ibáñez, S. and P. Alvareza (2005). Development of a system of indicators for sustainable port management. *Marine Pollution Bulletin,* 50, 1,649-1,660.
[4] Schrooten, L., Vlieger, I., Panis, L., Styns, K. and R. Torfs (2008). Inventory and forecasting of maritime emissions in the Belgian sea territory, an activity-based emission model. *Atmospheric Environment,* 42, pages 667-676.
[5] Wang, C., Corbett, J. and J. Firestone (2008). Improving spatial representation of global ship emissions inventories. *Environmental Science and Technology,* 42, pages 193-199.
[6] Meyer, P., Maes, F. and A. Volckaer (2008). Emissions from international shipping in the Belgian part of the North Sea and the Belgian seaports. *Atmospheric Environment,* 42, pages 196–206.
[7] Han, C. (2010). Strategies to reduce air pollution in the shipping industry. *The Asian Journal of Shipping and Logistics,* 26, pages 7-30.
[8] Villalba, G. and Gemechu, E. (2011). Estimating GHG emissions of marine ports—the case of Barcelona. *Energy Policy,* pages 39, 1,363–1,368.
[9] Fitzgerald, W., Howitt, O. and I. Smith. (2011). Greenhouse gas emissions from the international maritime transport of New Zealand's imports and export. *Energy Policy,* 39, pages 1,521–1,531.
[10] Liao, C., Tseng, P., Cullinane, K. and C. Lu (2010). The impact of an emerging port on the carbon dioxide emissions of inland container transport: an empirical study of Taipei port. *Energy Policy,* 38, pages 5,251-5,257.
[11] Gibbs, D., Muller, P., Mangan, J. and C. Lalwani (2014). The role of sea ports in end-to-end maritime transport chain emissions. *Energy Policy,* 64, pages 337-348.
[12] Chang, Y., Song, Y. and Y. Roh (2013). Assessing greenhouse gas emissions from port vessel operations at the Port of Incheon. Transportation Research Part D. *Transport and Environment,* 25, pages 1-4.
[13] Greenhouse gas protocol. www.ghgprotocol.org/standards.
[14] World Ports Climate Initiative (2010). Carbon footprinting for ports, Guidance document.
[15] IVL (2004). Methodology for calculating emissions from ships: update on emission factors. Prepared by IVL Swedish Environmental Research Institute for the Swedish Environmental Protection Agency.
[16] Port of Los Angeles (2012). Inventory of air emissions.
[17] European Commission Directorate General Environment (2005, August). Service contract on ship emissions: assignment, abatement and market-based instruments. Task 2—General report. http://ec.europa.eu/environment/air/pdf/task2_general.pdf, accessed 5 July 2017.

Chapter 18

Renewable Energy Based Smart Microgrids—A Pathway to Green Port Development

Atulya Misra, Gayathri Venkataramani, Senthilkumar Gowrishankar, Elayaperumal Ayyasam and Velraj Ramalingam

Ports are an industry that accounts for 3% of the global greenhouse gas emissions. Sustainable initiatives and zero net energy goals are driving the use of renewable energy sources, including solar power, wind, and other alternate source systems. The main objective of this chapter is to reduce greenhouse gas emissions inside the port by integrating suitable renewable energy technologies in an efficient manner using micro grids. Clean energy-based, direct current (DC) microgrids are considered as a revolutionary power solution. In several development sectors, they are becoming increasingly attractive to researchers since they generate less greenhouse gases (GHGs), have lower operating costs, and offer flexibility. This chapter highlights and details the benefits of implementing renewable-based DC microgrids with the suitable energy storage technologies for sustainable energy management of ports.

INTRODUCTION

Global shipping is normally powered by standalone diesel generators for electricity supply. Shipping and port facilities are affected considerably by the cost of electricity generation. The use of renewable energy (RE) resources in the shipping industry is advantageous in reducing CO_2 emissions and reducing dependence on the fossil fuels. However, the intermittency associated with RE and large increases in its share of total power generation can result in problems with power generation, distribution and demand, contributing to electric grid instability. The magnitude of these problems can be reduced by developing large numbers of microgrids with energy storage capabilities. Microgrids are small electricity grids that can operate either independently or be connected to larger utility grids. These could have individual capability to manage supply and demand if these microgrids were integrated with the primary electrical grids. Then smart microgrids would enable consumers to connect to the primary energy delivery networks.

Several researches are developing on board, renewable-based microgrid systems for maritime applications. Corredor et al. developed a novel hybrid propulsion system for an internal combustion engine (ICE) coupled with an electric motor powered by batteries through an electric DC bus device in a parallel configuration system, increasing a boat's travelled distance to fuel consumption ratio [1]. Parameters including boat size and engine hybrid configuration were simulated. Strunz et al. proposed a DC microgrid with RE systems [2]. Further, a new method to quantify the uncertainties affiliated with the forecast of aggregated wind and photovoltaic (PV)-based power generation was developed and used to quantify the energy reserve of battery energy storage systems. Microgrid systems can operate in autonomous mode, supporting an uninterruptable power supply (UPS) when connections to the main grid are unavailable. Jayant Kumar discussed the importance of developing smart microgrids in port owned applications [3]. A new microgrid system has been suggested for Philadelphia Navy Yard Alstom. Krkoleva et al. developed a pilot microgrid for a rural location [4]. Their microgrid encompassed a part of a farm's existing low voltage grid and included sample loads plus a generator. Electrical loads within the microgrid could be supplied either by the grid or by electricity produced by renewable sources.

Cherry et al. stated that there is a strong correlation between electrification and rising human development, but in developing countries access to electricity is often unreliable, unavailable, or unobtainable, especially from a centralized electric grid [5]. They developed the concept of using a biogas digester to supply an internal combustion engine with fuel to generate electricity for portable energy storage devices (PESDs) for decentralized electrification. Hebener et al. examined the aspects of terrestrial microgrids and ship power systems [6]. They stated that the balancing strategy is an effective tool to improve system-level stability on finite inertia power systems, an important consideration in early-stage microgrid and ship power system designs. Gerry et al. optimized the distributed energy resources (DER) that includes distribution generation (solar PV array, wind energy, hydro and biogas fuelled generators) with battery backup [7]. They adopted an approach based on mathematical modeling for each component of DERs. Optimization was performed using HOMER (a software tool) to determine the economic feasibility of DERs and ensure reliable power supply to load demand and minimal cost. The methodology for overcoming several technical challenges in RE systems for isolated applications was also considered. Mariam et al. performed a technical and economic analysis of micro generation for the development of community based microgrid systems [8]. They performed this analysis for wind power supplied micro generation systems for communities that lacked RE feed-in-tariff policies.

This chapter next considers the integration of renewable based power generation and introduces the concept of a smart microgrid for the Port of Chennai, India. This study is useful to analyze the feasibility of RE deployment within the community context while minimizing environmental pollution. The proposed DC microgrid is a model for the development of greener ports and their facilities.

ENERGY CONSUMPTION PATTERN AT THE PORT OF CHENNAI

The Port of Chennai, formerly known as Madras Port, is the fourth largest port in India in terms of throughput. It is situated on the Coramandel coast and has a handling capacity of 86 million tonnes (Mt) annually. The port has 24 berths for cargo handling, distributed along three docks and has an entrance channel of seven kilometers in length. The port's electricity is provided by the Tamil Nadu Generation and Distribution Corporation (TANGEDCO).

The port uses enormous amounts of electricity and diesel for the intermodal transportation of goods and for providing essential services. It uses diesel fuel for trucks, generators, rail locomotives, cranes and mooring launches. The diesel operated cranes are located in container terminals 1 and 2. The port has five tug boats for maneuvering merchant vessels to berth, two dredgers for maintaining the depth of the quay, four mooring launches, 8 Nos. of 700 HP Locos (railway locomotives) and 2 Nos. of 1,400 HP Locos. The transmission of power to the locos makes use of an alternator. The diesel consumption by port owned vehicles, cranes operated at container terminal 1, 2, and other equipment total 6.3 million liters annually. Of this total, 59.2 % of the diesel consumption is used by cranes, 25.5% by tug boats and 15.3% by other equipment. The diesel consumption by the port's vehicles and equipment are illustrated in Figure 18-1.

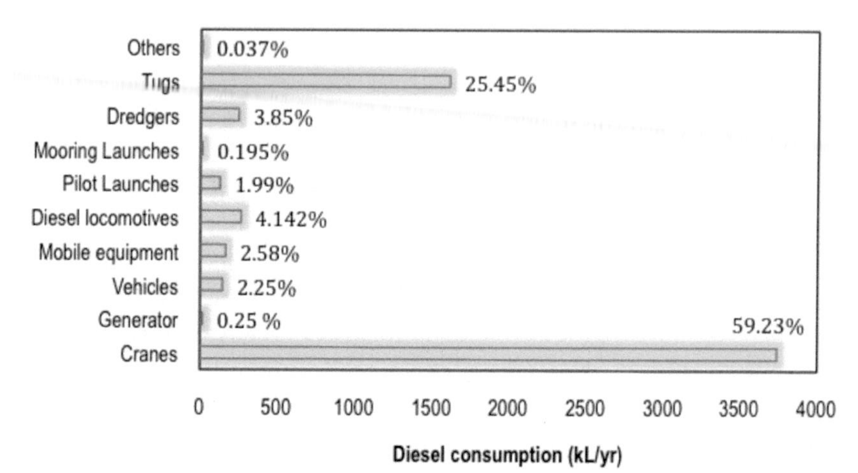

Figure 18-1. Diesel consumption by port-owned vehicles and equipment.

The major electrical energy consumers in Port of Chennai are: 1) port owned equipment including cranes, pumps, reefer (refrigerated vessels) containers; 2) private operated cranes for container movement, transfer cranes for intermodal transportation of containers; and 3) housing colony. The electrical energy consumers in the Port of Chennai are shown in Figure 18-2.

At the port, electrical energy is used for lighting, air conditioning for buildings, and to power cargo handling equipment. Electric cranes, located on the quays, are used mainly for handling dry bulk cargo and occasionally for dredging along berths to salvage cargo that has slipped or spilled along the berth during handling. The total electricity consumption of the Port of Chennai includes port-owned equipment, port tenants and user operated equipment, and the electrical energy consumed by housing colony. The total electrical usage is 25.0 GWh per year and is allocated as shown in Table 18-1. The electricity consumption by the port owned equipment is 5.65 GWh/year (22.6% of the total electrical energy consumption). The energy consumption by the port tenants and other users due to operation of cranes for container movement and transfer cranes for intermodal transportation of containers in container terminals 1 and 2 accounted for 16.55 GWh/year. The electrical en-

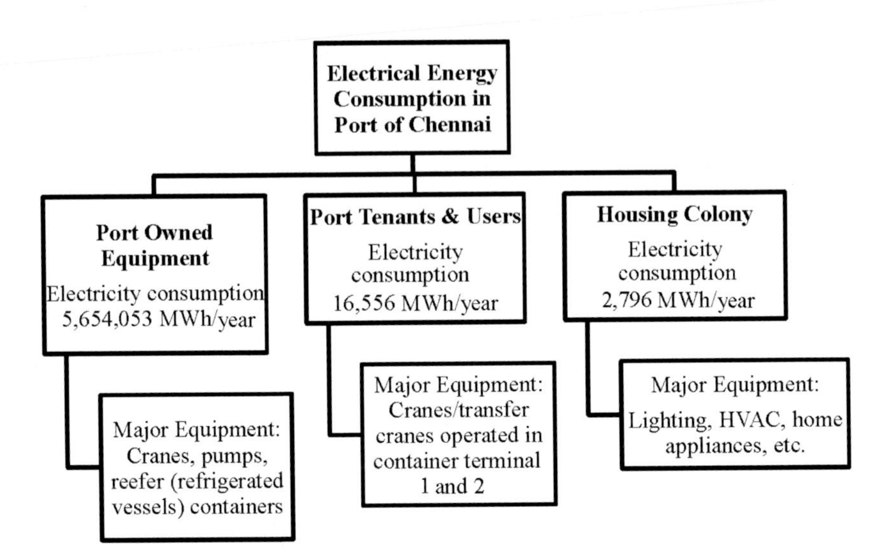

Figure 18-2. Major electrical energy consumers at Port of Chennai.

Table 18-1. Major electrical consumption equipment/utilities at Port of Chennai.

Equipment	Electrical consumption (MWh/yr.)
Port owned motors for crane operation	2,092
Port owned pumps	2,035
Port owned lighting, HVAC & other loads	1,527
Container terminal 1	8,297
Container terminal 2	6,511
Other port users	1,748
Housing colony	2,796
Total	25,006

ergy consumed in the housing colony due to lighting, air conditioning loads and other home appliances accounted for 2.8 GWh annually.

RENEWABLE ENERGY TECHNOLGIES FOR PORT ACTIVITES

The process of developing a strategic plan to match the proposed renewable power generation for a port's electricity demand involves technical, organizational and port policy analysis. This process involves fully integrating the port within the host community to achieve the goal of being "Green Port." One option for the Chennai Port is to harness sea breezes to generate electricity by installing wind turbines, the sun to generate power using PV systems and produce biogas from fishing harbor wastes. Electricity produced from these RE resources would support the development of a smart microgrid, using the electricity efficiently for various on-site applications. Next is a summary of the proposed power generation from renewable sources.

Power Generation Using Solar PV

India is endowed with rich solar energy resources due to its location in the Earth's equatorial areas. The daily average solar energy incidence in India varies from 4 to 7 kWh/m^2 and is roughly 5.6 kWh/m^2 in Chennai. There are about 300 clear and sunny days in a typical year. The capacity utilization factor (CUF), as evaluated by Equation 18-1 is 19% in Tamil Nadu. For example, a one kWp solar installation would be able to generate about 4.56 units (4.56 kWh) of electricity daily.

$$\text{CUF} = (\text{actual energy from the plant}) / [\text{plant capacity (KWp)} \times 24 \times 365] \quad (18\text{-}1)$$

The average potential roof area requirement for a typical 1 kWp solar PV power plant is approximately 9.5 m^2 of shade-free area. Port of Chennai, with a roof-top area (warehouse, transit shed, administrative blocks and housing colony) of 82,142 m^2, would be able to develop a 8.65 MW solar PV power plant, generating close to 14.4 million units (14.4 GWh) of electricity per annum. This clean source of electrical energy would reduce the GHG emissions by 16,262 tonnes annually. Jurong Port, Singapore [1,5] announced its installation of solar panels in more than 95,000 square meters of warehouse roof space, which would make it among the world's largest port-based solar facilities. Port of Chennai is considering the installation of solar PV systems using the rooftop space available on buildings. A minimum of 5 MW is recommended for the efficient installation, which would generate 8.32 million units (8.32 GWh) of electricity annually.

Wind Energy

Among the types of renewable energy sources available, power generation using wind energy is the most cost effective. The cost of generation is comparable with the other conventional sources of generation. Tamil Nadu, the southern state in India where the Port of Chennai is located, has three prominent passes with high wind potentials during the southwest monsoon. Installing wind turbine generators (WTGs) at the port can capture the wind energy from coastal breezes to drive the generators and produce electric power. WTGs during required rotational speeds produce electricity that is delivered to batteries and electrical loads in fishing boats using an AC to DC converter and a battery charger. Offshore wind development is an ideal way for modern green ports to increase revenue and gain a competitive advantage.

Port of Chennai plans to develop offshore wind energy for the following reasons:

- Wind energy is abundant and available.
- WTGs occupy minimum space.
- Being environmentally neutral, it complies with air emission regulations.
- Modern wind energy technology is closely associated with port technology and operations.
- Ports are an ideal location for both land-based and offshore farms.
- From financial and commercial standpoint, ports are ideal consumers of renewable energy, which is compatible with maritime and port energy generating technologies.

Continuing efforts towards its green port initiative, the Chennai port trust proposes to develop a wind farm project with an initial capacity of 6.5 MW. Total annual net energy generation from the wind farm will be about 14.6 million units (14.6 GWh), which will account for about 58% of the port's annual electrical energy requirements.

Energy Generation Using Fishing Harbor Waste

The Port of Chennai intends to identify the potential for biogas production from fishery wastes. Fish waste was determined to be a potential substrate for biogas production. It has potential as a source of high valued organic carbon for methane production.

The energy content of biogas is directly related to methane concentrations. Biogas production from fish waste is about 1,279 m^3 per tonne of total solids (TS). Biogas produced using fish waste has high methane concentrations of about 71%. A typical cubic meter of methane has a calorific value of 10 kWh, while carbon dioxide has zero. In the case of biogas produced from fish waste, the calorific value is 7.1 kWh/Nm^3. This biogas can be used as a fuel in stationary engines and 30% to 40% of the energy in the fuel will be used to produce electricity while the remaining energy becomes heat.

The fishing harbor managed by the Port of Chennai generates large volumes of fish waste (1,200-1,300 metric tons annually). This waste is disposed by the corporation of Chennai. Using a 30% conversion efficiency, the Port of Chennai can generate 3.8 million kWh of electricity per annum with the accumulated fish wastes, producing nearly 15% of the total electricity consumption by the Port of Chennai. To enhance the conversion efficiency from the biogas, combined heat and power generation is proposed as shown in Figure 18-3. The heat generated can be utilized to generate potable water with a multi-effect evaporator, meeting the port's entire potable water requirements.

Considering the scenario depicted in Figure 18-3 and its benefits, the port is proposing the installation of a biogas power system with two biogas generators (one 200 kW and one 300 kW), to reduce the potential of part load operation. This system would generate 3.5 GWh of electricity annually.

Microgrid Using Renewables with Energy Storage

Sustainable renewable energy can be provided with a DC microgrid to minimize the energy needed to operate cranes and other equipment at the Port of Chennai. Most renewable energy resources generate direct current electricity. Microgrids are decentralized electric grids that combine clusters of loads and parallel distributed generation systems in a local area. Local DC microgrids are being used for data and telecommunications centers. There is a potentially large market for DC microgrids in the developing world. The benefits they offer to the world's underserved regions continue to grow as RE power generation in developing countries increases. Distributed RE based DC microgrids can reduce distribution and transmission requirements. Other advantages of DC microgrids support consumer needs, local utilities and their communities. These include enhanced power performance, reduced overall power consumption, lowered greenhouse gases and pollutant emissions, increased service quality and local reliability.

There is a growing realization that energy storage (ES) systems will be key technologies in future electricity transmission networks, particularly those with heavy dependence on RE resources. Energy storage systems

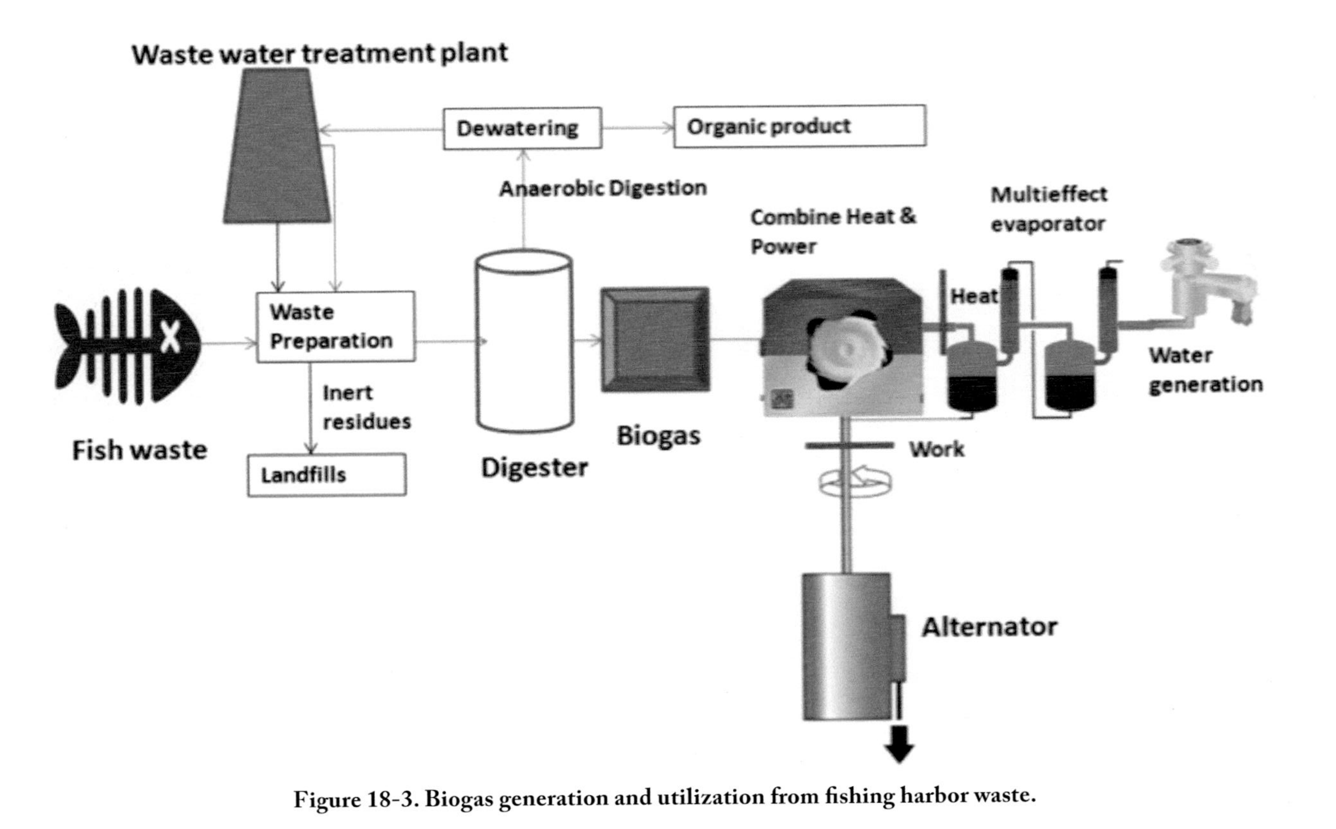

Figure 18-3. Biogas generation and utilization from fishing harbor waste.

can: 1) enable a match between supply and demand; 2) replace inefficient auxiliary power production; 3) ensure grid stability with a diversified energy supply and increased levels of renewable penetration; 4) ensure security of supply; and 5) facilitate distributed generation. A diverse range of electrical and thermal energy storage technologies exist, differentiated by power and energy density, physical size, cost, charge and discharge time periods and market readiness. For medium to large-scale electrical storage requirements involving timescales on the order of hours, mechanical and electrochemical ES technologies are the most viable, but no single technology is capable of fulfilling all of the required roles. The energy storage technologies being developed include advanced batteries, compressed air energy storage, fuel cells and others to store intermittent renewable energy resources. Integration of energy storage technologies into generation and distribution networks requires an understanding of the potential benefits, risks, and conformity with network operational rules. It requires an optimal mix of thermal and electrical storage to achieve the multiple objectives of CO_2 emissions reductions, cost minimization, safety and reliability.

Demonstrating the use of electric mobility in ports and port owned areas is increasingly important. Electric boats are more efficient than boats with diesel engines. An ideal 100% clean energy port is achievable if all boats, vehicles and machinery operating in the port used electrical power.

As handling cargo in harbors and ports is typically performed using cranes and vehicles, efforts are underway to make them more environmentally friendly. Regardless, most of these types of vehicles are diesel driven but could be equipped with electric battery operated systems. A major portion of the energy used in the Port of Chennai is for operating cranes and heavy machinery. Container cranes that lift and lower containers consume the most electrical energy in port terminals. While lifting the container, the current required depends upon the applied torque. While the crane is lowered, it is possible to generate electricity and store it in batteries.

PROPOSED SYSTEM

A ship's power system is microgrid. It contains electrical generation, distribution and loads. At sea, it is both isolated and self-sufficient. In a similar manner, a DC microgrid is proposed for Port of Chennai (see Figure 18-4), with power generation from RE sources.

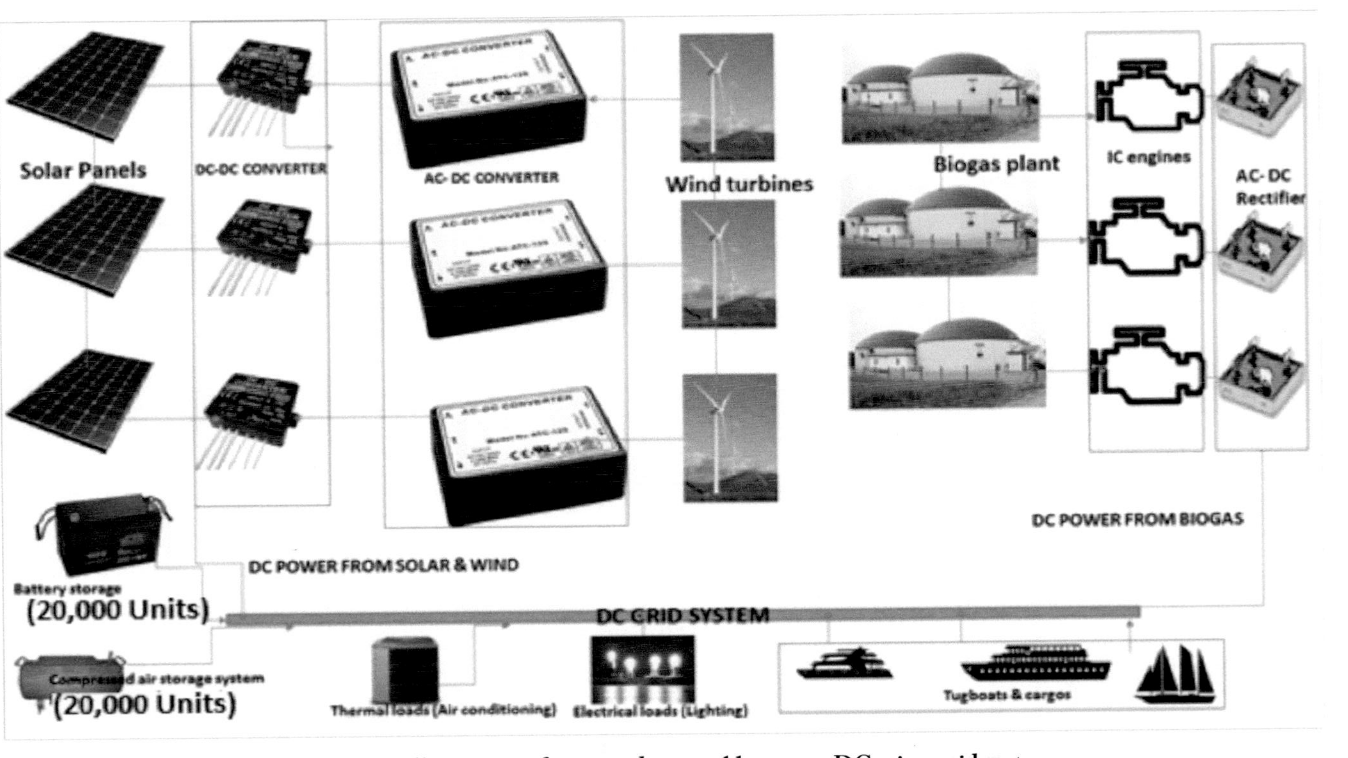

Figure 18-4. Illustration of proposed renewable energy DC microgrid system.

The port plans to meet its entire electrical energy requirement with RE using a microgrid. The power required for the electrical distribution system is obtainable using a 5 MW solar PV system (8.32 GWh/year), a 6.5 MW of wind power system (14.6 GWh/year), and 500 kW of biogas generation (3.5 GWh/year). The power generated from these systems is fed to the DC microgrid after conditioning the generated electrical power. A new DC bus system and transmission lines will be installed based on microgrid requirements.

Energy storage devices, including batteries and super capacitors, are being considered for use by port vehicles and cranes to reduce power fluctuations that result in higher costs and energy consumption. The average daily electricity consumption is about 70,000 kWh. To store approximately 60% of the total daily electricity requirements, a battery storage and compressed air energy storage system will be installed. The battery storage system will be designed to store 20,000 units of electricity and provide a buffering source for electricity to operate cranes and other port owned vehicles. The battery will be sized for a capacity of 72 volt, 630 amp hours to provide a continuous energy supply of 20,000 units of electricity. A new compressed air energy storage system (CAES) is suggested to store nearly 20,000 units of electricity using a 10,000 m^3 storage tank. It is also suggested that the stationary cranes and hoists operated in the port may be replaced by pneumatic cranes using the stored compressed from the CAES.

CONCLUSIONS

Generating renewable on-site power at ports can significantly reduce pollution, improve public opinion of the ports, and reduce their energy expenses. The RE technologies suggested in this chapter are useful in planning effective CO_2 mitigation strategies to incorporate technological advancement and management policies. Microgrids provide an opportunity to allow ports to meet their electrical energy requirements.

An assessment, accounting and estimate of energy consumption from the individual systems for the Port of Chennai, India are offered in this study. The port has the potential to implement renewable energy technologies with energy storage by deploying a microgrid. The suggested direct current microgrid uses solar power, wind turbine generators, and biogas generation to enable the port to provide most of its electrical energy needs.

Energy storage systems, including conventional batteries and compressed air, can provide 60% of the port's total daily electricity requirements. The information in this chapter can assist policy makers in developing mandatory measures for certain systems and accelerate the growth of environmentally friendly renewable energy and technologies that encourage green port development.

Acknowledgement

The authors acknowledge the administration and the technical staffs of the Port of Chennai for financial support and the assistance rendered while collecting data for this study.

References

[1] Corredor, L., Baracaldo, L., Jaramillo, J., Gutiérrez, D. and Jiménez (2012, March). A comprehensive energy analysis of a hybrid motorization for small/medium boats. International conference on renewable energies and power quality (ICREPQ'12). Santiago de Compostela, Spain.

[2] Strunz, K., Abbasi, E. and D. Huu (2014). DC microgrid for wind and solar power integration. *IEEE Journal of Power Electronics*, 2, pages 115-126.

[3] Jayant, K. and Alstom (2015, July). Microgrid system for Philadelphia Navy Yard, NREL—Advanced grid technology workshop series. http://www.nrel.gov/esi/pdfs/agct_day3_kumar.pdf.

[4] Krkoleva, A., Taseska, V., Markovska, N., Taleski, R. and V. Borozan (2010). Microgrids: The Agria test location. *Thermal Science*, 14, pages 747-75.

[5] Cherry, C., Rios, M., McCord, A., Sarah, S. and G. Venkatramanan (2016). Portable electrification using biogas systems. *Procedia Engineering*, 78, pages 317- 326. doi: 10.1016/j.proeng.2014.07.073.

[6] Hebner, R., Uriarte, F., Kwasinski, A., Gattozzi, A., Estes, H., Anwar, A., Cairoli, P., Dougal, R., Feng, X., Chou, H., Thomas, L., Pipattanasomporn, M., Rahman, S., Katiraei, F., Steurer, M., Aruque, M., Rios, M., Ramos, G., Mousavi, M. and T. McCoy (2016). Technical cross-fertilization between terrestrial microgrids and ship power systems, *Journal of Modern Power Systems and Clean Energy*, 4, pages 161- 179. doi: 10.1007/s40565-015-0108-0.

[7] Gerry and Sonia (2013). Optimal rural microgrid energy management using HOMER. *International Journal of Innovations in Engineering and Technology*, 2, pages 113-118.

[8] Mariam, L., Basu, M. and M. Conlon (2013). Community microgrid based on micro-wind generation system. IEE Conference on Power & Engineering (UPEC).

Chapter 19

Carbon Capture and Geologic Storage in India's Power Sector

Udayan Singh and Garima Singh

India is a major developing country with ambitious developmental goals. Future development is expected to increase energy demands and subsequently greenhouse gas (GHG) emissions. Carbon capture and storage (CCS) is one of the GHG mitigation strategies that might be adopted in this context. This chapter summarizes the scope of deployment of CCS in India's power sector. It also offers perspectives with regard to CO_2 capture technologies *vis-à-vis* Indian power plants. The potential geologic CO_2 storage sites and their resulting storage capabilities are discussed with references to the Indian research work being performed.

This chapter reviews the scope of CO_2 capture and storage in India's power sector which is largely dominated by coal. It considers how coal-based power generation is expected to increase in the future, the potential role of CCS, and the various perspectives of capture and storage strategies. This is followed by a discussion of the economic and regulatory aspects of the CCS technology, the two largest non-technical deterrents to implementing CCS. Finally, recommendations are offered regarding improving CCS technologies and policies in India. The major theme is CO_2 capture from the power sector. Applications in other sectors, such as the fertilizer industry are similar. The Jagdishpur fertilizer plant, as one example, has been performing CO_2 capture for a considerable time.

This chapter considers the Indian perspective, summarizes and reviews past CCS work in India, and offers suggestions that include progressive ideas about how technologies and policies can advance CCS in India. We conclude that CCS is an important transition technology to minimize GHG emissions while technologies develop to enable future deployment of renewable energy sources.

INTRODUCTION

India is a major developing economy which ranks fourth in the world in terms of energy consumption [1]. India must produce more energy to meet its growing demands since no country has increased its Human Development Index without a corresponding increase in per-capita energy consumption [2]. However, a large portion of India's primary energy supply is reliant upon fossil fuels which are sources of atmospheric greenhouse gases. India's CO_2 emissions are expected to increase in the future in order to meet its developmental challenges.

The problem of climate change is a major concern for India considering its developmental ambitions and natural climatic vulnerabilities. These include a long coastline, tropical rainforests, assorted biodiversity and the Himalayan mountain range. Climate change mitigation is one of the key issues faced by India [3]. The Maplecroft Index [4] suggests that India is the second most vulnerable country with regards to climate change. In a recent review by Sathaye and Shukla [5], it has been predicted that regions growing demographically and economically will need to be engaged in climatic change mitigation. Economic growth in India is strongly linked to energy consumption. If we consider the economic growth trajectory of India from 2001 to 2007, we find that this period's increase in the growth rate of the real gross domestic product from 5.2% to 9.6% was accompanied by an increase in the growth rate of CO_2 emissions from 0.7% to 8.6% [6]. While the relationships are not strongly linear, there is certainly a strong co-relation between GHG emissions and growth rate.

India has a history of environmental protection policies, beginning with the Water Act of 1974 [7] to the National Action Plan on Climate Change announced in 2008 [8,9]. India is a signatory to the Kyoto Protocol and has ratified the United Nations Framework Convention on Climate Change (UNFCCC). India is a founding member of the Carbon Sequestration Leadership Forum (CSLF) organized by the U.S. Department of Energy and is a past vice chair for this group [9]. Thus, India has been proactive in the formulation and implementation of policies related to climatic change mitigation.

India's total GHG emissions in 2007 were 1,831,647 gigatons of CO_2-equivalent of which 1,388,307 (~75%) gigatons of CO_2-equivalent were from the energy sector [10]. A large portion of India's GHG emissions come from large point sources (LPSs) such as thermal power plants, the steel industry, the cement industry and petroleum refining. Figure 19-1 provides a graphical representation of projected trends and contributions of all-India CO_2 emissions and the contributions of LPSs. As shown, CO_2 emissions will continue to increase and the contribution of LPSs will be roughly 70% [11]. LPSs offer a manageable way to control CO_2 emissions as a large amount of emissions can be mitigated while controlling a smaller number of sources.

Of the LPSs, coal-fired power plants are the most prominent contributors. In 2020, roughly 47% of India's CO_2 emissions are projected to be from coal-fired power plants [11]. The rate of CO_2 emitted per unit of electricity is much higher for coal-based power generation than for oil and natural gas based ones [12,13]. This includes emissions from both pulverized coal (PC) and integrated gasification combined cycle (IGCC) plants. For these reasons, coal-fired power plants are the most important sources from the perspective of CO_2 emission reductions.

Carbon capture and storage, also referred to as carbon capture storage and utilization (CCSU), encompasses industrial utilization of CO_2

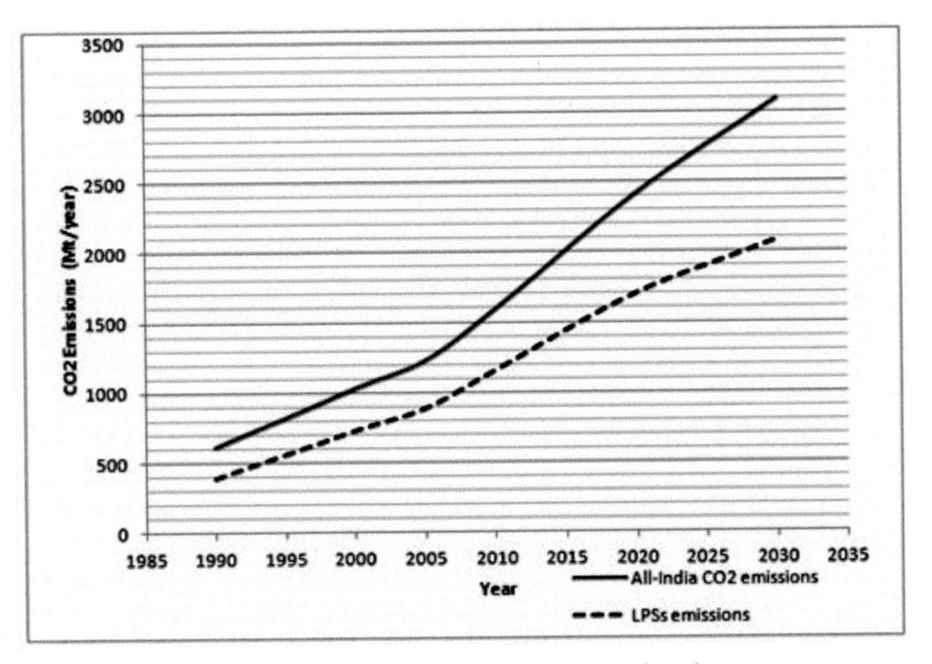

Figure 19-1. All-India CO2 emissions and CO2 emissions from large point sources (past and projected trend). Graph plotted from referenced data [11].

along with geologic storage. It is a technology that involves the capture of CO_2 from LPSs flue gases, its compression, transport and subsequent storage in geologic reservoirs. These reservoirs might include coal seams, saline aquifers, basalt formations and depleted hydrocarbon reservoirs. The reader may refer to IPCC (2005), DOE (1999) and Haszeldine (2009) for excellent reviews on this subject. Since coal-fired power plants are the largest emitters of CO_2 among all the other LPSs, it is imperative to judge the prospects of CCS in such plants. The government of India is taking steps towards energy security with the development of ultra mega power plants (UMPPs) with generation capacities on orders of gigawatts [17]. Government expenditures on climate change adaptation in India already exceed 2.6% of GDP, indicating that the government intends to work towards greenhouse gas (GHG) emission mitigation [8]. Development of renewable energy sources have been hampered as they are location dependent, have high establishment costs and non-competitive pricing [13]. To mitigate rising GHG emissions, CCS is considered to be a tool for India's power sector.

India's Power Sector and CCS Deployment

India's power sector is currently dominated by coal. At 210 GW, India has the world's fifth largest electricity generation sector, with plans to add 76 GW from 2012 to 2017 and 93 GW more from 2017 to 2022 [18]. Coal accounts for more than 50% of total electrical power generation. It is predicted that this domination of coal will increase in the future [19,20]. Integrated economic, energy and environmental modeling for India suggests that roughly 150,000 MW of coal-based power could be added by 2030. This is a factor in the projected increase in CO_2 emissions by 2.5 times during this period [21]. Given the increase in coal-based power generation, it is likely that CCS technologies will be used by power plants [11]. If renewable power becomes less expensive in the future, yet remains costlier than the conventional fossil fuel power generation, CCS could allow India to continue using coal in a more climate friendly manner. CCS will thus be a form of energy security for India [11]. For instance, the current predicted cost of concentrated solar power in India is about \$230(U.S.)/MWh [22], compared to the \$110(U.S.)/MWh cost of coal-fired power with CCS [23].

Thus, we must determine by what margins CCS will increase given the planned increases in coal-based power generation. The India Energy Security

Scenarios 2047 model, developed by the Planning Commission, Government of India, provided estimates in this regard [24]. Their model indicated four levels of deployment of CCS technology in Indian power plants:

- Least effort scenario (Level 1)—Involves a capacity addition of 10 GW to the year 2052. In this scenario, the installation of CCS-based plants will begin in approximately 2030, after which the growth of such plants will be very slow.
- Determined effort scenario (Level 2)—CCS-based capacity additions will reach roughly 40 GW by the year 2052. For this scenario, CO_2 capture based plants must begin functioning by 2022.
- Aggressive effort scenario (Level 3)—Involves CCS-based capacity additions of 88 GW by 2052 and requires such plants to be operational by 2017. This is the same level of CCS deployment found in the International Energy Agency (IEA) roadmap on CCS [25].
- Heroic effort scenario (Level 4)—Results in cumulative capacity additions of 100 GW of CCS-based power in India. This scenario occurs if the IEA Global Vision on CCS technology is followed [25].

Based on the previous data and the study by Singh and Rao [26], we can predict the extent of resulting CO_2 reductions from CCS. Assuming a super-critical boiler of 660 MW as characterized in this study, there will be a reduction of 0.78 kg/kWh using CO_2 capture. Using this data, we obtain results as shown in Figure 19-2.

The role that CCS will play in India will be determined by several factors. This includes cost of the technology, the attitude of the government and other stakeholders, and the development and pricing of other low carbon energy initiatives. As indicated in Figure 19-2, almost 200 million metric tons could be mitigated in the Indian power sector assuming a Level 3 scenario and nearly 230 million metric tons could be mitigated assuming a Level 4 scenario by 2052. Since the current CO_2 emissions are 665.4 million tons [27], this technology can serve as a useful tool to mitigate CO_2 from the largest emission source. This is in accordance with an International Energy Agency (IEA) assessment [28], which suggests that CCS may be instrumental in meeting the 19% CO_2 reduction necessary to achieve the less than 2°C (3.6°F) climate change mitigation.

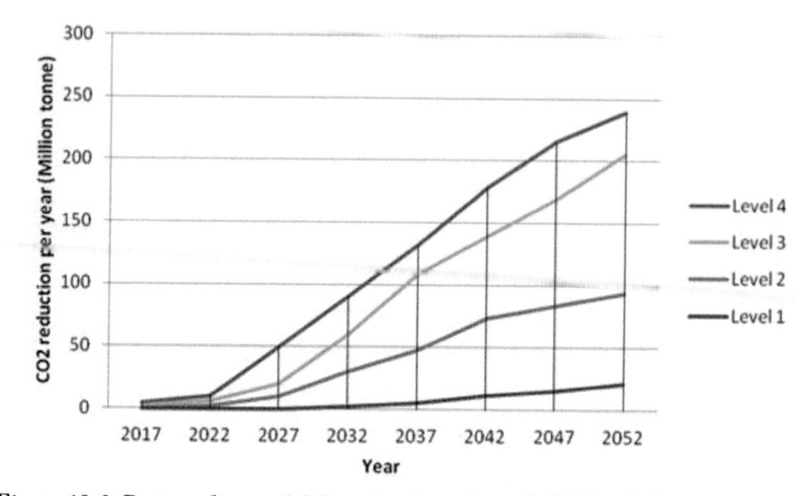

Figure 19-2. Projected annual CO_2 reductions through CCS in Indian power plants from 2017 to 2052. This graph has been plotted by multiplying the projected CCS installations in India [24] with the data on CO_2 reductions from Indian coal-fired power plants [26].

CO_2 Capture Perspectives for India

The first major step in the CCS technology is to capture CO_2 from LPSs, or in this case, thermal power plants. There are two basic considerations to be studied. The first is the type of capture technology to be used and the second is the implications of base plant power generation or those lacking CO_2 capture. For a plant with CO_2 capture, the two factors are related (i.e., the plant conditions govern which capture technology should be used). For capture, one assessment studied the energy, exergy, and environmental (3E) aspects for a plant with CCS compared to the same plant without CCS. Hasan et al. [29] have suggested a modeling framework for the CCS chain to arrive at the most cost-effective configuration while dealing with several sensitivities. These sensitivities include the type of material for carbon capture, process optimization, capture technologies and deployment of CO_2 utilization. A similar framework may be applied to India to suggest an optimized CCS chain for cost-effective CCS and CCUS deployment.

The first consideration in the base plant or the plant without CO_2 capture is the type of plant. Does it use combustion (pulverized coal plant) or gasification (IGCC)? When dealing with pulverized coal (PC) plants, the type of boiler plays a crucial role. It must initially be determined whether it is sub-critical (SubC), super-critical (SuperC) or ultra-super-critical

(USC). For carbon capture, the capture technology can be post-combustion, pre-combustion or oxyfuel based capture. Post-combustion capture has several sub-divisions including sorbent based (amine and ammonia) and non-sorbent based (membrane).

Only a few studies have been performed for thermodynamic as well as environmental comparisons of Indian PC plants with and without CCS. Numerous studies report that supercritical technology is well-suited for power plants with CCS [30]. A 2007 study of the risks of moving to more advanced steam conditions for the UMPPs being commissioned by the Indian government was undertaken by Mott MacDonald with funding from the UK Foreign and Commonwealth Office [17]. It considered factors such as environmental conditions in India, limited local manufacturing capability and a lack of experience with deploying the technology.

If CCS is to be tested on a reasonable scale, it is expected that the future UMPPs would be carbon capture and storage ready (CCSR). According to Kapila and Gibbins, a CCSR facility is a large-scale industrial or power source of CO_2 which is intended to be retrofitted with CCS technology when the necessary regulatory and economic drivers are in place [31]. Designing a plant to be carbon capture ready could lead to CCS implementation at a minimal cost. The factors to consider if it is economically feasible to make an older power plant CCSR include installed capacity, remaining lifetime, load factors, cost of retrofitting and the required rate of return. For a new power plant factors include cost of CCSR investment and expected plant lifetime [32].

Karmakar and Kolar simulated a 500 MW sub-critical plant operated by the National Thermal Power Corporation (NTPC) [33]. They concluded that the net plant efficiency drops by 8.3% to 11.2%, while the net plant exergy drops by 7.6% to 10.7%. They also suggested that the net plant efficiency improves when using low ash coal instead of high ash coal (most Indian coals are high ash). They also stated that the amount of CO_2 avoided varies between 0.68 and 0.70 kg per kilowatt-hour of electricity generated.

In a similar study, Singh and Rao simulated a 500 MW sub-critical and a 660 MW super-critical PC plant [34]. While Karmakar and Kolar studied only amine based capture, this study analyzed amine, ammonia, membrane and oxyfuel based capture [33]. They also studied the implications of using an auxiliary natural gas boiler. They calculated the energy penalty to be between 39% and 46% using CCS for the super-critical plant. They also predicted that using an auxiliary natural gas boiler would not be beneficial in India. Singh and Rao, in another paper, studied the implications of CCS on power plant emissions and resource use [26]. They concluded that CO_2 emissions would be reduced by 0.78 kg/kWh for a 660 MW super-critical plant. They also stated that for the sub-critical unit, the increase in coal use is between 46% and 52% while for the super-critical unit the increase is between 40% and 48%. Water consumption almost doubles for amine and membrane-based capture. The increase is more than 150% using ammonia-based capture and approximately 50% using oxyfuel technology.

These studies suggest a clear energy penalty as well as penalties on the use of resources. Plants with higher boiler efficiencies using better quality coals are more suited to CO_2 capture. This is not beneficial for Indian plants as they have lower efficiencies [35]. Furthermore, Indian coals generally have lower heating values and higher ash contents [36].

International studies, such as Rubin et al. suggest that IGCC plants are more suited to CO_2 capture than PC plants [37]. However, India currently lags in IGCC technology. Initial studies of the economics of power generation with CO_2 capture have suggested that the costs of producing electricity from coal with CO_2 capture could be similar for IGCC and state-of-the-art pulverized coal-fired power plants with CO_2 capture [38]. The Indian company Bharat Heavy Electricals, Limited (BHEL) held research trials in 1989 on IGCC in a pilot plant with a capacity of 6.2 MW [39].

While a few experimental and modeling initiatives have been performed with IGCC, none has dealt with CCS. In 1991, Iyengar and Haque [40] studied performance of high-ash Indian coals gasified by a fluidized bed reactor (FBR) at the Central Fuel Research Institute (CFRI) in Dhanbad. Similarly, Krishnudu [41] studied the performance of moving-bed gasification on Indian coals at the Indian Institute of Chemical Technology in Hyderabad (IICT). Some Indian indigenous coal samples were also tested for IGCC at the Gas Research Institute in the U.S. [39]. Goel made a comparative study of the IGCC and oxyfuel options in the Indian context [42]. More recently, Singh et al. [43] and Ajilkumar et al. [44] developed mathematical models for gasification of Indian coals in a moving bed and tubular gasifiers respectively. However, despite discussions concerning IGCC for over three decades, it has not yet come to fruition in India [45].

Lessons for CO_2 capture from power plants may be drawn from industrial CO_2 sources. The Jagdishpur fertilizer plant operated by the Indo

Gulf Corporation has had success with CO_2 capture. This plant has been instrumental in capturing 9,131 Mt CO_2 and establishing a precedent for the power sector [46].

CO_2 STORAGE SITES

After capturing CO_2, the question arises as to whether there are adequate storage sites. India is a large country with varied landforms. Options for potential geological storage for CO_2 include basalt formations, saline aquifers, hydrocarbon reserves, coal seams and use in enhanced oil recovery. We next consider each of these in a detail.

Basalt Formations

Basalt is a volcanic rock composed of metal silicates. Deccan Basalts in Southern and Central India are composed of 48 flows, covering 5,000 km^2 (1,930 $miles^2$) and are one of the world's largest flood eruptions [47,48]. The Rajmahal traps which are smaller than the Deccan traps also have substantial storage potential [49]. Tectonically, these traps are considered to be stable (see Figure 19-3). These provide solid storage for CO_2 with long-term integrity as they form mineral carbonates upon reaction with CO_2. Basalt formations have high storage capacities but have not been widely researched, perhaps because the IPCC Special Report on CCS has stated that this technology is not yet mature [14]. Nevertheless, if storage in basalt formations is developed, it might be a highly attractive option for India. McGrail et al. [50] mapped the locations of India's power plants and their proximity to basalt formations and concluded that 26% of the former lay in proximity to the latter. They also compared India's Deccan basalts spectroscopically with Columbia River basalts and identified similarities. Thus, international experience in CO_2 storage in basalt formations may help India obtain a suitable understanding of this alternative.

India's Hyderabad based National Geophysical Research Institute in collaboration with U.S. Pacific Northwest National Laboratory (PNNL) are involved in a joint project entitled "Demonstration of Capture, Injection and Geologic Sequestration of CO_2 in Basalt Formations of India" [51]. This is both a research and demonstration initiative endorsed by the CSLF. It involves assessing selected basalt storage areas and injecting 2,000 tons of CO_2.

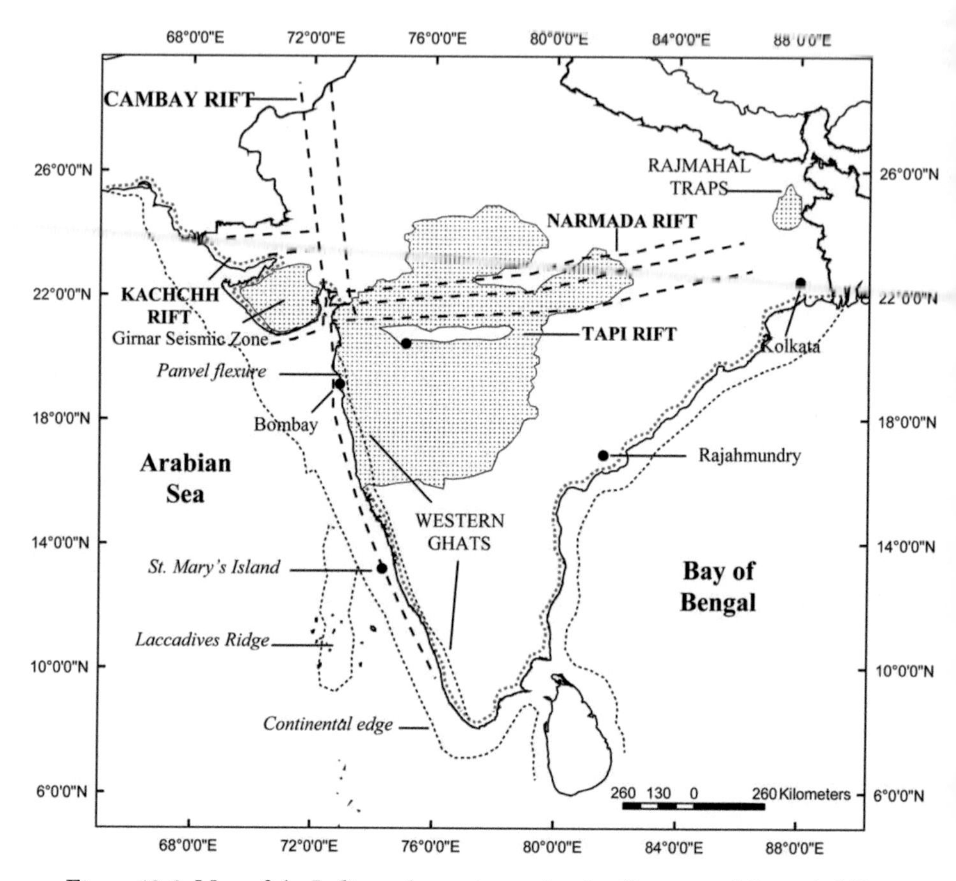

Figure 19-3. Map of the Indian sub-continent showing Deccan and Rajmahal Trap Province [11].

Saline Aquifers

Saline aquifers are also a very good storage option due to their ubiquity [52]. Many locations in Rajasthan and southern Haryana have electrical conductivity (EC) values of ground water greater than 10,000 s/cm at 25°C, making the water non-potable [53]. The Central Ground Water Board and the Geological Survey of India have established that saline aquifers exist at depths greater than 300 meters (984 ft) in the Ganga river basin. These are in the states of Rajasthan, Uttar Pradesh, Haryana and Punjab [53]. Research under the IEAGHG program [54] divided these basins into three parts:

- Good potential basins—Assam Basin, Assam-Arakan fold belt, Mahanadi basin (deep water part), Krishna-Godavari Basin, Cauvery Basin, Mumbai Basin, Cambay Basin, Barmer Basin and Jaisalmer Basin.
- Fair potential basins—Mahanadi Basin, Bikaner-Nagaur Basin and Kutch Basin.
- Limited potential basins (areas with uncertain storage potential)—Ganga Basin (and Punjab Shelf), Bengal Basin (Indian part), Vindhyan Basin, Cuddapah Basin, Chhatisgarh Basin, Konkan-Kerala Basin, Narmada Basin, Saurashtra Basin, Rajmahal Basin, Pranhita-Godavari Basin, South Rewa Basin, Satpura Basin and Damodar Valley Basins.

Coal Seams

Coal has a porous structure which contains adsorbed methane. When CO_2 is injected into coal, methane is displaced since coal has a higher affinity for CO_2 than for methane. Enhanced coal bed methane recovery (CO_2-ECBM) expedites the conventional coal bed methane recovery process. Methane is a fuel source that can supplement India's natural gas reserves.

The coal seams which have a good potential for CO_2 storage are divided into three categories, unmineable coal beds in well-delineated coalfields, grey area coal beds and concealed coal beds [55]. Table 19-1 provides a detailed categorization of coal beds.

ECBM is considered to be a good prospect for initial plants as the high costs of CCS are supplemented by the methane extraction. Vishal et al. have considered numerical modeling for CO_2 storage in India coal beds. They initially studied the permeability of Indian coal seams and identified subsequent implications for CO_2 storage [56]. They also modeled Gondwana coal seams in India as CBM reservoirs substituted for CO_2 sequestration [57]. In a more recent study, they have analyzed the influence of sorption time in the CO_2-ECBM process in Indian coals using coupled numerical simulation [58]. Researchers from the state funded CSIR-Central Institute of Mining and Fuel Research based at Dhanbad have studied CBM and ECBM [59,60].

Oil and Gas reservoirs

India is heavily dependent on imported oil and natural gas. As only 27% of oil-in-place is normally recovered, recovering the balance would

Table 19-1. Categories of coal beds in India [55].

Category of coal beds	*Grade of coal*	*Candidates /Basins*
Unmineable Coalbeds in explored areas	Power Grade coal	Singrauli MandRaigarh Talcher Godavari
Grey Areas Coalbeds	Coking coal	Jharia East Bokaro Sohagpur South Karanpura
	Superior non coking coal	Raniganj South Karanpura
	Power grade coal	Talcher
Concealed Coalfields	Tertiary age coal	Cambay basin Barmer Sanchor basin
	Power grade coal	West Bengal Gangetic Plain Birbhum DomraPanagarh Wardha Valley Extension Kamptee basin Extension

boost India's energy sector. The Oil and Natural Gas Corporation Limited, India's petroleum giant is exploring the possibilities of enhanced oil recovery (CO_2-EOR), which is the acceleration of oil recovery using CO_2 injection. EOR projects have been approved in both the Ankaleshwar (Gujarat) and Rajasthan oilfields. The possibility of CCS for disposal of acid gas at Uran is also being explored [61].

CO_2 Storage Potential

A major issue for implementation of the CCS technology is accurate assessment of CO_2 storage potential in various geological formations. A first order estimate by Singh et al. suggested a storage potential of 360 Gt of CO_2 in saline aquifers, 200 Gt of CO_2 in basalt formations, 5 Gt in coal seams and 7 Gt in oil and gas fields [55]. These total 572 Gt of CO_2. While encouraging, another study by the IEAGHG [63] regards this total as highly exaggerated and estimates total storage potential to be only 142 Gt of CO_2 for good, fair and limited quality areas. The major differences are that the latter study excluded basalt formations and included only 0.345 Gt of CO_2 storage potential for coal seams. This is due to an assumption that much of India's coal would be mined instead of being used for CO_2 storage. Table 19-2 compiles the results of these two studies and another by Dooley et al. [64,65].

Table 19-2. Overview of estimates for theoretical storage capacity in India [65].

	Dooley et al. 2005[64]	*Singh et al. 2006[62]*	*Halloway et al. 2008[63]*		
Oil fields	—	7	*Good, fair and limited quality*	*Good and fair quality 10.0 - 1.1*	*Good quality*
Gas fields	2			2.7-3.5	
Aquifers	102	360	138	59	43
Coal Streams	2	5		0.345	
Basalts	—	200		—	
Total	104	572	142	63	47

Interestingly, Narain states that CO_2 storage sites are not constrained by geography or geology [66]. Doig meanwhile suggests that Indian geologic formations offer inadequate CO_2 sequestration capability [67]. It is clear that estimates of CO_2 storage potential vary widely and that progress will be hindered until reliable estimates of storage capacities are available. To reduce discrepancies, a standard methodology needs to be used to estimate storage potential, such as one suggested by Bachu et al. [68]. Figure 19-4 provides locations of India's CO_2 emission sources and potential storage area.

The data calculated by the researchers mentioned above consider theoretical storage capacity, which is based on the physical limits that the geological systems can accept [68]. When a range of technical and geological cut-offs are applied to the theoretical storage limit, it provides a more realistic storage limit [69,70]. Technical, legal, regulatory, economic and geological barriers limit the practical CO_2 storage capacity. By matching

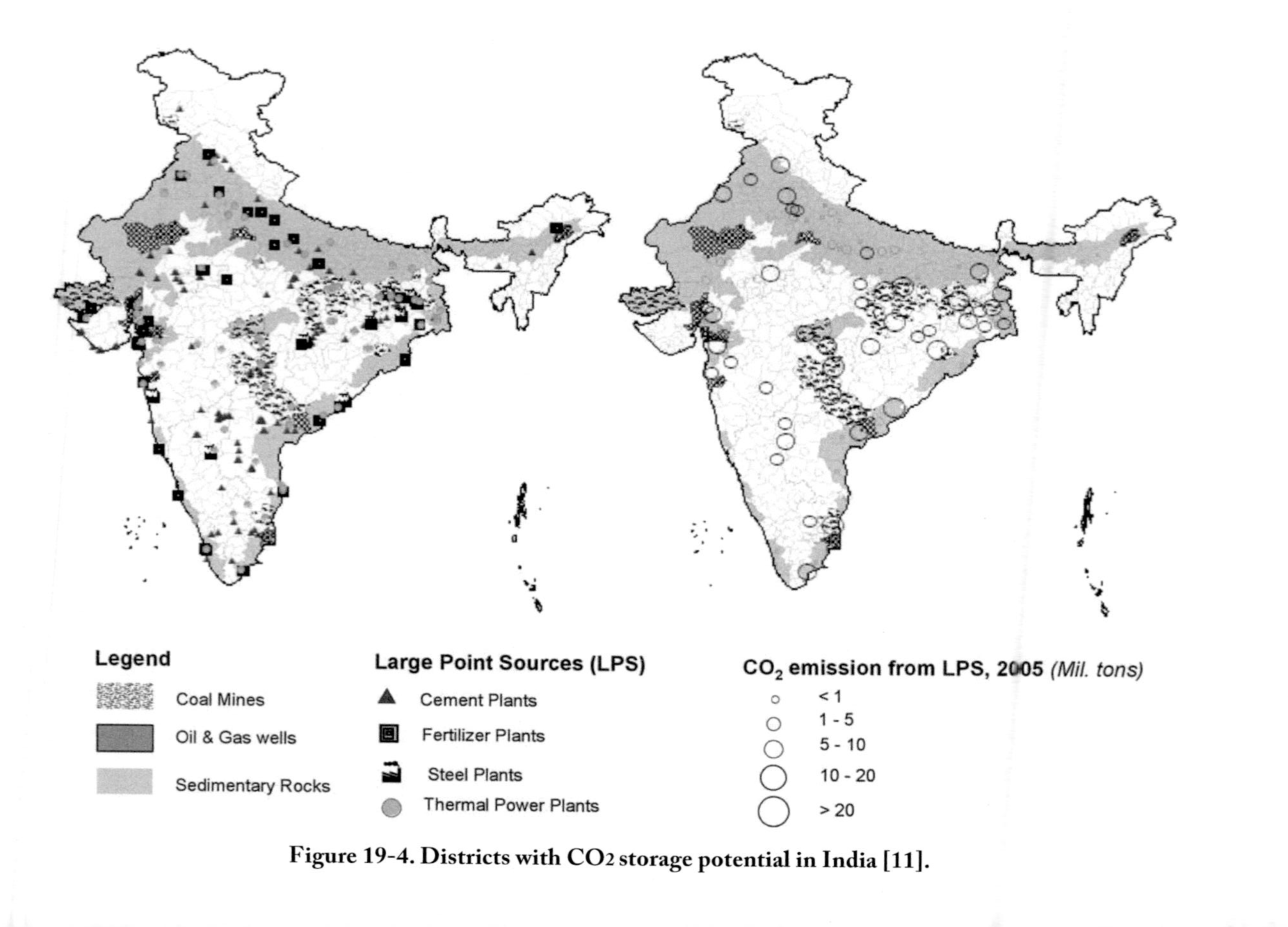

Figure 19-4. Districts with CO_2 storage potential in India [11].

the three CO_2 sources with the sinks, we can define the matched storage capacity [68]. Source sink matching by Beck et al. mapped a number of industrial sector LPSs to CO_2 sinks [71]. Specific to the power sector, Jain et al. used a graphical information system (GIS) for source sink matching for eastern India considering seven major power plants and four sinks [72].

ECONOMIC AND POLICY CONSIDERATIONS

Reducing GHG emissions by any means is costly. Carbon capture and storage is often criticized as an expensive technology. Based on expert assessments, the cost for the total chain of capture, transportation and compression for India is expected to be roughly $50 to $60 (U.S.) per metric ton of CO_2. The various components of the costs are capture (45%), compression (20%), transportation, injection and monitoring (35%) [11]. Reliable economic estimates are not available for CCS in India [69]. This hampers policy-making.

Attempts have been made to quantify the economic penalty for Indian power plants. Mott MacDonald estimated the cost of CO_2 abatement by CCS to be between $35 and $42 (U.S.) for each metric ton of CO_2 avoided [17]. The TERI scooping study on CCS stated that the cost of electricity with CCS will increase by 47% for the UMPPs [61]. Viebahn et al. estimated that the cost of electricity would increase by 45% to 51% for CCS in India [65]. A recent study by Rao and Kumar, estimated the incremental cost of electricity due to CCS between $39(U.S.)/MWh and $47(U.S.)/MWh and the cost of CO_2 avoidance between $48 and $58 (U.S.) per metric ton of CO_2 avoided [23]. This study was based on retrofit installation of CCS in four Indian power plants. This is substantially less than the estimated cost in the international power sector, which ranges from $72 to $114 (U.S.) for each metric ton of CO_2 avoided [73-76]. One reason why the cost of avoidance in India could be lower is due to the lower price of coal which is regulated domestically.

The studies referenced have not followed the standard costing methodology suggested by the Global CCS Institute. In this methodology, guidelines have been established to avoid ambiguity in the results as CCS cost estimates have several pitfalls [77]. The assumptions made for the study must be clearly specified per the Global CCS Institute guidelines [78, 79].

Future studies should follow a probabilistic methodology rather than a deterministic methodology based only on studying nominal values of just a few plants. Also, our belief is that some of the assumptions made in the aforesaid studies are not appropriate. In the study by Rao and Kumar, the reduction in gross size due to solvent regeneration was not considered [23]. Such improvements are needed in future studies to obtain more reliable estimates of CCS implementation costs in India.

CO_2-ECBM and CO_2-EOR are expected to be tested earlier due to their greater cost benefits compared to storage in basalt formations and saline aquifers. Initially, it is expected that CCS will be implemented for thermal power plants which are near storage areas as this offers lower transportation costs [11]. It is also envisaged that capture from coal-fired thermal power plants will be initiated before oil and natural gas fired ones. Tax incentives for emission reductions implemented by the Norwegian Sleipner project will help in the progress of CCS projects [13]. Factors that are likely to have roles in determining the economics of CCS in India include the price of fuel and imposition of carbon taxes. The policies of India's government and support by financial and regulatory agencies will be key determinants for the success of CCS technology in India. This includes issues such as the debt-to-equity ratio for the power plants and interest rates on loans which lead to differences in the costs of the plants. Technical parameters and new developments in CO_2 capture mechanisms also affect the CCS costs substantially [80, 81].

The Indian government's policies toward CCS can be either major enablers or deterrents to implementing this technology. India's government has been very pro-active in mitigating climatic change. However, whether or not the mitigation will include CCS depends on future policies and governmental regulations.

In India, research on carbon sequestration began with the support of the industry and the government [42]. The Indian government continues to look favorably upon carbon sequestration technologies including biological and terrestrial storage. The Department of Science and Technology, initiated a national program on carbon sequestration with one of its thrusts being carbon capture process development [9]. However, programs like these remain restricted to research and development (R&D). There is a consensus in developing nations that they should test this technology and then it can later be adopted in emerging economies [82]. It is desirable to develop local technical, economic and regulatory capabilities, so that CCS technologies can be more easily adopted in India. One assessment states that the Indian government has adopted a cautious stance towards CCS for the following reasons [83]:

- CCS leads to substantial energy penalties, whereas the principal aim of the government is to ensure maximum power generation.
- CCS leads to a higher economic penalty, whereas the government wants a low electricity price.
- CCS is not yet commercially viable.

Public sector industries have shown a positive approach towards CCS with two of the largest companies in the Indian power sector, NTPC and BHEL, being involved in R&D projects. This contradicts Kapila and Haszeldine [32] who state that CCS in India may progress in the same way as the information technology (IT) sector, which benefited from investments made by private enterprises. There is a consensus among several stakeholders that developed countries will have to demonstrate CCS at a commercial scale prior to commercial–scale development in India [84].

For development of CCS in India, there are also national and international laws that need to be considered. These relate to long-term monitoring of stored CO_2, ownership issues, site-selection and remedial responsibilities in case of accidents [85]. The reader may refer to TERI who mentioned specific laws that need to be considered when proposing CCS-specific legislation in India [61].

CONCLUSIONS AND RECOMMENDATIONS

There is a strong agreement that India needs to focus on climatic change mitigation. India's government has already provided a strong mandate towards mitigation efforts. Since coal is expected to dominate India's energy mix, there is a need for clean coal development. The prime minister of India recently stated that there are vast opportunities for development of clean coal technology (CCT) in India and invited foreign investments [86]. Yet the question remains whether the thrust on CCT will focus on improving power generation efficiency or radical reductions in emissions when CCS is deployed.

CCS will require additional research before implementation. While work has been successful in determining plant level results and storage capacity, there is a need for less ambiguity. This may be achieved using standard assessment methodologies. Adoption of CCS in India would create major challenges including reductions in energy efficiency and higher electricity costs. Reductions in efficiency may be offset (at least partially) by the following means:

- Development of IGCC technology in India as IGCC plants are more suited to CO_2 capture than PC plants.
- Higher capacity installations of super-critical and ultra-super-critical boilers.
- A major breakthrough may be achieved by research and development of membrane based capture in India. Singh and Rao [26] state that the energy penalty for membrane-based capture is less at lower capture efficiency. Research can focus upon making membrane-based capture more economically feasible and efficient. Improved blending and beneficiation of coal as better coal quality leads to more economic and efficient CO_2 capture. Plants which use imported coal should be among the first to employ CCS [26].
- Indigenous development of capture and compression equipment would lower costs. Current estimates are based on using equipment found in developed countries. Since Indian boilers are less expensive than their western counterparts, radically lower capital costs are possible if similar R&D is undertaken for capture and compression equipment.

The following must be considered regarding economic policies towards CCS:

- Currently, CCS is not part of the Clean Development Mechanism (CDM). If CCS can be brought within the ambit of CDM, India could benefit from its use [87].
- Financial and regulatory bodies need to support the technology for CCS to be successful. If lower taxes or interest rates are available for plants with CCS than for plants lacking CCS, then such favorable regulation would help CCS become established. Also, the availability of higher loans guarantees, production credits, purchasing agreements

or other policy instruments, through government interventions [37] would help create a positive investment opportunity for CCS in India.

- Long-term carbon pricing will enhance the feasibility of CCS. However, the Indian government has not shown interest in placing a tax on CO_2.
- Additional revenues from CCS, such as ECBM and enhanced oil recovery (EOR), will help reduce incremental electricity costs due to CCS. However, India's ECBM and EOR prospects are not favorable.
- Foreign funding should be attracted for R&D in CCS. Such funding can be from other governments or international institutions such as the Asian Development Bank.

There is also the need to strengthen the CCS knowledge base by organizing workshops and courses for college students, as public perception plays a major role in CCS deployment [13]. Political will and bureaucratic acceptance of the technology are crucial. At least two foreign studies on CCS in India have deemed that the Indian governmental bureaucracy has too many roadblocks to innovations in technology and financing [35,89]. Kapila and Haszeldine stated that the coalition structure of the Indian government was a deterrent towards innovation [35]. It remains uncertain whether CCS will become an important policy area for India's government.

While considerable research has been performed in various areas for CCS in India, there is a need for the studies to be integrated. Singh [13] and Jayanthu et al. [85] have vouched for a national mission project on CCS involving academic institutions (IITs, IIMs, NITs and other universities), research laboratories (NGRI, CIMFR, IIP, NCL) and industries (NTPC, CIL, BHEL, private sector industries). Funding for such a project might be financed by industry or foreign sources. Interdisciplinary research is the key to lowering the costs of CCS. As Rubin et al. have indicated, there have been considerable decreases in costs from lessons learned from the power-plant based technologies with emission controls [90,91].

We conclude by reiterating the tone of Maroto-Valer that CCS is an important transition technology to minimize GHG emissions while developing technologies for deployment of renewable energy sources [92]. A larger number of abatement options will help ensure fewer difficulties in the mitigation of climate change [88].

Acknowledgements

Our thanks to Professor Anand B. Rao, IIT Bombay and Dr. Ajay K. Singh, CSIR-CIMFR, for their guidance in the subject.

References

[1] U.S. Energy Information Administration —India. www.eia.gov/country/country-data.cfm?fips=IN, accessed 26 June 2014.
[2] Ghosh, P. (2009). Climate change: Is India a solution to the problem or a problem to the solution? In: Climate change: Perspectives from India. UNDP India. Pages 17-36.
[3] Singh, U., Agrawal, A., Chhangani, M. and A. Milton. (2014). Climate change and India. In: International Undergraduate Summit on Global Climate Change.
[4] Maplecroft. Big economies of the future—most at risk from climate change. http://maplecroft.com/about/news/ccvi.html, accessed, 25 December 2014.
[5] Sathaye, J. and P. Shukla (2013). Methods and models for costing carbon mitigation. *Annual Review of Environment and Resources,* 38, pages 137-168.
[6] Govindaraju, V. and F. Tang (2013). The dynamic links between CO_2 emissions, economic growth and coal consumption in China and India. *Applied Energy,* 104, pages 310-318.
[7] MoEF. The Water (Prevention and Control of Pollution) Act. http://www.moef.nic.in/legis/water/wat1.html, accessed 13 November 2015.
[8] PMCCC (2009). National action plan on Climate Change, Government of India.
[9] Goel, M. (2010). Perspectives in CO_2 sequestration technology and an awareness programme. In: CO_2 sequestration technologies for clean energy (Eds. Qasim, S. and M. Goel). Daya Publishing House. Pages 23-40.
[10] Sharma, S., Choudhury, A., Sarkar, P., Biswas, S., Singh, A., Dadhich, P., Singh, A., Majumdar, S., Bhatia, A., Mohini, M., Kumar, R., Jha, C., Murthy, M., Ravindranath, N., Bhattacharya, J., Karthik, M., Bhattacharya, S. and R. Chauhan. (2011). Greenhouse gas inventory estimates for India. *Current Science,* 101, pages 405-415.
[11] Garg, A. and P. Shukla, P. (2009). Coal and energy security for India: Role of carbon dioxide (CO_2) capture and storage (CCS). *Energy,* 34, pages 1,032-1,041.
[12] Johnsson, F. (2011). Greenhouse gases. *Science and Technology,* 1, pages 119-133.
[13] Singh, U. (2013). Carbon capture and storage: An effective way to mitigate global warming. *Current Science,* 105, pages 914-922.
[14] IPCC (2005). Special report on carbon dioxide capture and storage. Cambridge University Press.
[15] DOE (1999). Carbon sequestration: State of the nation. U.S. Department of Energy.
[16] Haszeldine, R. (2009). Carbon capture and storage: how green can black be? *Science,* 325, pages 1,647-1,652.
[17] Mott Macdonald (2008). CO_2 capture ready UMPPs, India. British High Commission at India.
[18] Prayas (2011). Thermal power plants on the anvil: implications and need for rationalization. Pune: Prayas Energy Group.
[19] Chikattur, A., Sagar, A. and T. Sankar (2009). Sustainable development of the Indian coal sector. *Energy,* 34, pages 942-953.
[20] Chikattur, A. and D. Sagar (2009). Rethinking India's coal power technology trajectory. *Economic and Political Weekly,* 44, pages 53-58.
[21] Nair, R., Shukla, P., Kapshe, M., Garg, A. and A. Rana (2003). Analysis of long-term energy and carbon emission scenarios for India. *Mitigation and Adaptation Strategies for Global Change,* 8, pages 53-69.
[22] Krishnamurthy, P., Mishra, S. and R. Banerjee (2012). An analysis of costs of parabolic

trough technology in India. *Energy Policy*, 48, pages 407-419.
[23] Rao, A. and P. Kumar. (2014). Cost implications of carbon capture and storage for the coal power plants in India. *Energy Procedia*, 54, pages 431-438.
[24] Planning Commission (2014). India energy security scenarios 2047. Government of India: New Delhi.
[25] International Energy Agency 2009). Technology roadmap—carbon capture and storage.
[26] Singh, U. and A. Rao (2014). Estimating the environmental implications of implementing carbon capture and storage in Indian coal power plants. In: 2014 International Conference on Advances in Green Energy. Pages 226-232.
[27] Guttikunda, S. and P. Jawahar (2014). Atmospheric emissions and pollution from the coal-fired thermal power plants in India. *Atmospheric Environment*, 92, 449-460.
[28] International Energy Agency (2010). Energy technology perspectives 2010—scenario and strategies for 2050.
[29] Hasan, M., Boukouvala, F., First, E. and C. Floudas (2015). A multi-scale framework for CO_2 capture, utilization and sequestration: CCUS and CCU. Computers and Chemical Engineering. http://dx.doi.org/10.1016/j.compchemeng.2015.04.034.
[30] Odeh, N. and T. Cockerill, T. (2008). Life cycle GHG assessment of fossil fuel power plants with carbon capture and storage. *Energy Policy*, 36, pages 367-380.
[31] Kapila, R. and J. Gibbins. (2010). Getting India ready for carbon capture and storage. *Energy Manager*, pages 10-13.
[32] Li, J., Liang, X. and T. Cockerill (2011). Getting ready for carbon capture and storage through a 'CCS (carbon capture and storage) ready hub': a case study of Shenzhen city in Guangdong province, China. *Energy*, 36, pages 5,916-5,924.
[33] Karmakar, S. and A. Kolar, A. (2013). Thermodynamic analysis of high-ash coal-fired power plant with carbon capture and storage. *International Journal of Energy Research*, 37, pages 522-534.
[34] Singh, U. and A. Rao (2014). Prospects of carbon capture and storage (CCS) for new coal power plants in India. In: Proceedings of the 1st National Conference on Advances in Thermal Engineering. Indian School of Mines. Pages 165-174.
[35] Kapila, R. and R. Haszeldine, R. (2009). Opportunities in India for carbon capture and storage as a form of climate change mitigation. *Energy Procedia*, 1, pages 4,527-4,234.
[36] Chandra, A. and H. Chandra (2006). Impact of Indian and imported coal on Indian thermal power plants. *Journal of Scientific and Industrial Research*, 63, pages 156-162.
[37] Rubin, E., Chen, C. and A. Rao (2007). Cost and performance of fossil fuel power plants with CO_2 capture and storage. *Energy Policy*, 35, pages 4,444-4,454.
[38] Davison, J. (2007). Performance and costs of power plants with capture and storage of CO_2. Pages 1,163-1,176.
[39] Principal Scientific Advisor to the Government of India (2005). Development of the integrated gasification combined cycle (IGCC) technology as suited to power generation using Indian coals.
[40] Iyengar, R. and R. Haque (1991). Gasification of high-ash Indian coals for power generation. *Fuel Processing Technology*, 27, pages 247-262.
[41] Krishnudu, T., Madhusudhan, B., Reddy, S., Rao, K. and R. Vaidyeswaran (1989). Moving-bed pressure gasification of some Indian coals. *Fuel Processing Technology*, 23, pages 233-256.
[42] Goel, M. (2009). Recent approaches in CO_2 fixation research in India and future perspectives towards zero emission coal based power generation. *Current Science*, 97, pages 1,625-1,633.
[43] Singh, N., Raghavan, V. and T. Sundararajan (2014) Mathematical modeling of gasification of high-ash Indian coals in moving bed gasification system. *International Journal of Energy Research*, 38, pages 737-754.

[44] Ajilkumar, A., Sundararajan, T. and U. Shet (2009). Numerical modeling of a steam-assisted tubular coal gasifier. *International Journal of Thermal Sciences,* 48, pages 308-321.
[45] Shahi, R. (2010). Carbon capture and storage technology: a possible long-term solution to climate change challenge. In: CO_2 sequestration technologies for clean energy (Eds. Qasim, S. and M. Goel). Daya Publishing House. Pages 13-22.
[46] IEAGHG. The Indo Gulf Fertilizer Company Plant, India. http://ieaghg.org/rdd/gmap/project_specific.php?project_id=42#top, accessed 25 December 2014.
[47] Eldholm, O. and M. Coffin (2000). Large igneous provinces and plate tectonics. In: The history and dynamics of global plate motions. *Geophysics Monograph Series*, 121, AGU, pages 309-326.
[48] Tiwari, V., Rao, M. and C. Mishra (2000). Density in homogeneities beneath Deccan Volcanic Province, India as derived from gravity data. *Geodynamics*, 31, pages 1-17.
[49] Charan, S., Kumar, B. and R. Singh (2008). Evaluation of basalt formation in India for storage of CO_2. In: International Workshop Carbon Capture and Storage in the Power Sector. New Delhi.
[50] McGrail, B., Schaef, H., Ho, A., Chien, Y., Dooley, J. and C. Davidson (2006). Potential for carbon dioxide sequestration in flood basalts. *Journal of Geophysical Research*, 111, page B12201.
[51] IEAGHG. Demonstration of capture, injection and geologic sequestration of CO_2 in basalt formations of India. IEAGHG RD&D Database. http://ieaghg.org/rdd/gmap/project_specific.php?project_id=157#top, accessed 25 December 2014.
[52] Bachu, S., Gunter, W. and E. Perkins (1994). Aquifer disposal of CO_2: Hydrodynamic and mineral trapping. *Energy Conversion and Management,* 35, pages 269-279.
[53] Central Ground Water Board. Ground water quality—. www.cgwb.gov.in/GW_quality.html, accessed 26 June 2014.
[54] Energy Information Administration (2008). A regional assessment of the potential for CO_2 storage in the Indian subcontinent. GHG R&D Program.
[55] Singh, A. (2008). R&D challenges for CO_2 storage in coal seams. In: Carbon capture and storage: R&D technologies for a sustainable energy future (eds. Goel, M., Kumar, B. and S. Charan). *Alpha Science International*, pages 139-149.
[56] Vishal, V., Ranjith, P. and T. Singh (2013). CO_2 permeability of Indian coals: Implications for carbon sequestration. *International Journal of Coal Geology,* 105, pages 36-47.
[57] Vishal, V., Singh, L., Pradhan, S., Singh, T. and P. Ranjith (2013). Numerical modeling of Gondwana coal seams in India as coalbed methane reservoirs substituted for carbon dioxide sequestration. *Energy,* 49, pages 384-394.
[58] Vishal, V., Singh, T. and P. Ranjith (2015). Influence of sorption time in CO_2-ECBM process in Indian coals using coupled numerical simulation. *Fuel*, 139, pages 51-58.
[59] Singh, A. (2010) Opportunities for extraction and utilization of coal mine methane and enhanced coal bed methane recovery in India. In: CO_2 sequestration technologies for clean energy (Eds. Qasim, S. and M. Goel). Daya Publishing House. Pages 131-142.
[60] Mendhe, V. (2010). Coal bed methane: prospects and challenges. In: *CO_2 Sequestration Technologies for Clean Energy* (Eds. Qasim, S. and M. Goel). Daya Publishing House. Pages 143-154.
[61] The Energy and Resources Institute (2013). India CCS scoping study: final report.
[62] Singh, A., Mendhe, V. and A. Garg (2006). CO_2 sequestration potential of geologic formations in India. In: Proceedings of the 8th International Conference on Greenhouse Gas Control Technologies. Elsevier.
[63] Holloway, S., Garg, A., Kapshe, M., Deshpande, A., Pracha, A., Khan, S., Mahmood, M., Singh, T., Kirk, K. and J. Gale (2009). An assessment of the CO_2 storage potential of the Indian subcontinent. *Energy Procedia,* 1, pages 2,607-2,613.
[64] Dooley, J., Kim, S., Edmonds, J., Friedmann, S. and M. Wise (2005). A first-order global

geological CO_2-storage potential supply curve and its application in a global integrated assessment model. In: Proceedings of the 7th International Conference on Greenhouse Gas Control Technologies. Elsevier.

[65] Viebahn, P., Vallentin, D. and S. Holler (2014). Prospects of carbon capture and storage (CCS) in India's power sector—an integrated assessment. *Applied Energy,* 117, pages 62-75.

[66] Narain, M. (2007). Pathways to adoption of carbon capture and sequestration in India: technologies and policies. MS Thesis. Massachusetts Institute of Technology.

[67] Doig, A. (2009). Capturing India's carbon: the UK's role in delivering low-carbon technology to India. Christian Aid.

[68] Bachu, S., Bonijoly, D., Bradshaw, J., Burruss, R., Holloway, S., Mathiassen, O. and N. Christensen (2007). CO_2 storage capacity estimation: methodology and gaps. *International Journal of Greenhouse Gas Control,* 1, pages 430-443.

[69] Carbon Sequestration Leadership Forum. (2005). A taskforce for review and development of standards with regards to storage capacity measurement.

[70] Bradshaw, J., Bachu, S., Bonijoly, D., Burruss, R., Holloway, S., Christensen, N. and O. Mathiassen (2007). CO_2 storage capacity estimation: issues and development of standards. *International Journal of Greenhouse Gas Control,* 1, pages 62-68.

[71] Beck, R., Price, Y., Friedmann, S., Wilder, L. and L. Neher (2013). Mapping highly cost-effective carbon capture and storage opportunities in India. *Journal of Environmental Protection,* 4, pages 1,088-1,098.

[72] Jain, P., Pathak, K. and S. Tripathy (2013). Possible source-sink matching for CO_2 sequestration in Eastern India. *Energy Procedia,* 1, 3,233-3,241.

[73] U.S. Department of Energy's National Energy Technology Laboratory (2007). Cost and performance baseline for fossil energy plants. Pittsburg, Pennsylvania.

[74] U.S. Department of Energy's National Energy Technology Laboratory (2010). Cost and performance baseline for fossil energy plants. Vol. 1: bituminous coal and natural gas to electricity. Pittsburg, Pennsylvania.

[75] DOE (2010). Report of the interagency task force on carbon capture and storage. Interagency Task Force on Carbon Capture and Storage. Washington, D.C.

[76] Hasan, M., Boukouvala, F., First, E. and C. Floudas (2014). Nationwide, regional, and statewide CO_2 capture, utilization, and sequestration supply chain network optimization. *Industrial and Engineering Chemistry Research,* 53, pages 7,489-7,506.

[77] Rubin, E. (2012). Understanding the pitfalls of CCS cost estimates. *International Journal of Greenhouse Gas Control,* 10, pages 181-190.

[78] Global CCS Institute (2013). Toward a common method of cost estimation for CO_2 capture and storage at fossil fuel power plants—a white paper.

[79] Rubin, E., Short, C., Booras, G., Davison, J., Ekstrom, C., Matuszewski, M. and S. McCoy (2013). A proposed methodology for CO_2 capture and storage cost estimates. *International Journal of Greenhouse Gas Control,* 17, pages 488-503.

[80] Hasan, M., Baliban, R., Elia, J. and C. Floudas (2012). Modeling, simulation, and optimization of postcombustion CO_2 capture for variable feed concentration and flow rate. 1. Chemical absorption and membrane processes. *Industrial and Engineering Chemistry Research,* 51, pages 15,642-15,664.

[81] Hasan, M., Baliban, R., Elia, J. and C. Floudas (2012) Modeling, simulation, and optimization of postcombustion CO_2 capture for variable feed concentration and flow rate. 2. Pressure swing adsorption and vacuum swing adsorption processes. *Industrial and Engineering Chemistry Research,* 51, pages 15,665-15,682.

[82] Centre for Science and Environment (2009). Climate change: politics and facts.

[83] Wuppertal Institute for Climate, Environment and Energy (2012). Prospects of carbon capture and storage technologies (CCS) in emerging economies. Final Report to the

German Federal Ministry for the Environment, Nature Conservation and Nuclear Safety (BMU).

[84] Kapila, R., Chalmers, H., Haszeldine, S. and M. Leach (2011). CCS prospects in India: results from an expert stakeholder survey. *Energy Procedia,* 4, pages 6,280-6,287.

[85] Jayanthu, S., Singh, U., Madan, M. and Y. Rao (2013). Methane recovery from unmineable coal seams through carbon sequestration—a critical appraisal. In: International Conference on Coal and Energy: Technological Advances and Future Challenges. BESU Shibpur. Pages 59-71.

[86] Mohan, V. (16 November 2014). PM Narendra Modi makes strong pitch for clean energy at G20 summit. *The Times of India.*

[87] Verma, V. (2010). Should carbon be priced? Open round table discussion. In: CO_2 sequestration technologies for clean energy (Eds. Qasim, S. and M. Goel). Daya Publishing House. Pages 183-202.

[88] Herzog, H. (2001). What future for carbon capture and sequestration? *Environmental Science and Technology,* 35, 148-153.

[89] Massachusetts Institute of Technology (2007). The future of coal.

[90] Rubin, E., Yeh, S., Antes, M., Berkenpas, M. and J. Davison (2006). Estimating future costs of CO_2 capture systems using historical experience curves. In: Proceedings of the 8th International Conference on Greenhouse Gas Control Technologies. Elsevier.

[91] Rubin, E., Yeh, S., Antes, M., Berkenpas. M. and J. Davison (2007). Use of experience curves to estimate the future cost of power plants with CO_2 capture. *International Journal of Greenhouse Gas Control,* 1, pages 188-197.

[92] Maroto-Valer, M. (2011). Why carbon capture and storage? Greenhouse gases. *Science and Technology,* 1, pages 3-4.

Chapter 20

Renewable Energy in India—Barriers to Wind Energy

Sanjeev H. Kulkarni and T.R. Anil

As the world moves toward greater development and growth, it is imperative to appreciate the looming ramifications of environmental degradation and ecological imbalances that are caused by atmospheric carbon emissions. The increased negative effects of fossil fuels on the environment has forced many countries, especially the developed ones, to use renewable energy sources. Recently, renewable energy systems have emerged as a clean and easy-to-maintain alternative to the use of diesel engines for rural electrification. Among them wind energy, biomass energy systems and solar photovoltaic (PV) offer capacity and proven technologies to provide continuous and reliable energy. Wind energy, the fastest developing energy source, is renewable and environment friendly. Systems that convert wind energy to electricity have developed rapidly.

Many clean and energy-efficient technologies contribute to sustainable development and energy security in developing economies. However, in practice these technologies are rarely used. Barriers exist that prevent sustainable, energy-efficient technologies from being more widely utilized and having greater market penetration and diffusion. It is absolutely necessary to overcome these barriers. This chapter aims to identify and rank the barriers to greater diffusion of renewable energy systems particularly wind generation. The barriers identified include lack of knowledge about energy sources, higher investment costs, preferences for grid extension projects, lack of arrangements for long-term operation and maintenance, absence of a certification systems for equipment, and lack of financial instruments for renewable energy entrepreneurs.

Our study identifies and ranks the barriers for the diffusion of wind energy in three wind farm clusters in the southern Indian state of Karnataka based on the perceptions and judgements of the various stake holders.

Five main barrier groups are considered and their dimensions are recognized before ranking them based on four different criteria using multicriteria decision making. While all barrier groups have significant weights, barriers related to policy, organizatinal form and awareness were found to be the primary barriers. This provides a promising domain for regulatory and political policy interventions to enhance the implementation of wind energy. The results provide evidence of how consumers receive wind energy information and make decisions using their analytical capabilities.

INTRODUCTION

At the dawning of the twenty first century, the world experienced growing demand for energy, increased environmental pollution and depleting energy sources. Human society faces two great challenges: the transition towards more sustainable development, and the eradication of poverty. The influential Brundtland report (World Commission on Environment and Development: Our Common Future) defined sustainable development based on the ideal of meeting the needs of the present without compromising the ability of future generations to also meet their needs [1,2]. Eradicating poverty requires the cooperation of industrialized and developing countries. A primary element of eradicating poverty is greater economic development of rural areas in developing countries [3].

Climate change is considered to be among the most serious threats to the world's environmental sustainability. Most scientists agree that the earth's climate is being affected by atmospheric greenhouse gases (GHGs) released by human activities. Since the main economic sector contributing to GHG emissions is the energy sector, transitioning to a more sustainable developmental model must include reducing the use of non-renewable primary energy sources [3]. This calls for increasing the use of renewable energy, which offers other positive consequences including decentralizing electricity generation, reducing external energy dependence, diversifying energy resources and creating employment.

The fundamental characteristic of a sustainable energy system is its ability to deliver required services without exhausting resources. The primary step to designing such a system is to use existing energy resources efficiently and increase the use of renewable energy (RE) resources. Thus, a shift from non-renewable to renewable energy generation technologies is a top priority in efforts to transition to more sustainable energy systems [4]. There exists a direct relationship between development and energy consumption, yet energy production needs to be increased [5]. There is a large potential for renewable energy technologies such as solar PV, wind energy, biomass energy and micro hydro to meet the needs of rural areas in India [6]. Renewables directly contribute to improved quality of life and reduced environmental pollution [7].

Energy has received widespread attention due to the burgeoning demand from the emerging economies, geopolitical factors and excessive volatility in international crude oil prices [8]. India imports 79% of its petroleum. Relying heavily on imported fuels, it is difficult for India to sustain the use of imported fossil fuels [9]. India has been increasingly reliant on imported coal, creating the necessity of considering domestic sources of energy [10]. Regardless, the growth of renewables in India over the last five years has been impressive. RE-based technologies reduce GHG emissions by displacing energy production from the combustion of fossil fuels which emit large quantities of CO_2. Renewable energy reduces India's carbon footprint [10]. However, for India to continue on its path of economic growth, there are issues which must resolved with improved governance [9].

Power Scenario in India

About 1.3 billion people in the world (or about 1 in 5) lacked access to electricity in 2010; the challenge of providing reliable and cost-effective electrical services remains one of the world's major challenges [11]. In 2012, despite a slowing global economy, India's electricity demand continued to rise. India's electricity demand is projected to more than triple between 2005 and 2030. In the recently released national electricity plan (2012), the Central Electricity Authority projected the need for 350-360 GW of total generation capacity by 2022 [12]. Despite major capacity additions over recent decades, power supplies struggle to match demand.

By the beginning of 2015 India had an installed power generation capacity of over 280,000 MW yet many plants are facing shortages of fuel supplies lowering production. About 53 million of the country's homes lack electricity and many industries depend on diesel generators to meet their electricity requirements [13]. Though the government has announced it would improve fuel supplies to natural gas fired plants, it needs to develop plans and compliment them with relief measures since many plants have struggled due to lack of fuel [14]. The sources of installed power in India are shown in Figure 20-1. Of the total installed capacity of 282,023 MW, 70% is from thermal sources (coal, oil and gas). The rest includes hydro (15%) and other renewables (13%).

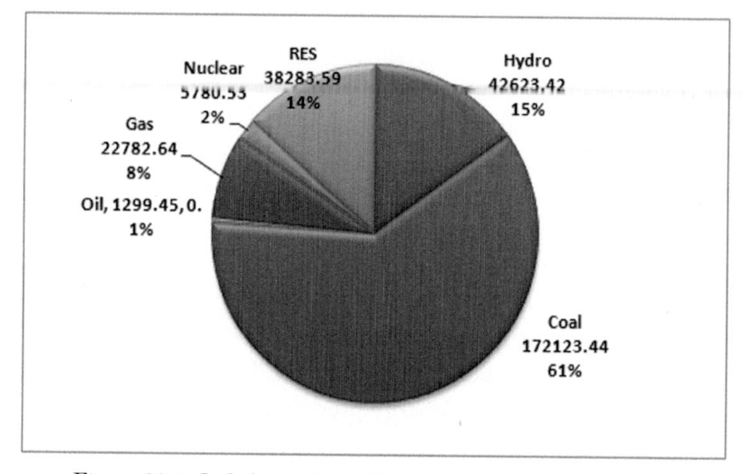

Figure 20-1. India's 2015 installed power capacity (MW) [15].

Promising continuous electricity supplies to all households by 2019, the government also sees an investment potential of up to 250 billion Rs (approximately 4 billion U.S. dollars) for power generation, transmission, distribution and coal mining. India's total power consumption would double to 2 trillion units by 2019 [15]. Though the majority of investment funds would be provided by the private sector, part of the total would come from the government. The government is moving forward with plans for renewables. It hopes to increase solar power generation by 1,000,000 MW by 2022. It also aims to double the installed capacity of wind generation to more than 40,000 MW by 2019 [16]. Wind energy constitutes a major portion of the installed capacity (excluding rooftop solar). Table 20-1 shows the potential and installed renewable energy in India.

THE NEED FOR RENEWABLES AND THEIR POWER SECTOR PENETRATION

Industrialization, urbanization, population growth, economic growth, greater per capita consumption of electricity, depletion of coal reserves, in-

Table 20-1. Potential and installed capacity of renewables in India, November 2015 [13, 15].

Sector	Potential (MW)	Installed (MW)	% Achieved	% of Total Installed
Wind energy	102,788	24,759.3	24.1	64.7
Solar energy	100,000	4,684.7	4.7	12.2
Small hydro power (up to 25 MW)	20,000	4,161.2	20.8	10.9
Biomass and bagasse cogeneration	23,700	4,550.6	19.2	11.9
Waste to energy	3,880	106.6	2.7	0.3
Tidal/wave	Tidal: 8,000-9,000 Wave: 40,000	none	-	-
OTEC	180,000 MW	none	-	-
Geothermal	10,000 MW	none	-	-
Total	270,368	38,261	14.2	100 %

creasing imports of coal, crude oil and other energy sources and the rising concerns about climate change have placed India in a difficult position. Like many developing countries, India must balance economic development and environmental sustainability. One of the primary challenges for India is to alter its present energy mix which is dominated by coal, including a greater share of cleaner and sustainable energy sources. The total renewable energy potential from various sources in India is 270,368 MW [13]. As of November 2015, the total installed capacity from renewable energy, both grid-interactive and off-grid or captive power, was 38,261 MW. Thus, the untapped market potential for renewable energy in India is 232,107 MW [15].

> *"...Why not consider RE to be the main occupant of the "house" and then work out the rest of the system around RE, because RE is the future?"*

This remains a key and critical question [17]. For over 100 years, India's conventional fossil-fueled power plants were the core of power generation systems. Those systems had particular engineering and technical characteristics. For decades, operating and governance institutions were created, designed and operated to support systems with those characteristics. But renewables are different. Capturing the benefits of renewables as "the main occupant of the house" will require institutions to be reengineered, policies redefined, power grids and systems retuned and old habits replaced with new ones.

India's RE program is the biggest and most extensive among the developing countries [15]. Though India's renewable resources are abundant, the output of wind and solar photovoltaic is variable and subject to uncertainty. To capture the benefits, India would need to raise more capital and become more comfortable with managing the variability and uncertainty of RE generation [17].

Wind Energy for Power Generation

The use of fossil fuels increases the emission of pollutants including SO_x, NO_x, and carbon monoxide that have detrimental effects on the environment. Hence, the use of alternate energy sources such as wind, solar and hydrogen is gaining importance [18]. Wind power has proven to be a very effective source of energy due to technological richness, infrastructure and relative cost attractiveness. Renewable energy and especially wind energy does not emit atmospheric CO_2—thus it protects us from global warming.

Wind turbine generators (WTGs) have been used for over 100 years to generate electricity [19]. The use of WTGs for renewable energy has become one of the most viable sources of power generation due to its lucrative economics and ecofriendly impacts [20]. Many companies, institutions, organizations and researchers have reported that wind turbines with higher efficiencies are needed to fulfill energy mandates [21]. Though countries lacking natural reservoirs of fossil fuel stocks need to be careful with their use of alternate energy sources, wind turbine technologies have improved during the last three decades [22]. Developments in the technology of WTGs such as power electronics, aerodynamics and mechanical drive train design have made them a more efficient way of producing electrical energy [23,24].

Since 2010, Asia has led the growth of wind energy, as wind energy installations in the region have outstripped both North America and Europe. While China and India have been the main drivers of growth, the projected investments in wind projects in the rest of Asia are expected to exceed $50 billion (U.S.) between 2012 and 2020 [25]. According to the report by the Intergovernmental Panel on Climatic Change (IPCC), wind energy will have a major role in RE electricity generation by 2050 [26]. Its contribution will be about 80% of the world's energy demand.

Progress of Wind Energy in India

Wind energy is increasingly accepted in India as a major complementary energy source for securing a sustainable and clean energy future. [27]. Demonstration wind energy projects were started in 1985. Five wind farms were developed in 1986 with a generation capacity of 3.3 MW. By 1989, India had 10 MW of total installed wind capacity, all from governmental demonstration projects [28]. The first commercial wind power generation projcct started in Tamil Nadu in 1990. By 1992, many wind turbines were installed in coastal areas of Tamil Nadu, Gujarat, Maharashtra and Orissa. Afterwards, wind power production in India began to grow, mostly by using land-based systems.

Asia's largest and tallest wind turbine was built in India in 2004. Installed wind energy capacity in India increased from 41.3 MW in 1992 to 22,465 MW by the end of 2014 [29]. Among the Indian states, Tamil Nadu experienced the largest growth in wind power energy development and was producing 5,867.2 MW by 2011 and 7,254 MW at the end of 2014 [30]. Other states achieved significant wind power growth by the end of 2014—Maharashtra 4,024.7 MW, Gujarat 3,405.7 MW and Karnataka 2,331.3 MW. India has the wind energy potential to produce 49,130 MW at the 50m height and 102,788 MW at the 80m height above the ground level. By November 2015, India had a total installed wind power generation capacity of 24,759.3 MW [30].

India's renewable energy policies which predominately support onshore systems, need improvement. European countries, most notably the UK and Germany, have adopted effective offshore policies [24]. Offshore wind power policies should be developed suitable to India's situation.

Wind energy initially generated interest from the scientific perspective rather than for its potential of meeting increasing demand for electricity. It is gaining market share because of its usefulness for decentralized energy generation and distributed generation. The growing gap in electricity demand and supply, greater environmental awareness and the attention to decentralized energy supplies made wind energy important to meeting India's growing demand for energy [31]. In India's 12th plan period from 2012-17, renewable energy capacity was envisaged by the Central Electricity Authority to grow to 32 GW [32]. Nearly 20 GW was to come from wind and 10 GW from solar. India's new government plans to scale up installations to almost 140 GW by 2020—100 GW from solar and 40 GW from wind. This outlook presents new challenges. According to industry estimates, if all barriers to wind farm development were resolved, nearly 40 GW (6-7 GW annually) of wind power installations could be possible by 2020 [33]. These capacity additions necessitate a detailed and critical examination of the barriers faced by renewables.

Karnataka: Renewable Energy Scenario and Rural Electrification

Karnataka has about 30,000 MW of estimated RE potential, making it one of the country's top five RE-rich states. Table 20-2 shows the sources of installed electrical generation capacity in Karnataka which totals about 15.0 GW with 5.1 GW or 30% from RE sources [34]. The state agency, responsible for RE development under the purview of the energy department is the Karnataka Renewable Energy Development, Ltd. (KREDL). As a facilitator between industry, finance, government and technical experts, KREDL works to increase the deployment of RE in the state.

Table 20-2. Karnataka: source of installed capacity [34].

Sl. No	Source	Capacity in Nov 2015 (MW)
1.	Hydro	3,773
2.	Thermal	27,20
3.	Diesel	108
4.	CGS	2,169
5.	NCE Source	5,082
6.	IPP	1,200
	Total	15,052

The biggest challenge for utility-scale projects in Karnataka has been project implementation (i.e., progressing from allocation to actual commissioning). Although 60% (**19,200.4** MW) of the state's RE potential has been 'allocated' by KREDL, only about 15% (**5,082** MW) has been commissioned [35,36]. Table 20-3 indicates the available and allotted potentials plus the commissioned capacity of the types of renewables in Karnataka.

The allotted capacity of wind energy generation to various groups equaled 13,245 MW as of 2015 with a commissioned capacity of only 2,686 MW (20.3%). A total of 2,623 MW was cancelled and the balance allotted capacity remaining to be commissioned stands at 7,935 MW [34].

The cumulative progress of wind energy installations for major Indian states is shown in Figure 20-2. The installed capacity for most of the states has doubled over last six years.

Table 20-3. Achievements in the renewable energy sector, November 2015 [37].

RE Source	Available Potential (MW)	Allotted Capacity (MW)	Commissioned Capacity (MW)
Wind	13,983	13,245	2,686
Small hydro	3,000	2,956	785
Cogeneration	2,000	1,677	1,176
Biomass and waste to energy	1,135	400	113
Solar	10,000	1,100	84
Others	-	-	249
Total	30,118	19,200	5,082

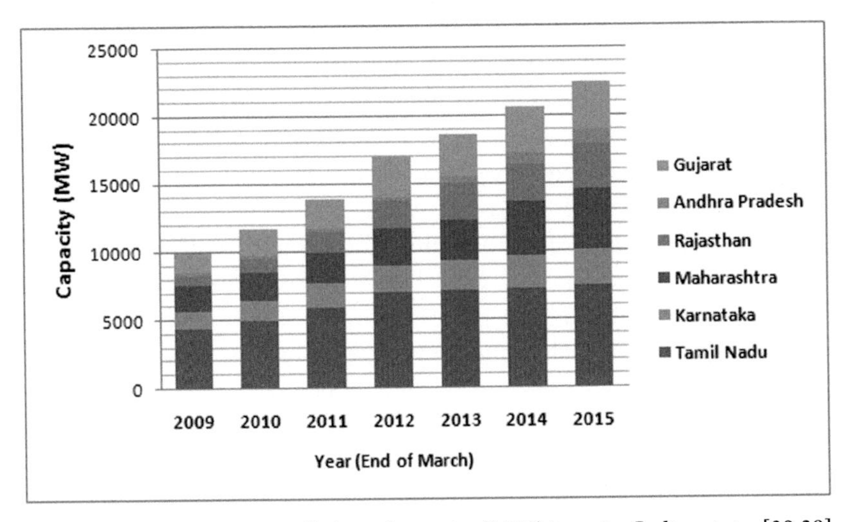

Figure 20-2. Cumulative installed wind capacity (MW) in major Indian states [38,39].

Electricity Poverty in Rural Areas (Karnataka)

Access to quality electricity services continues to be a major issue in the state of Karnataka, similar to rest of the country. Electricity access is not only essential for households, but is also important basic infrastructure for hospitals, schools and industries. Supported by several years of governmental schemes, electrification in Karnataka has increased between 2001 and 2011 [36]. Almost all (99.95%) of villages in Karnataka are now electrified (Table 20-4) as defined by the national rural electrification program, Rajiv Gandhi Grameen Vidyutikaran Yojana (RGGVY). A village is deemed electrified if public places like schools, Panchayat offices, health facilities, dispensaries, community centers and 10% of all households are electrified. While the village might be grid connected, this status does not guarantee that all households are.

Table 20-4: Village electrification in Karnataka [32,36].

Total Inhabited Villages as per 2011 census	Villages Electrified as on 31-03-2014		Un-electrified Villages (March 2015)
	Number	Percentage	
27,481	274,684	99.95	13

Even with a high electrification achievement, the 2011 census mentions that 8.6% of the households in Karnataka use kerosene for their lighting, implying that several households lack access to electricity even in officially electrified villages [11,37]. There remains a gap in rural and urban access to electricity, with 96.4% of urban households using electricity as a source of lighting compared to 86.7% in rural areas.

There are a number of issues with electricity access in rural areas. Electrified villages have erratic and unreliable supplies. Secondly, there are remote hamlets, less than 100 households, which are not electrified nor included in the RGGVY plan [41]. Finally, the cost of delivering power from the grid, which includes generation, transmission, and distribution of electricity from a centralized coal thermal power plant is variable.

Kerosene is used in about 12% of the rural households for lighting compared to approximately 3% for urban households [40]. Kerosene combustion emits toxic fumes, causing eye and respiratory ailments and higher associated health costs [32]. Access to electricity improves health and reduces these costs.

Karnataka is home to the ecologically sensitive Western Ghats. For remote hamlets in these and similar locations, it is technically and economically infeasible to extend the electric grid. Small-scale renewable energy technologies (RETs) can be economically attractive compared to electrical grid extension and provide electricity to these rural areas [32]. Studies suggest that RE resources can meet a substantial portion of the energy demand with currently available technologies [35,36]. This may become a reality only after issues that prevent greater market penetration of RETs are properly addressed. While the government is promoting RETs, renewables have failed to emerge as prominent competitors to conventional energy technologies. This implies that there are barriers to implementing renewable technologies. For RETs to successfully penetrate the power sector markets, the involvement of various stakeholders in the identification of barriers is extremely important [42]. The identification and perceptions of barriers and barrier-removal measures vary across the stakeholders. Unless these are addressed, implemented policies may not function as intended.

IDENTIFYING AND ASSESSING BARRIERS

The need to enact policies supporting renewables is often attributed to a variety of "barriers" or conditions that prevent greater investments. The barriers create economic, regulatory, or institutional disadvantages for developing renewable energy relative to other forms of energy supply. Barriers include subsidies for conventional forms of energy, high initial capital costs coupled with lack of fuel-price risk assessment, imperfect capital markets, lack of skills or information, poor market acceptance, technology prejudice, financial risks and uncertainties, high transactions costs, and a variety of regulatory and institutional factors [43]. Barriers clearly exist that prevent energy-efficient technologies from being more widely utilized. The major barriers are different for renewable energy technologies. Policy barriers are the key barriers which have given rise to sets of related barriers such as financial and institutional barriers. High investment costs, lack of performance guarantees and lack of access to information are other key barriers [6].

Meyers [44] and UNFCCC [45] outline the following typical types of barriers:

- Institutional—lack of legal and regulatory frameworks, limited institutional capacity and excessive bureaucratic procedures.
- Political—instability, government intervention in domestic markets (for example, subsidies), corruption and lack of civil society.
- Technological—lack of infrastructure, lack of technical standards and institutions for supporting the standards, low technical capabilities of firms and lack of a technology knowledge base.
- Economic—economic instability, inflation, poor macroeconomic conditions and disturbed or non-transparent markets.
- Information—lack of technical data, financial information, and demonstrated capabilities.

- Financial—lack of investment capital and financing instruments.
- Cultural—particular consumer preferences and social biases.
- General—insufficient intellectual property protection and unclear arbitration procedures [46].

Identifying Barriers to Wind Energy Development

Barriers can be defined as factors that inhibit or hinder technology transfer or adoption. This study considers wind energy generation systems (WEGS) as de-centralized renewable energy generation technologies (DCRE). The analysis of barriers and their rankings was performed by completing a comprehensive review of published academic literature and technical reports on energy systems and listing the major barriers noted as hindering the diffusion of renewable energy technologies. The full list of relevant barriers was then refined to a shorter list. Finally, the barriers for implementation of wind energy were categorized into five broad groups. The details of capacity and number of WTGs at the study cluster are provided in Table 20-5.

Table 20-5. Details of the wind energy cluster area for study [34].

Capacity of WTGs in kW												Total no. of WTGs	Total capacity MW
250	400	500	800	850	1,250	1,500	1,600	1,650	1,700	2,000	2,100		
Number of WTGs													
14	5	4	178	31	8	10	20	5	29	9	5	318	319.3

The Belgaum district wind farm cluster consists of 318 WTGs with a total installed capacity of 319.3 MW. Their installed capacities range from 250 kW to 2 MW. The district of Belagavi is among the top five districts with the highest potential for wind energy generation.

Our study's objectives were to: 1) identify the barriers to WEGS adoption; 2) specify the major factors that need to be addressed from a policy perspective; and 3) assess the significance of removing the barriers. For the analysis, a questionnaire-based research survey was conducted. Opinions and judgments were collected from wide variety of experts and stakeholders. All respondents were knowledgeable about the power sector and were familiar with RE generation technologies and the barriers hindering their widespread diffusion and adoption. The stakeholders included representatives from households, industrial firms, wind energy developers and policy development. The survey first gauged the respondents' perceptions as to the most significant barriers from a list of major factors (e.g., awareness, information, cost, policy regulations, etc.) and then delved into each factor to determine how it posed obstacles to WEGS.

Information on the state of the wind energy technologies, various policies, and data on factors effecting the penetration of wind energy were collected. The wind energy farms in Belagavi district cluster (Chikkodi, Saundatti, Belagavi and Ramdurg) in the state of Karnataka were selected for the study. Policy makers in the state's energy department, Karnataka State Electricity Board (SEB), KREDL, and energy researchers and practitioners, were consulted to obtain their views on barriers to the diffusion of wind energy technologies. The survey was conducted during the years 2014-2015. The information collected was used for designing the survey's questionnaires. The questionnaires were structured to elicit information on the perceived benefits of WECS, awareness about their costs and benefits, and queries related to barriers such as lack of sufficient information, high initial cost, low electricity tariff, lack of incentives, maintenance problems, and lack of a suitable agency to deal with problems. The respondents were asked to identify the barriers and rank their importance on a given scale. Rankings were normalized and the weighted ranking was determined for each barrier. The barriers to development of RE in India are presented in Table 20-6. Most are specific to a technology, policy, site or region [4,42,47].

The major barriers vary for the different renewable energy technologies. Policy barriers appear to be key barriers, giving rise to a group of related barriers such as financial and institutional barriers. Other key barriers are high investment costs, lack of guaranteed performance and lack of access to information. Finally, the barriers were grouped into five broad categories: 1) policy, political and regulatory barriers (PPRB); 2) technological and geographical barriers (TAGB); 3) economic and financial barriers (EAFB); 4) organizational and institutional barriers (OAIB); and 5) knowledge, information and awareness barriers (KIAB). A brief explanation of these barriers is presented next.

Policy, political and regulatory barriers: There are no sufficient government regulations or incentives to stimulate the adoption of RETs by businesses or industries. Policy support from governmental bodies is lacking in terms of tax benefits, subsidies, depreciation, power distribution, interconnection standards, licensing requirements, environmental and

Table 20-6. Classification of energy barriers [12].

Barrier	Description
Policy and regulatory barriers	Policy framework for RE
	Provision of accelerated depreciation for wind developers
	Regulatory framework for promotion of RE
Institutional barriers	Inter-institutional coordination
	Single window clearance system
Fiscal and financial barriers	Budgetary constraints
	Financing of RE projects
Market-related barriers	Level playing field and market for RE
	Inadequate market prices
	Transmission network
	High equipment costs
	Inputs for RE plants
	Absence of serious developers
Technological barriers	Technology risk
	Absence of minimum standards
	R&D and manufacturing capabilities
	Lack of local technology
Information barriers	Lack of skilled manpower
	Lack of information and awareness

social considerations. Others include: a) lack of explicit national policy at end-use level; b) incomplete transition to cost-based electric tariffs for most residential and some industrial customers; c) poor availability of credit for RETs; and d) lack of modern management skills in energy development agencies. These contribute to many renewable energy technologies in India remaining in the development stage.

Technological and geographical barriers: These include demonstration of project, geographic externalities, wind resource assessment, intermittency or dispersed nature of potential, back up of technology, mismatching load duration curve and grid instability. Wind energy technologies are considered proven and low risk. However, there may be site specific issues which create technical application risks and provide rational reasons for project rejection. Other barriers include: a) lack of standardization in system components resulting from the wide range in design features and technical standards; b) absence of long-term policy instruments that resulted in manufacturing, servicing and maintenance difficulties for wind turbines; c) mismatches between locally manufactured components and imported parts; d) reliability of the overall system and absence of effective servicing and maintenance networks; e) inadequate user-training; and f) lack of co-ordination among research groups, academic institutions and the wind industry.

Economic and financial barriers: These include payback period requirements, financing or funding mechanisms, incentives, discount rates, uncertain economic benefits, market demand, nature of competition and transmission costs. Lack of adequate financial resources has been a chronic problem for establishing wind energy projects. The Indian Renewable Energy Development Agency, Ltd. (IREDA) has a crucial role in supporting the country's wind energy projects. There is a need to create financial institutions who will support wind projects.

Organizational and institutional barriers: These include institutional arrangements through nodal agencies, the importance of the technology, disregarding the options for energy management, small markets, unfair competition, internalization of generating costs, lack of knowledge, lack of infrastructure, and lack of local association activities. Institutional barriers constitute a real constraint, not only to the development of RE sources such as wind but also to their wider dissemination. RE technologies that are relevant to developing countries like India are available. While improvements may be required in individual cases, especially to reduce production costs, the hardware for harnessing wind energy is known and reliable.

Knowledge, information and awareness barriers: These include lack of awareness of concepts, key terms, media exposure, limited databases, availability of training, partial knowledge about the technologies and their benefits and government policies. It is believed that the technologies are not adopted due to lack of information or customer knowledge, or a lack of confidence in obtaining reliable information. India lacks public capacity for disseminating information. There is hardly any consumer knowledge acquisition capability (software or hardware) that is readily available and easily accessible. Given these circumstances, information collection and processing, consumes time and resources which can be difficult for small businesses. Table 20-7 presents a social, technological, economic and political (STEP) framework for factors that strongly influence India's wind power development [48].

Considering the parameters in Table 20-7, an analysis of the survey data was performed. The data provided valuable insights into stakeholders' awareness about RETs and barriers to their adoption and diffusion. The opinions of policy analysts about the actions needed to remove the barriers were also revealed [4,42].

Table 20-7. Framework of factors influencing wind power development [48].

Social	Technical	Economic	Political
NIMBY concerns (Not in my Back Yard)	Stochastic nature of wind power	Externalities not internalized	Political conflict over optimal electricity mix
Level of civic activism	Multi-stakeholder grid management	Other competing alternative technologies	Level of fossil fuel industry opposition
Geographic hurdles	Logistical "bother"	Subsidies to traditional technologies	Diffused alternative energy support
Market information asymmetry	Distance to grid	Insufficient renewable energy subsidies	Energy efficiency initiatives prioritized
Social complacency	Inadequate R&D to improve storage	Long-term fossil fuel purchase commitments	Complacency regarding CO_2 reductions
Electricity price sensitivities, concerns over community impact	Underestimated potential	Market players lack investment incentives; government budget limitations; national advantage in other energy resources	Vertically integrated utility monopoly; weak adjoining grid coordination; lack of R&D support

Overview of AHP Methodology and Barrier Analysis

The barriers for implementing wind energy are broadly categorized into five groups. The ranking of barriers is based on four different criteria namely: 1) barrier removal impact on technology adoption (BRITA); 2) financial difficulty in removal of barriers (FDROB); 3) barrier removal impact on socio-environmental benefits (BRISEB); and 4) barrier removal impact on techno-economic performance (BRITEP). All criteria were measured using Thomas Saaty's nine-point scale as shown in Table 20-8. The weights for each of the criteria were based on their importance as perceived by the respondents.

The foundation of an analytic hierarchy process (AHP) is a set of axioms which carefully delimits the scope of the problem environment. It is based on a well-defined mathematical structure of consistent matrices and their associated right Eigen vector's ability to generate true or approximate

Table 20-8. Saaty's scale for pair-wise comparison [49].

Numerical Ranking	Definition of Verbal Judgment
1	Equal importance
3	Moderate importance
5	Essential or strong importance
7	Very strong importance
9	Extreme importance
2,4,6,8	Intermediate values (between two adjacent judgments)
1/2, 1/3...1/9	Reciprocals

weights [49]. The AHP methodology compares criteria, or alternatives with respect to a criterion, in a natural pair-wise mode [49]. The three steps of the AHP methodology are: 1) identifying barriers and structuring a hierarchy prioritization model; 2) constructing a questionnaire and collecting data; and 3) determining the normalized weights for each barrier category and each specific barrier. Opinions from different stakeholders were collected through carefully designed questionnaires and then synthesized and analyzed using the AHP. Consistency checks for the pair-wise comparison matrix were performed by calculating the consistency ratio which should be less than 10 [50].

1. Calculate the Eigen vector or relative weights and λ_{max} for each matrix of order n.
2. Compute the consistency index for each matrix of order n by: $CI = (\lambda_{max} - n)/(n-1)$
3. The consistency ratio is then calculated using the formula: $CR = CI/RI$

Table 20-12 presents the pair-wise comparisons of the selected four criteria with respect to the objective. Values indicated in Table 20-12 represent pair wise comparisons in the selected clusters which were obtained by calculating the geometric means of all individual pair-wise comparisons of (XX) participants from the wind farm clusters. It also shows the priority vector (weights of each criterion) in the clusters computed by a process of averaging over the normalized columns as per the AHP methodology. The priority column in Table 20-9 is obtained by dividing each matrix element by the sum of respective column elements (normalization of column) and then by calculating the arithmetic mean of each row.

From Table 20-9 it can be inferred that the various stakeholders give maximum priority to BRITEP and BRISEB. It is obvious that the stake holders insist on the advancement of the technical and economic performance of wind farms. Improving socio-environmental benefits and outcomes after the technical benefits is also important. FDROB scores a reasonable priority ahead BRITA which has the least weight. Considering a very low weight for BRITA, it is ascertained that the stakeholders are not overly concerned about barrier removal impact on technology adoption and are more concerned about improving the technical and economic performance. Improvements in technical and economic performance benefit wind farm development.

Table 20-9. AHP- pair-wise comparison of criteria.

CRITERIA	FDROB	BRITA	BRISEB	BRITEP	PRIORITY
FDROB	1.00	1.25	0.80	0.22	0.132
BRITA	0.80	1.00	0.40	0.18	0.114
BRISEB	1.25	2.50	1.00	0.33	0.163
BRITEP	4.55	5.56	3.00	1.00	0.592

Next the relative weights for individual barrier groups under each of the four criteria were obtained. For this purpose, the pair-wise comparisons of five barriers with respect to each of the four criteria were performed using AHP. The corresponding results of AHP analysis of barriers under the criteria (FDROB) are provided in Table 20-10. Similar to the previous step, column normalization followed by computing the arithmetic mean of each row yields the respective weights of the barriers under a particular criterion. This above mentioned procedure was extended to the remaining criteria to obtain the relative weights.

Table 20-10. Pair-wise comparision of barriers with respect to criteria.

Barrier removal impact on Technology Adoption (BRITA)						
Barrier	EAFB	PPRB	OAIB	KAIB	TAGB	FRW
EAFB	1.00	1.59	5.26	1.16	1.35	0.293
PPRB	0.63	1.00	4.55	2.04	0.96	0.255
OAIB	0.19	0.22	1.00	0.65	0.51	0.076
KAIB	0.86	0.49	1.55	1.00	2.44	0.210
TAGB	0.74	1.04	1.96	0.41	1.00	0.166
CR = 7.63%						

Table 20-11 presents the values for the weight of the criteria under each barrier, the priorities of the criteria, and the final composite weights for each barrier group based on the given rankings. To determine the final composite weight for each barrier group in the clusters, these local weights are multiplied by criteria priority and then aggregated. From these results it can be opined that PPRB leads OAIB and TAGB with a marginal difference followed by EAFB and KAIB ranked last.

In Table 20-11 it can be observed that, given the criterion FDROB, the PPRB barrier group has the maximum priority. This needs to be addressed for implementation of WTGs. In the same critrion, PPRB is followed by TAGB which is the second most important barrier. Again both EAFB and AIOB obtained lower values. The lowest ranking barrier group

Table 20-11. Composite weight and barrier ranking in wind farm cluster.

Barrier Group	Criterion Weights				Composite Weight - Final	Rank
	FDROB	BRITA	BRISEB	BRITEP		
EAFB	0.120	0.293	0.112	0.062		4.00
PPRB	0.530	0.255	0.497	0.466	0.132	1.00
OAIB	0.096	0.076	0.213	0.254	0.114	2.00
KAIB	0.027	0.210	0.045	0.061	0.163	5.00
TAGB	0.227	0.166	0.133	0.156	0.592	3.00

in this category was KAIB.

Similarly, given the weights of barriers related to criterion BRITA, the wind farm has ascribed a maximum value to EAFB. PPRB and KAIB obtained values that are close to each other. The barrier group of OAIB had less importance with respect to the criterion followed by TAGB. Considering BRISEB and BRITEP, the wind farm clusters judged PPRB the highest, ahead of all other barriers. OAIB and TAGB were second and third with respect to both criterion BRISEB and BRITEP with KAIB obtaining the least weight.

Barriers and Ranking by Weighted Average Score

Each respondent from the different sectors was given a detailed examination of the context of the issues that are involved together with the list of barriers. They were asked to indicate the importance of each barrier to them on a five-point scale (Table 20-12). The '1' on the scale indicated 'extremely important' (indicating maximum impact of removing a barrier on adoption of technology), 3 on the scale indicated 'important' (average), and '5' indicated the 'least important' (least impact of removing a barrier on adoption of technology) [4].

Table 20-12. The five-point scale for ranking [4].

Scale	1	2	3	4	5
Importance of barrier	Extremely important	Moderately important	Important	Less important	Least important

The ability to make qualitative distinctions is represented by attributes, such as equal, weak and strong or otherwise stated, rejection, indifference and acceptance. Each of these can be subdivided into low, medium and high indicating nine scales of distinction. The weighted average score for each barrier was determined by using normalized weights. The barriers were ranked according to their weighted average scores. These final ranks indicate the relative importance of the barriers from the stakeholders' perspective. The weights given to the scale of importance (1-5) of the stakeholders were 5/15 points (to importance 1), 4/15, 3/15, 2/15 and 1/15 (to importance 5), the total of weights being 1. A negative response was assigned a 0 weight, indicating that the barrier is not at all important to the respondent. These weights were multiplied by the number of responses for each barrier and the weighted averages were calculated. The barriers were then ranked based on these weighted averages [4,42].

Barrier Rankings: Wind Energy Developers

Information-related issues emerged in the category of most important barriers. Economic barriers were of major concern to the wind energy developers which indicated difficulties in accessing financing. The regulatory barriers resulted in problems concerning land acquisition and problems in obtaining clearances whereas the lack of infrastructure added to the cost. Ever changing governmental policies created uncertainty and increases in associated costs. In addition, economic barriers were perceived to result from the high cost of development, inadequate incentives, and delays in receiving funds from governmental agencies. Changing depreciation methodologies were also seen as a factor causing increased wind power costs. Also, the prices paid by utilities for power was quite low. Failure of the existing institutions to deliver the results needed is evident from this analysis. Various organizations including governmental agencies were considered to be insensitive to the needs of developers.

Overall Barrier Ranking in Wind Farm Clusters

For the wind farms, policy, political and regulatory barriers (PPRB) are ranked first followed by organizational and institutional barriers (OAIB), technological and geographical barriers (TAGB), economic and financial barriers (EAFB), and knowledge awareness and information barriers (KAIB) in that order. It is clearly established that PPRB is the single greatest barrier group affecting wind energy technology (WET) implementation. This ranking identifies the following outcomes:

- No barrier group is negligible. Their weights support the assumption of our study that all considered barrier groups are relevant and significant in the wind farm clusters.

Table 20-13. Ranking of barriers.

Barrier	**Ranks and responses (number)**					**Weighted average**
	1	2	3	4	5	
Difficulty in getting clearances	2	1	1	2	3	0.133
Poor Infrastructure facilities	3	2	3	2	1	0.206
Inadequate incentives	3	2	2	2	1	0.189
Hurdles in land acquisition	4	4	3	2	0	0.272
Inconsistency in government policies	0	1	1	2	2	0.072
Long delay in receipt of payments	0	1	1	1	2	0.061
Miscellaneous and others (local)	0	1	1	1	3	0.067

Table 20-14. Ranking of barriers groups.

Barrier	**Ranks and responses (number)**					**Weighted Average**
	1	2	3	4	5	
Economic and financial barriers	1	2	0	1	4	0.106
Policy, political and regulatory barriers	6	4	5	3	1	0.378
Organizational and institutional barriers	2	4	5	1	2	0.250
Knowledge, information and awareness barriers	1	0	1	3	2	0.089
Technological and geographical barriers	2	2	1	4	3	0.178
	12	12	12	12	12	1.000

- PPRB and OAIB are the most important brriers that need to be considered to enhance the implementation of wind energy.
- The least weight for KAIB indicates that there is considerable awareness and information about WETs. These concerns are not significantly hindering WET implementation.
- A relatively low weight for EAFB underscores the point that investors are ready and willing to invest in WET provided the appropriate policies and institutional networks are provided.

The implementation of wind energy initiatives must address the policy and regulatory aspects of wind energy power plants. There are multiple reasons for the highest ranking of the PPRB barrier group in the wind energy clusters. First, most stake holders feel that obtaining approvals from government agencies or organizations is very difficult. Secondly, requiring that developers sell their electricity to the State Electricity Board (SEB) through power purchasing agreements (PPAs) is a major policy hurdle. Recently, government orders for allotment and enhancement were issued subject to this condition. This condition should be removed for captive or third party sales, since it should be the investor's decision as whether or not to sell power to the SEB or to opt for wheeling and banking. Thirdly, policies for fixing the deadlines for project completion, capacity enhancements, and repowering of wind farms are needed for effective implementation of WETs.

Apart from PPRB, addressing the OAIB and the TAGB are also important. It is generally felt that any policy and regulatory initiatives will not be able to effectively address WET implementation unless the critical problems of technology, geography, finance, institutional arrangements and organizational issues are addressed on a prioritized basis. EAFB is fourth in the ranking, indicating that stake holders do not rate this barrier highly. In other words, investors are wiling to invest in wind technology provided other hurdles are cleared. The ranking also shows the awareness and information barrier is not posing a serious challenge in WET implementation.

SUMMARY OF RESULTS

The objective of this study was to prioritize the five barrier groups based on all four criteria using the AHP. An initial ranking of barriers based on each individual criterion indicates the need for a multi criterion approach. Based on the 'impact of barrier removal' alone, EAFB has the highest ranking for wind farms while PPRB and KAIB ranked second and third for wind energy implementation. Further, OAIB was the lowest ranked barrier. For 'financial difficulty in barrier removal of barrier' as a criterion for ranking, the PPRB received the top ranking, while TAGB and EAFB ranked second and third rank for wind energy implementation. Further, OAIB ranked fourth while KAIB ranked last. Considering the potential of 'barrier removal impact on techno-economic performance' PPRB again tops the ranking. It was followed by OAIB and TAGB, while both EAFB and KAIB were equally strong and shared the lowest ranking. OAIB again ranked second. With the slight difference between TAGB and EAFB, the former ranked third while the later ranked fourth. KAIB had the lowest ranking as a barrier.

This approach was based on individual criterion while ignoring others. It is inadequate for prioritizing the barriers since it does not provide a holistic view. For example, the barrier KAIB, which obtained the last rank using three criteria, has considerable strength in terms of the impact of barrier removal on adoption of technology and must not be ignored. Therefore, prioritizing the barriers by considering the effects of all the influencing factors simultaneously is improtrant. It is for this reason that a multicriteria decision tool (the AHP) was adopted in this study. The results of the weighed averages present a similar scenario.

Policy recommendations: The state of Karnataka is viewed as the key enabler for promoting renewable and energy-efficient technologies. There is an urgent need for a public policy to invest in these solutions. A transition from fossil-based conventional fuels to renewables-based energy systems would have to rely largely on successful development and diffusion of these technologies. The RETs could become increasingly competitive through cost reductions resulting from technological and organizational development. Different technologies vary widely in their technological maturity, commercial status and integration aspects. Policies aimed at accelerating renewable energy must be sensitive to these differences.

Key issues related to faster diffusion of RETs include a strong need to improve the reliability of technologies and introduce consumer-desired features (in terms of services and financial commitments) in the designs and sales packages. Including renewable energy strategies in development programs will promote decentralized applications. Renewable energy strategies should be included in the energy sector regulatory framework. Governmental policies should encourage more private participation and industry collaboration in research and development (R&D) for rapid commercialization of RETs. Most renewables are not yet competitive with fossil technologies, especially for power generation purposes. Further commercialization will demand intense R&D efforts. Renewables need to gain the confidence of developers, customers, planners and financiers.

The lack of knowledge about the technologies and their applications is the most important barrier to the use of renewables. The absence of a reliable offer for renewable energy equipment and the frequent lack of arrangements for the long-term management and operation of renewable energy projects are barriers to the penetration of renewables in rural markets. Both barriers limit the long-term sustainability of renewable energy projects.

- Two basic measures have been proposed to guarantee the long-term sustainability of renewable energy projects: 1) training for operation and maintenance (O&M) technicians; and 2) standardization and certification of renewable energy equipment to warrant its quality.
- A solid management model is the key for the long-term sustainability of rural electrification projects. The participation of local people in O&M tasks, and a tariff system that covers the costs associated to those tasks, are basic requirements for a sound management model.
- Crucial elements for the sustainability of rural electrification projects are the quality, reliability, and warranty of the equipment installed, and the service provided by the equipment supplier. An assessment of the local characteristics of the project—including local energy needs, available energy resources, capacity and willingness to pay—is important for the success of any rural electrification project.

CONCLUSIONS

The enormous potential of renewable energy resources is sufficient to meet the world's demand for energy many times over. Renewables can reduce local and global atmospheric emissions, enhance diversity in energy supply markets and have potential to contribute to long-term sustainable energy supplies. Many countries including India have established national targets for the long-term development of renewables and are integrating clean energy into national regulatory frameworks. Communities, individual consumers and investors are also actively contributing to and participating in renewable energy development plans. India, despite being a pioneer in the Asian region in formulating and implementing innovative policies for promoting renewable energy technologies, has experienced only slow to moderate growth in the use of alternative technologies. This is mainly due to the presence of variable barriers to the promotion of renewable technologies. Among them, policy-related barriers appear to be the major ones, particularly the financial and institutional barriers. Therefore, an emphasis on supportive policy initiatives by the government is essential for overcoming barriers to the promotion of renewables in India.

RETs have potential to provide commercially attractive options to meet specific needs for energy services, particularly in rural areas. They create new employment opportunities, and offer opportunities to manufacture equipment locally. To achieve this goal, a number of barriers must be overcome to increase the market penetration of renewables. A formal survey was used to analyze the major barriers to wider adoption of wind energy technologies. The results of the expert survey can be summarized as follows: the dominant barriers to wider adoption of wind power are financial or infrastructure hurdles, institutional constraints and deficiencies in government policy.

Currently, about 9.6% of every 100,000 households in rural Karnataka lack access to electricity even though 99.5% of villages are electrified. Several households already in electrified villages are not grid connected and those that have grid access are plagued with poor electrical quality and supply availability. The state needs to encourage the growth of small-scale rural electrification projects by making clear, comprehensive guidelines for the market-based implementation of RE projects in un-electrified and under-served areas.

India's renewable resources are abundant, but the output of wind and solar photovoltaic is variable, and in the case of wind subject to uncertainty. To capture the benefits, India would need to raise the necessary capital, and become comfortable with managing the variability and uncertainty of renewable energy generation. To help policymakers identify the new approaches, a stakeholder-driven analysis of the barriers to the rapid deployment of RE was performed. The resulting process and its findings have relevance. In the present scenario, the government of India has enhanced its aspirations multifold—amending them from 20 GW of solar power (by 2022) to 100 GW (by 2019) and adding 15 GW of wind power (during 2012-17) to an additional 40 GW (by 2019).

Acknowledgements

The authors wish to thank those who have contributed their valuable time to respond to our questionnaire. The authors express their gratitude to the staff at the wind sites for sharing information and technical data. The authors also wish to thank the Visvesvaraya Technological University (VTU) Belagavi, the management of KLS and the principal of KLSGIT for allowing us to perform this research.

References

[1] Frauke, U. (2008). Sustainable energy for developing countries: modeling transitions to renewable and clean energy in rapidly developing countries. Thesis, ISBN 978-90-367-3703-6.

[2] Vilma, R. (2007). Community stakeholder management in wind energy development projects: a planning approach. Thesis.

[3] Ricardo, F. (2003, February). Removal of barriers to the use of renewable energy sources for rural electrification in Chile. Thesis, Engineering Systems Division. Massachusetts Institute of Technology.

[4] Sudhakara, R. and P. Balachandra (2007, November). Commercialization of sustainable energy technologies, Indira Gandhi Institute of Development Research. Mumbai, India. WP-2007-018

[5] Bilgili, M. and B. Sahin (2010). Electrical power plants and electricity generation in Turkey. *Energy Sources*, part B, 5, pages 81–92.

[6] Balachandra, P., Rao, K., Dasappa, S. and N. Ravindranath (2005). Ranking of barriers and strategies for promoting bioenergy technologies in India. Centre for Sustainable Technologies, Indian Institute of Science, Bangalore, India and the Sustainable Energy and Environment Division, UNDP. New Delhi, India.

[7] Usha, R. and V. Kishore (2009). Wind power technology diffusion analysis in selected states of India. *Renewable Energy*, 34(4), pages 983–988.

[8] Narula, K. (2014). Is sustainable energy security of India increasing or decreasing? *International Journal of Sustainable Energy*, 33(6), pages 1,054–1,075.

[9] Krithika, P. and S. Mahajan (2014, March). Governance of renewable energy in India: issues and challenges. The Energy and Resources Institute TERI-NFA Working Paper Series No.14.

[10] Reenergizing India (2014). Policy, regulatory and financial initiatives to augment renewable energy deployment in India: Climate Parliament, IDAM Infrastructure Advisory Private Ltd. New Delhi, India.

[11] Bhattacharyya, S. (2012). Energy access programmes and sustainable development: a critical review and analysis. Centre for Energy, Petroleum and Mineral Law and Policy, University of Dundee.

[12] Planning Commission, Government of India (2013). Twelfth five-year plan, India 2012-2017, sustainable growth, economics and social factors. Sage Publications India, Ltd. New Delhi, India.

[13] GREEN (2014, June). Global Renewable Energy Summit. India renewable energy status report. Green summit. Bangalore, India.
[14] Rural Electric Corporation (2013). www.recindia.co.in.
[15] Ministry of New and Renewable Energy, Government of India (2015). National offshore wind energy policy. mnre.gov.in.
[16] Ministry of Power, Central Electricity Authority (2015, July). Executive summary. New Delhi, India.
[17] National Institution for Transforming India, Government of India (2015, February). Aayog: report on India's renewable electricity roadmap 2030. Toward accelerated renewable electricity deployment. Executive summary. New Delhi, India.
[18] Demirbas, A. (2001). Biomass and the other renewable and sustainable energy options for Turkey in twenty-first century. *Energy Sources*, 23, pages 177–187.
[19] Gokcek, M. and M. Sedar (2009). Evaluation of electricity and energy cost of wind energy conversion systems in central Turkey. *Applied Energy*, 86, pages 2,731–2,739.
[20] Hansen, A., Iov, D., Blaabjerg, F. and L. Hansen (2004). Review of contemporary wind turbine concepts and their market penetration. *Journal of Wind Engineering*, 28, pages 247–263.
[21] Hansen, A., Sorensen D., Iov, P., and F. Blaabjerg (2006). Centralized power control of wind farm with doubly fed induction generators. *Journal of Renewable Energy*, 31, 935–951.
[22] Sahin, B. and M. Bilgili (2009). Wind characteristics and energy potential in Belen-Hatav, Turkey. *International Journal of Green Energy*, 6, pages 157–172.
[23] Gopalkrishan, N. and K. Thyangarajan (1999). Optimization studies on integrated wind energy systems. *Renewable Energy*, 16, pages 940–943.
[24] Ravindra S. and Y. Mahajan (2015). Review of wind energy development and policy in India. *Energy Technology and Policy*, 2(1), pages 122-132.
[25] ADB (2012). Wind energy future in Asia, a compendium of wind energy resource maps, project data and analysis for 17 countries in Asia and the Pacific. ADB-GWEC report.
[26] Intergovernmental Panel on Climate Change (2014). IPCC-2014: Working group II contribution to the fifth assessment report, summary for policy makers. Cambridge University Press. Cambridge, UK.
[27] Sangroya, D. and J. Nayak. (2015). Development of wind Energy in India. *International Journal of Renewable Energy Research,* 5(1).
[28] Haselip, J., Nygaard, I., Hansen, U. and E. Ackom (2011, November). Diffusion of renewable energy technologies case studies of enabling frameworks in developing countries technology transfer. Perspectives Series UNEP.
[29] Global Wind Energy Council (2014). VDMA/BWE: The German wind industry takes a breather. Brussels, Belgium.
[30] IWP (2015, February). *Indian Wind Power*, 1(2). http://www.indianwindpower.com/pdf/.
[31] Jagadeesh, A. (2000). Wind energy development in Tamil Nadu and Andhra Pradesh, India, institutional dynamics and barriers—a case study. *Energy Policy*, 28, pages 157 -168.
[32] Central Electricity Authority, Government of India (2014, February). All India installed capacity (MW). MNRE, physical progress.
[33] Jami, H. and C. Mehra (2014, May). Barriers to accelerating wind energy in India. Central electricity regulatory commission.
[34] KREDL-Karnataka Renewable Energy Development Ltr. www.kredlinfo.org.
[35] C-Step (2014). Roadmap for energy storage technologies in India. Bangalore: Center for Science, Technology and Policy.
[36] Re-Energizing Karnataka (2014). An assessment of renewable energy policies, challenges and opportunities—climate parliament—policy brief. Center for Science, Technology and Policy.

[37] Government of Karnataka, Department of Energy. www.gokenergy.in.

[38] InWEA (2014). Wind energy India: tariffs/regulation regime. Indian Wind Energy Association. http://www.inwea.org/tariffs.htm.

[39] INWEA (2015). http://www.inwea.org/installedcapacity.htm.

[40] Ministry of Statistics and Program Implementation, Government of India (2015). Annual Report. New Delhi, India.

[41] Nouni, M., Mullick, S. and T. Kandpal (2009). Providing electricity access to remote areas in India: niche areas for decentralized electricity supply. *Renewable Energy*, 34, pages 430–434.

[42] Painuly. J. (2002). Barriers to renewable energy penetration: a framework for analysis. *Renewable Energy*, 24(2), pages 73–89.

[43] Beck, F. and E. Martinot (2004). *Renewable energy policies and barriers*. Renewable energy policy project. Global environment facility, forthcoming in *Encyclopedia of Energy*, Cleveland, C., editor. Academic Press/Elsevier Science.

[44] Meyers, S. (1998). Improving energy efficiency: strategies for supporting sustained market evolution in developing and transitioning countries. Document ref: LBNL-41460, Ernest Orlando Lawrence Berkeley National Laboratory.

[45] United Nations Framework Convention on Climate Change (1998). Kyoto Protocol. United Nations.

[46] Nguyen, N., Ha-Duong, M., Tranb, T., Shresthad, R. and F. Nadauda (2010). Barriers to the adoption of renewable and energy-efficient technologies in the Vietnamese power sector. Centre International de Recherche sur l'Environnement et le Développement. France.

[47] Infrastructure Development Finance Company, Ltd. (2010, February 10). Barriers to development of renewable energy in India and proposed recommendations. Discussion paper.

[48] Valentine, S. (2010). A STEP toward understanding wind power development policy barriers in advanced economies. *Renewable and Sustainable Energy Reviews*, 14, pages 2,796–2,807.

[49] Saaty, T. (1980). *The analytic hierarchy process: planning, priority setting, resource allocation*. USA: Mcgraw-Hill International Book Company.

[50] Govindan, K., Kaliyan, Kannan, D. and A. Haq (2014). Barriers analysis for green supply chain management implementation in Indian industries using analytic hierarchy process. *International Journal of Production Economics*, 147, pages 555–568.

Chapter 21

Industrial and Environmental Governance Efficiency in China's Urban Areas

Sufeng Wang, Ran Li, Jia Liu, Zhanglin Peng, Yu Bai, Sufeng Wang, Ran Li, Jia Liu, Xi'an, Shaanxi, China, Zhanglin Peng and Yu Bai

Industrial efficiency is important for the development of regional economic policies. Based on a network data envelopment analysis (DEA) methodology which considered undesirable outputs and links between sub-processes, we studied the overall industrial efficiency, pollution governance efficiency and industrial production efficiency of China's largest five urban agglomerations (Beijing-Tianjin-Hebei, Yangtze River Delta, Middle Reaches of Yangtze River, Pearl River Delta, and Chengdu-Chongqing) during 2000-2014. Our results show that:

1) The overall industrial efficiency grows in a wave form. Yangtze River Delta and Beijing-Tianjin-Hebei occupy the highest two positions in overall industrial efficiency. Environmental governance in Pearl River Delta is the most effective. Both overall industrial efficiency and environmental governance efficiency in Chengdu-Chongqing are at the lowest position.

2) The poor efficiency of environmental pollution governance is the key factor that limits the industrial efficiency of the five urban agglomerations. The sources of the inefficiencies of the pollution governance sub-process are the inefficiencies of desirable outputs.

Increasing the efficiency and technical levels of industrial pollution treatment is an important measure to improve the ecological environment of urban areas and the overall industry efficiency, which will ultimately promote more sustainable urban economic and environmental development.

INTRODUCTION

According to the Eleventh Five Year Plan of National Economic and Social Development of China (March 2006), urban agglomerations are the primary form of urbanization and have a leading role in promoting inter-regional cooperation. It was proposed in the 2010 China Development Report that three mega agglomerations, namely Beijing Tianjin Hebei, Yangtze River Delta and Pearl River Delta, would be given development priority. The Middle Reaches of the Yangtze River and Chengdu-Chongqing agglomerations were also approved by the State Council of China in April 2015 and April 2016, respectively. So far, there are five major urban agglomerations. Statistics indicate that between 2000 and 2014, the total gross regional product (GRP) of the top five urban agglomerations (TFUA) has increased to 5.9 trillion renminbi (RMB) yuan, growing at an annual average rate of 1.3% and accounting for 53% of China's GDP in 2014. This indicates that the TFUA are increasingly important for economic growth and development.

The rapid expansion of urban agglomerations has led to a series of unsustainable problems including urban resource shortages and environmental pollution. By the end of 2014, the amount of three types of industrial waste discharge from 91 cities in the TFUA were 9.4, 0.055 and 0.045 billion tons, respectively, accounting for 48.3%, 34.6% and 36.4% of China's total emissions. The total investment by the TFUA for industrial pollution treatment has increased 38.8% from 46.7 billion in 2000 to 64.9 billion RMB yuan in 2014. As leaders of Chinese economic growth, there are questions relating to industrial efficiency and environmental impacts. What is the level of the TFUA's overall industrial efficiency? How are laws of environmental governance efficiency evolving. What are the main sources of inefficiency? These questions need to be considered and analyzed. Their solutions reflect the practical significance of our study and are provided in this chapter.

Literature Review

The concept of efficiency comes from physics and can be traced to its introduction during the first industrial revolution. Färe et al. in 1989 explored the evaluation of environmental efficiency [1]. Since then, many studies on environmental efficiency emerged and different types of evaluation models and methods were introduced [2]. These included total factor productivity (TFP), the environmental performance index (EEI), life cycle assessments (LCA), stochastic frontier analysis (SFA), data envelopment analysis (DEA), sustainable value (SV), among others [3-8]. The DEA is a nonparametric method of operations research often used to estimate production frontiers and to empirically assess the efficiency of decision making systems. This widely used analysis methodology requires neither uniform index dimensions nor advance determination of indicator weights, and handles multi-inputs and outputs flexibly [9]. To clarify the development tract of DEA efficiency evaluation, Table 21-1 compares common DEA models.

Traditional DEA methodologies (i.e., the early Chames, Cooper and Rhodes model and its extensions) fail to consider undesirable outputs in the study of environmental efficiency; therefore, the results tend to deviate from the actual values. Numerous researchers and analysts have placed undesirable outputs into analytical frameworks to reflect the impact of resource and environmental constraints on industrial efficiency [10-12]. Most of these DEA models were radial and oriented, leaving redundant input and output indicators unanalyzed. To this end, Zhou et al., and Mahdiloo and Saen used a non-radial DEA method to estimate environmental performance [13-14]. Zhao and Song used a four-stage DEA technology to eliminate the external environmental impacts on efficiency values [15]. Shi combined Banker-Charnes-Cooper (BCC) and the stochastic frontier approach (SFA) to propose a three-stage DEA model, determining that scale inefficiency was the dominant factor restraining

Table 21-1. Evolution of DEA model for environmental efficiency evaluation.

Models	*Consider undesirable outputs*	*Radial / Non-radial*	*Oriented / Non-oriented*	*Consider slacks*	*Consider intermediate process*	*Consider links between processes*
Traditional DEA	N	N/A	N/A	N	N	N
Radial DEA	Y	Radial	Oriented	N	N	N
Non-radial DEA	Y	Non-radial	Non-oriented	Y	N	N
Multistage DEA	Y	Non-radial	Non-oriented	Y	N	N
SBM	Y	Non-radial	Non-oriented	Y	Y	N
Network DEA	Y	Non-radial	Non-oriented	Y	Y	Y

the control efficiency of industrial wastewater on provincial levels in China [16]. Using the modified three-stages bootstrapped DEA model, Liu et al. determined that governance efficiency in Chinese local governments exhibited a "wavy shape" and deteriorated [17].

Though these studies were non-radial and non-oriented, treating the evaluated systems as "black boxes," they failed to reflect the impact of the intermediate product. The slacks-based measure (SBM) approach effectively deals with this problem [18]. Song et al., Hadi-Vencheh et al., and Lan and Chen applied a SBM model to calculate the efficiency of environmental governance [19-21]. Castellet and Molinos-Senante emphasized the significant manpower and energy cost saving potentials in sewage treatment plants through a weighted relaxation measure model [22]. Nevertheless, the links between adjacent production processes were often ignored and underestimated the efficiency of environmental governance [23]. By dividing the whole process of decision making units (DMUs) into several sub-processes, the network DEA method produces more accurate results [24,25]. According to Lozano and Gutiérrez, the network DEA method obtains reliable results due to its higher discrimination capacity compared to one-process DEA methods [26].

Even with identical network structures, there will be a variety of network DEA models with varying conclusions that result from variables evaluated as DMUs and parameter settings (see Table 21-2). Given the limited data available, the existing network DEA literature mainly focuses on measuring provincial environmental efficiency, failing to distinguish both overall and governance efficiency among urban agglomerations in terms of efficiency level, evolution law, and the main sources of inefficiencies. Understanding this is the theoretical significance of our study.

Models and Indicators: The Network DEA Model

Consider n DMUs, which has K divisions (or sub-processes). In division k of DMU i (DMU_i), β_k desirable outputs $y^k{}_i = (y_1, y_2, \dots, y_{\beta k})$ $y^k{}_i = (y_1, y_2, \dots, y_{\beta k}) \in R+^{\beta k}$ and $y_k \gamma_k$ undesirable outputs $b^k{}_i = (b_1, b_2, \dots, b_{Yk}) \in R+^{Yk}$ are produced by using α_k inputs $x^k{}_i = (x_1, x_2, \dots, x_{\alpha k}) \in R_+{}^{ak}$. $\tau(k,h)$ representing both the intermediate outputs of division k and the intermediate inputs of the division h. The number of intermediate products is represented by $\delta(k,h)$. According to Tone and Tsutui [18], the possible production set of the network DEA $\{(x^k, y^k, b^k, \tau^{(k,h)}\}$ can be described as:

Table 21-2.
Representative research of environmental efficiency in China based on DEA.

Models / Methods	*Research object*	*Key points*	*Ref.*
DEA and Conditional Generalized Minimum Variance Method	Environmental pollution control efficiency in Henan province in 2000-2007	There is large redundancy in pollution control investment of Henan, which has a declining scale return.	Guo and Zheng (2009) [27]
Four-stages DEA and Bootstrap-DEA model	Environmental governance efficiency of China in 2010	The overall efficiency of the central and eastern was significantly better than the west, and the western scale efficiency is better.	Zhao and Song (2013) [15]
Network DEA and the two-sided panel Tobit model	Industrial governance efficiency in 1998-2010 in China	Environmental technology efficiency measured by traditional method underestimates the environmental governance efficiency.	Song et al. (2013) [19]
BCC and Tobit model	Environmental governance efficiency of China in 2003-2010	The local government's efficiency is very low; Fiscal decentralization and public awareness have significant negative impact on environmental treatment efficiency.	Zhang and Li (2014) [28]
Three-stage DEA model	Treatment efficiency of industrial water pollution of China in 2012	The Treatment efficiency of industrial water pollution is only 0.682, and the scale inefficiency is the essential element to hinder the improvement of treatment efficiency.	Tone (2001) [29]
ISBM model and ISBM-Luenberger productivity index	Ecological management efficiency in China from 2003 to 2012	The changes in technical progress and scale efficiency were the main driver for China's ecological management TFP changes.	Hou (2015)
Network DEA based on RAM model	Industrial production efficiency and environmental governance efficiency in China from 2001 to 2010	Industrial production efficiency is higher than environmental governance efficiency. The insufficiency and inefficiency of investment in pollution treatment is the main reason that resulting in low environmental governance efficiency.	Wang and Luo (2015) [30]
Super-SBM model	Efficiency of air pollution abatement in China between 2002 and 2011	There are small differences in efficiency of air pollution abatement between provinces.	Lan and Chen (2015) [21]
Modified DEA	Environmental spending efficiency of 29 provinces from China in 2007-2013	There seems more efficiency loss after excluding the exogenous and random factors, and efficiency score appears wave shape in the time period and is being worsen.	Liu et al. (2016) [17]
Super DEA-Malmquist model	Efficiency of industrial air pollution treatment in China during 2006-2013	The overall treatment efficiency of industrial air pollution is not high. Input redundancy and output insufficient exist at the same time in industrial sectors.	Fan and Jiang (2016) [31]

$$x^k \geq \Sigma^n{}_{t=1}\, x^k{}_i \lambda^k{}_i\, (k=1,2\ldots K) \quad y^k \geq \Sigma^n{}_{t=1}\, y^k{}_i \lambda^k{}_i\, (k=1,2\ldots K)$$

$$\tau^{(k,h)} = \Sigma^n{}_{i=1}\, \tau_t{}^{(k,h)} \lambda^k{}_t (\forall(k,h)) \quad \tau^{(k,h)} = \Sigma^n{}_{i=1}\, \tau^{(k,h)}{}_t \lambda^k{}_t (\forall(k,h)) \quad (21\text{-}1)$$

$$e\lambda^k=1(\forall k),\ \lambda^k \geq (\forall k)$$

where $e\lambda^k=1(\forall k)$ indicates the variable return scale (VRS).

Equation 21-1 becomes the constant return scale if this constraint is neglected. In addition, there are always two types of links between two divisions. First, the free link (the connection can be disposed freely while maintaining the cohesion between the inputs and outputs) with the equation expressed as:

$$\tau^{(k,h)}\, \lambda^h = \tau^{(k,h)}\, \lambda^k\, (\forall(k,h)) \quad (21\text{-}2)$$

Second, the fixed links (the connections remain unchanged) with the formula expressed as:

$$\tau_o{}^{(k,h)} = \tau^{(k,h)}\lambda^h,\ \tau_o{}^{(k,h)} = \tau^{(k,h)}\lambda^k\, (\forall(k,h)) \quad (21\text{-}3)$$

where the subscript *o* means "overall."

In accordance with the majority of relative literature, the free link is considered in the current study. Then, the network DEA model containing undesirable outputs can be formulated as:

$$\theta^* = \underset{\lambda^k, s^{k-}_{\alpha}, s^{k+}_{\beta}, s^{k-}_{\gamma}}{Min.} \frac{1 - \Sigma^k{}_{k=1}\, p_k[(1/\alpha_k)\ \Sigma^{\alpha x}{}_{\alpha=1}\, s^{k-}{}_{\alpha o}/x^k{}_{\alpha o}]}{1 - \Sigma^k{}_{k=1}\, p_k[1/(\beta_k + \gamma_k)]\Sigma^{\beta x}{}_{\beta=1}\, (s^{k+}{}_{\beta o}/y^k{}_{\beta o}) + \Sigma^{yx}{}_{y=1}(s^{k-}{}_{yo}/b^k{}_{yo})} \quad (21\text{-}4)$$

$$x^k{}_{\alpha o} = x^k{}_{\alpha o}\lambda^k + s^{k-}{}_{\alpha o}$$
$$y^k{}_{\beta o} = y^k{}_{\beta o}\lambda^k - s^{k+}{}_{\beta o}$$
$$b^k{}_{\gamma o} = b^k{}_{\gamma o}\lambda^k + s^{k-}{}_{\gamma o}$$
$$ep^k = 1,\ e\lambda^k = 1,\ p^k \geq 0,\ \lambda^k \geq 0,\ s^{k-}{}_{\alpha o} \geq 0,\ s^{k+}{}_{\beta o} \geq 0,\ s^{k-}{}_{\gamma o} \geq 0\ (\forall k)$$

where $s^{k-}{}_{\alpha o}$, $s^{k+}{}_{\beta o}$, $s^{k-}{}_{\gamma o}$ represent the slack vectors of inputs, desirable outputs and undesirable outputs respectively.

The objective function θ^* equals the unit when all of the slack variables are zero, indicating the most effective state of the DMU. The parameter p_k is the weight of division k, for the simplest situation, supposing p_k=1/K, implying the uniform weights of all divisions.

Further, we have

$$x^k = (x^k{}_1, x^k{}_2, \ldots, x^k{}_n) \in R^{\alpha_k \times n}$$
$$y^k = (y^k{}_1, y^k{}_2, \ldots, y^k{}_n) \in R^{\beta_k \times n}$$
$$b^k = (b^k{}_1, b^k{}_2, \ldots, b^k{}_n) \in R^{\gamma_k \times n} \quad (21\text{-}5)$$

Equation (21-4) can be solved by being transformed to linear programming according to Charnes and Cooper [32].

Models and Indicators: The Decomposition of Industrial Inefficiency

Based on the non-radial and non-oriented network DEA model above, we can obtain the efficiency of the overall industry and the divisions by using the input-output slack as follows:

$$\theta_o = \frac{1 - \Sigma^k{}_{k=1}\, p_k[(1/\alpha_k)\, \Sigma^{\alpha x}{}_{\alpha=1}\, s^{k-*}{}_{\alpha o}/x^k{}_{\alpha o}]}{1 - \Sigma^k{}_{k=1}\, p_k[1/(\beta_k + \gamma_k)]\Sigma^{\beta k}{}_{\beta=1} (s^{k+*}{}_{\beta o}/y^k{}_{\beta o}) + \Sigma^{yx}{}_{y=1}(s^{k-*}{}_{yo}/b^k{}_{yo})}$$

$$\theta_o = \frac{1 - (1/\alpha_k)\, \Sigma^{\alpha k}{}_{a=1}\, (s^{k-*}{}_{\alpha o}/x^k{}_{\alpha o})}{1 + [1/(\beta_k + \gamma_k)]\, [\Sigma^{\beta k}{}_{\beta=1}\, (s^{k+*}{}_{\beta o}/y^k{}_{\beta o}) + \Sigma^{yx}{}_{y=1}(s^{k-*}{}_{yo}/b^k{}_{yo})]} \quad (\forall k) \qquad (21\text{-}6)$$

where $s^{k-*}{}_{\alpha o}$, $s^{k+*}{}_{\beta o}$, $s^{k-*}{}_{yo}$ represents the optimal solution resulting from Equation (21-4).

Furthermore, the inefficiency can be decomposed in reference to Cooper et al. [33] as:

Input inefficiency $\theta_x = 1/\alpha_k\, [\Sigma^{\alpha k}{}_{a=1}\, (x^k{}_{\alpha o} - s^{k-*}{}_{\alpha o})/\, x^k{}_{\alpha o}]$ (21-7)

Desirable output inefficiency $\theta_y = [1/\beta_k\, [\Sigma^{\beta k}{}_{\beta=1}\, (y^k{}_{\beta o} + s^{k+*}{}_{\beta o})/y^k{}_{\beta o}]]^{-1}$ (21-8)

Undesirable output inefficiency $\theta_b = [1/\gamma_k\, [\Sigma^{\gamma k}{}_{\gamma=1}\, (b^k{}_{\gamma o} + s^{k-*}{}_{\gamma o})/b^k{}_{\gamma o}]]^{-1}$ (21-9)

In Equations (21-7, 21-8 and 21-9), the smaller the value of θ_x, θ_y, θ_b, the lower the efficiency of inputs and outputs. θ_x, θ_y, θ_b will reach the maximum value of 1 when $s^{k-*}_{\alpha o}$, $s^{k+*}_{\beta o}$, s^{k-*}_{yo} equals zero, which indicates that the input-output is the most efficient.

Indicators Design

Consider the industrial efficiency of a two stage production shown in Figure 21-1.

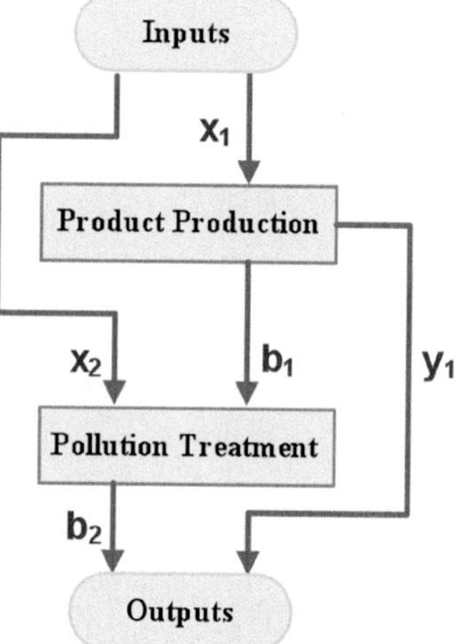

Figure 21-1. The two-stage industrial process.

In the first sub-process (product production), utilizing the input variables including labor L_p (represented by employment population of industry), capital K (represented by average annual net value of fixed assets, converted to year 2000 constant price according to the price index of fixed assets) and energy E (consumption of industrial energy), we obtain desirable output Y (gross industrial output, converted to year 2000 constant price according to the price index of industrial producer), and undesirable outputs (emissions of industrial wastewater W_p, industrial sulfur dioxide S_c, soot and dust F_p).

In the second sub-process (pollution treatment), input variables contain environmental protection staff L_c (the employment population of water conservancy, environment, and public management as proxy variables), investment in industrial pollution control I, and technology innovation in pollution control T (represented by scientific operating expenditure). Variables I and T were both converted to 2000 constant prices according to the industrial producer's price index. Only desirable outputs such as the amount of industrial wastewater treatment W_C, the removal of industrial sulfur dioxide S_c, the removal of industrial soot and dust F_c, urban green coverage area G_a, and the coverage rate G_r are considered. There is no undesirable output during this process.

As the keystone of this study is the industrial efficiency of 91 prefecture level cities in the TFUA (see Table 21-3), the industrial data of those cities in 2000-2014 are selected as the research sample. Data are mainly derived from the 2001-2015 *China City Statistical Yearbook*, the *Statistical Yearbook* of each city, and the *Statistical Bulletin of National Economic and Social Development*. Interpolation is used to deal with the missing data. The statistical summary of 15 variables is listed in Table 21-4.

Table 21-3. List of the 91 prefecture level cities in the TFUA.

Urban Agglomerations	*Cities*
Beijing-Tianjin-Hebei	Beijing, Tianjin, Shijiazhuang, Tangshan, Qinhuangdao, Handan, Xingtai, Baoding, Zhangjiakou, Chengde, Cangzhou, Langfang, and Hengshui.
Yangtze River Delta	Shanghai, Nanjing, Wuxi, Xuzhou, Changzhou, Suzhou, Nantong, Lianyungang, Huai'an, Yancheng, Yangzhou, Zhenjiang, Taizhou4, Suqian, Hangzhou, Ningbo, Wenzhou, Jiaxing, Huzhou, Shaoxing, Jinhua, Quzhou, Zhoushan, Taizhou2, Lishui, Hefei, Wuhu, Huainan, Ma'anshan, and Chuzhou.
Middle Reaches of Yangtze River	Wuhan, Huangshi, Yichang, Ezhou, Jingmen, Xiaogan, Jingzhou, Huanggang, Xianning, Changsha, Zhuzhou, Xiangtan, Hengyang, Yueyang, Changde, Yiyang, Loudi, Jingdezhen, Jiujiang, Xinyu, Yingtan, Ji'an, Yichun, Fuzhou, and Shangrao.
Pearl River Delta	Guangzhou, Shenzhen, Zhuhai, Foshan, Jiangmen, Zhaoqing, Huizhou, Dongguan, and Zhongshan.
Chengdu-Chongqing	Chongqing, Chengdu, Zigong, Deyang, Mianyang, Suining, Neijiang, Leshan, Nanchong, Meishan, Guang'an, Dazhou, and Ziyang.

During 2000-2014, there existed significant differences among the TFUA in gross industrial output, pollution emissions (taking industrial SO_2 emission as an example) and investment in environmental governance (see Figure 21-2).

Figure 21-2 conveys the following:

1) Considering relative levels of industrial production for the TFUA, pollution emissions and environmental governance investment for Beijing-Tianjin-Hebei and Pearl River Delta are ranked highest. Yangtze River Delta is ranked second, and the Middle Reaches of the Yangtze River rank lowest.

Table 21-4. The descriptive statistics of input-output variables.

Variables	*Mean*	*Standard Deviation*	*Minimum*	*Maximum*	*Count*
Employment population of industry (10^4 people)	24.8	34.0	1.5	279	1,365
Average annual net value of urban fixed assets (10^8 RMB Yuan)	377.7	638.2	16.5	5,501	1,365
Consumption of industrial energy (10^4 tons)	1,226.1	1709.2	10.4	15,382	1,365
Gross industrial output (10^8 RMB Yuan)	608.8	653.5	52.1	4,716	1,365
Emissions of industrial wastewater (10^4 tons)	12,326.3	13,824.2	232.0	91,260	1,365
Emissions of industrial sulfur dioxide (10^4 tons)	16.1	21.14	0.3	154	1,365
Emissions of industrial soot and dust (10^4 tons)	184.4	766.4	0.2	17,357	1,365
Employment population of water conservancy, environment and public management (10^4 people)	0.9	1.2	0.01	10	1,365
Investment in industrial pollution control (10^8 RMB Yuan)	6.1	17.2	0.02	178	1,365
Scientific operating expenditure (10^4 RMB Yuan)	3,337	1,0291	30	75,101	1,365
Amount of industrial wastewater treatment (10^4 tons)	11,208	12,925	57	88,072	1,365
Removal of industrial sulfur dioxide (10^4 tons)	8.8	16.2	0.0	145.0	1,365
Removal of industrial soot and dust (10^4 tons)	181.0	765.6	0.1	17,353	1,365
Urban green coverage area (ha)	5,872.2	9,204.7	16.0	83,729	1,365
Urban green coverage rate (%)	36.9	9.1	0.4	92.9	1,365

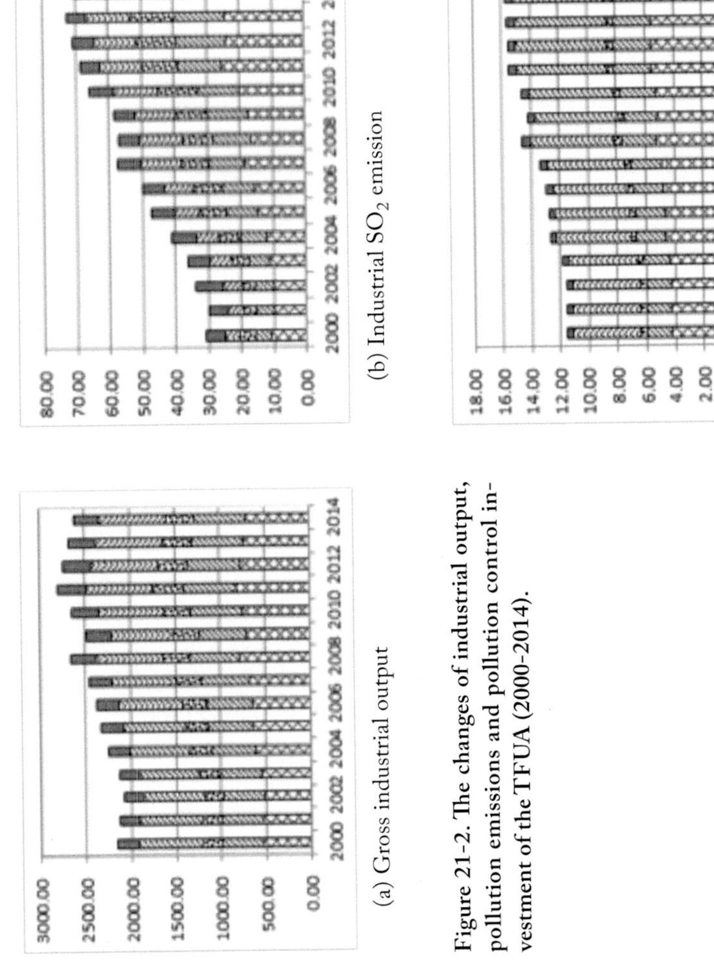

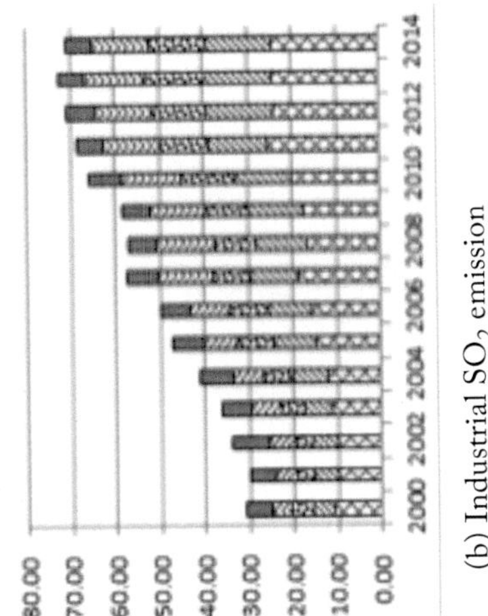

(a) Gross industrial output

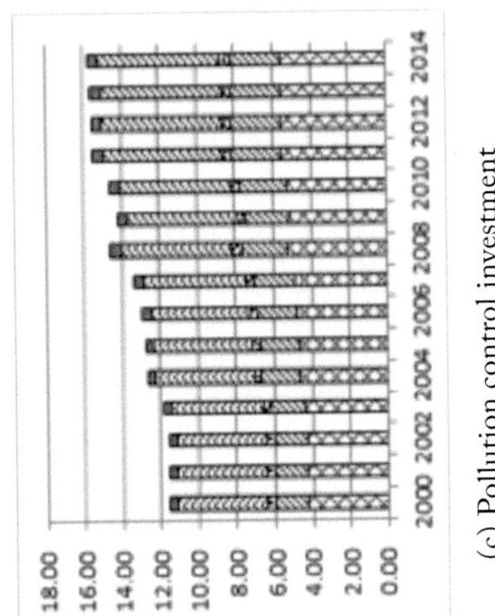

(b) Industrial SO_2 emission

(c) Pollution control investment

Figure 21-2. The changes of industrial output, pollution emissions and pollution control investment of the TFUA (2000-2014).

2) From the level of development, the trends of industrial growth and pollution emissions vary. As is shown in Figures 21-2(a) and 21-2(b), the gross industrial output between Beijing-Tianjin-Hebei and Pearl River Delta is similar, but industrial SO_2 emissions of the former are much higher than the latter. The growth rate of SO_2 emissions of Middle Reaches of Yangtze River is higher than that of Chengdu-Chongqing, although industrial development of the two agglomerations is similar.

3) The pollution emissions and investment in pollution control were disproportionate. Comparing Figures 21-2(b) with 21-2(c), given the comparable investment in environmental treatment, SO_2 emissions in Pearl River Delta were better controlled when compared to Beijing-Tianjin-Hebei. The Middle Reaches of Yangtze River produced similar SO_2 emissions at a much lower cost when compared with Yangtze River Delta and Pearl River Delta.

RESULTS

Temporal Differences in Industrial Efficiency

Considering the temporal dimension, industrial efficiency has evolved across urban agglomerations (see Figure 21-3). From 2000 to 2014, the total industrial efficiency for the TFUA and each agglomeration fluctuated within a narrow range. The industrial production efficiency showed an increasing trend after rapid development in the previous three years while the pollution control efficiency decreased sharply at first and then gradually flattened. For most of the 15-year period, pollution control efficiency moves between the overall industrial efficiency and industrial production efficiency in the TFUA with the exception of the Pearl River Delta. There the pollution control efficiency is higher than the other two types of efficiencies during almost the entire period.

Spatial Differences in Industrial Efficiency

Considering the spatial dimension, the distribution of industrial efficiency among the agglomerations is also variable (see Figure 21-4). The Yangtze River Delta has the highest overall efficiency, while Chengdu-Chongqing has the lowest. The rank of industrial production efficiency

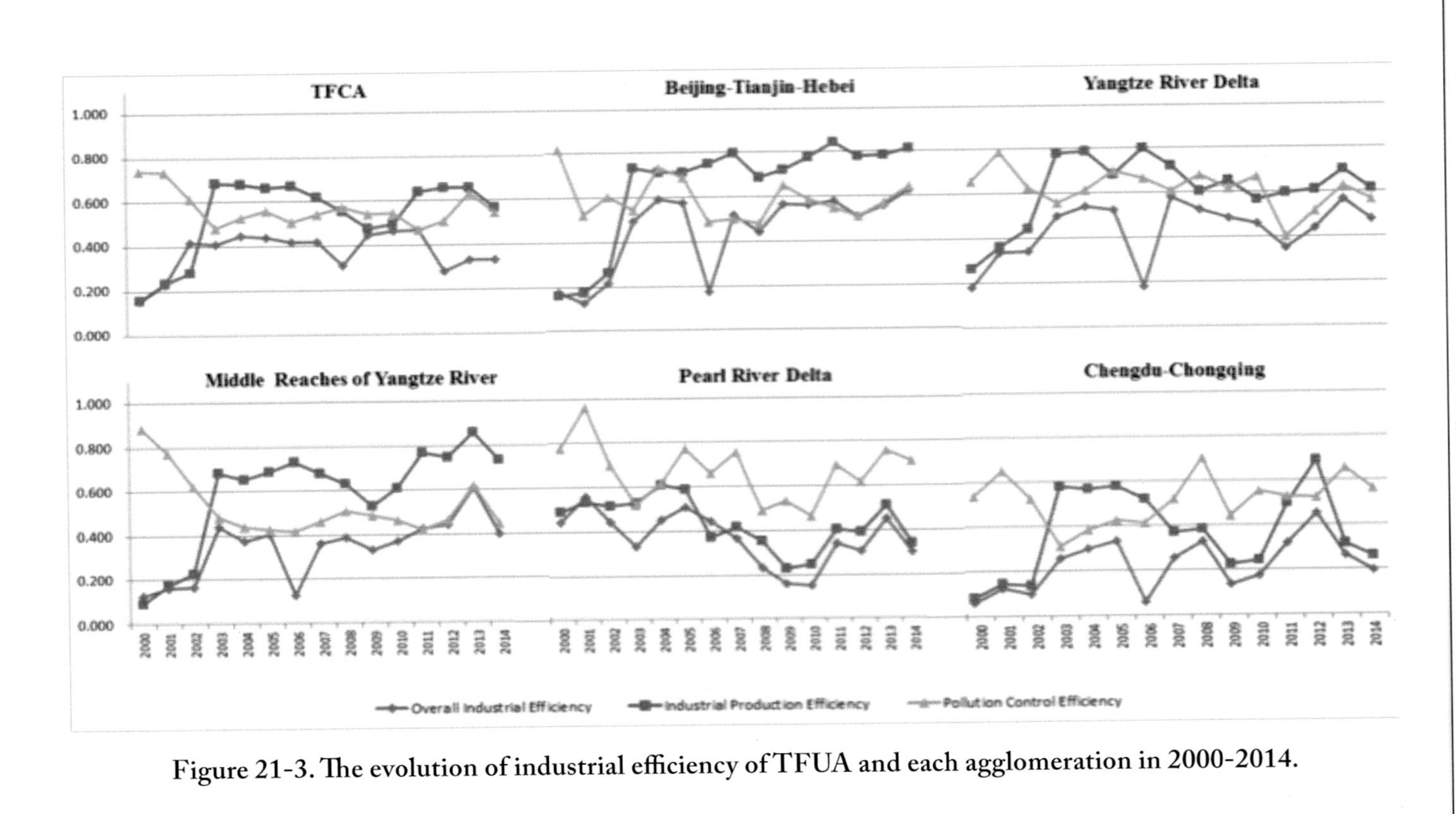

Figure 21-3. The evolution of industrial efficiency of TFUA and each agglomeration in 2000-2014.

for five urban agglomerations is similar to that of overall efficiency. Pearl River Delta leads in pollution control efficiency, followed by Yangtze River Delta and Beijing-Tianjin-Hebei. Chengdu-Chongqing ranks lowest in overall industrial efficiency and industrial production efficiency among the TFUA. The industrial efficiency of TFUA is proportionate to their economic development, which is consistent with the conclusions of previous research and shows the reliability of the model described by Equation (21-4).

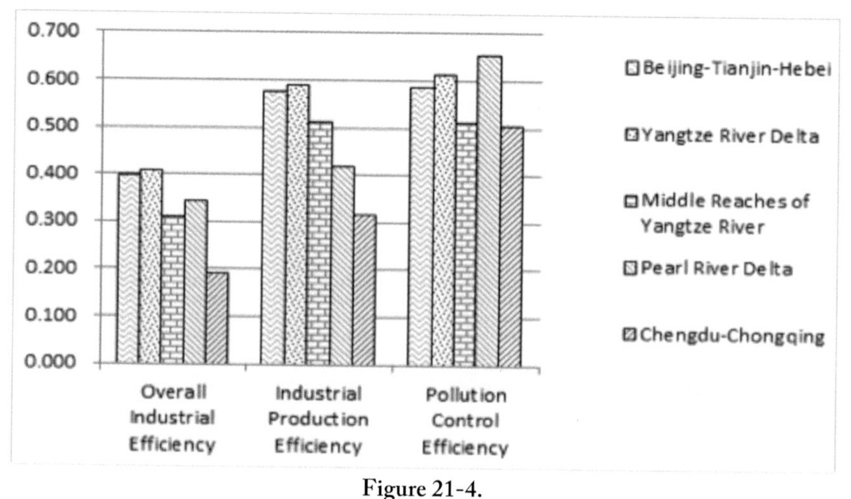

Figure 21-4.
TFUA industrial and pollution control efficiencies (2000-2014).

The spatial difference of industrial efficiency is reflected not only within each urban agglomeration but among them. Using Yangtze River Delta as an example (see Figure 21-5), the industrial efficiency for 30 cities in this region can be classified as three gradients.

Gradient 1—Urban agglomerations with the highest overall and sub-process efficiencies above 0.8 which include Shanghai, Hangzhou, Wuxi, Suzhou, Ningbo, Chuzhou and Zhoushan.

Gradient 2—Urban agglomerations with moderate efficiencies between 0.5 and 0.8 which include 10 cities from Shaoxing to Taizhou 2.

Gradient 3— The remaining urban agglomerations.

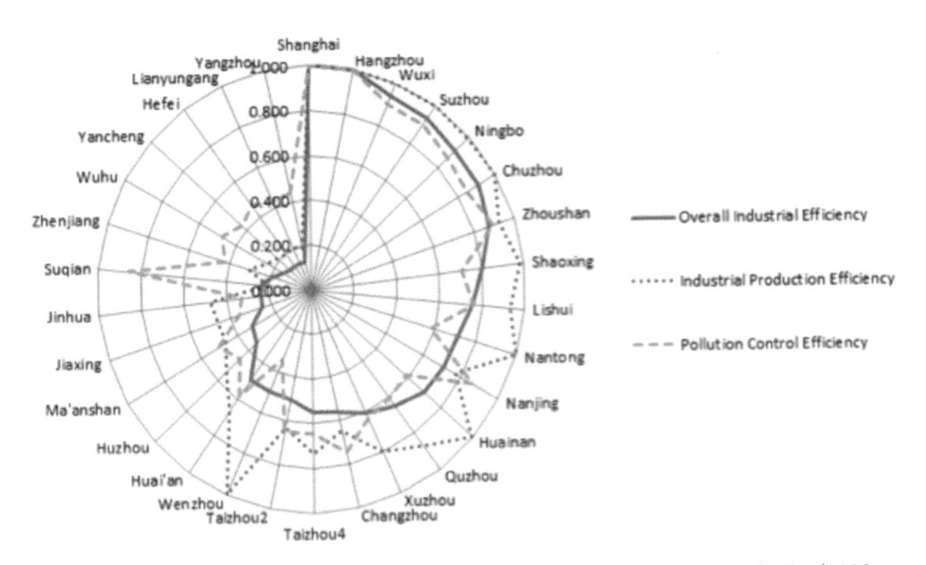

Figure 21-5. Comparison of industrial efficiency within Yangtze River Delta (2000-2014) decomposition of inefficiency.

The industrial efficiencies and economic levels of these cities are commensurate to a degree. While the industrial production efficiency in Wenzhou is very high, its pollution control efficiency is only 0.341, so it is classified as Gradient 3. Suqian is categorized in Gradient 3 due to its low overall efficiency and industrial production efficiency (only about 0.2), though its pollution control efficiency is high (0.859).

To explore the suppression of industrial efficiency, the decomposition of inefficiency was performed using Equations 21-7, 21-8 and 21-9, and the inefficiency in the overall process and the sub-processes was obtained (see Table 21-5).

The conclusions derived from the data in Table 21-5 are interesting. For the industrial production sub-process, undesirable output is most inefficient, indicating that pollution emissions during this stage have not been well controlled. For the pollution control sub-process without undesirable output, input efficiency is much higher than efficiency of desirable output, which suggests that desirable output (i.e., treatment of industrial wastes and urban greening) is inefficient. Finally, affected by the pollution control sub-process, the efficiency of desirable output through the overall industrial process is substantially lower than input efficiency—the lowest among three

Table 21-5. Industrial inefficiency of urban agglomerations (2000-2014).

		TFUA	*Beijing-Tianjin-Hebei*	*Yangtze River Delta*	*Middle Reaches of Yangtze River*	*Pearl River Delta*	*Chengdu-Chongqing*
Industrial production sub-process	Input Inefficiency	0.865	0.882	0.877	0.887	0.763	0.854
	Inefficiency of Desirable Output	0.996	1	0.997	0.993	1	0.993
	Inefficiency of Undesirable Output	0.546	0.644	0.618	0.562	0.495	0.39
Pollution control sub-process	Input Inefficiency	0.814	0.795	0.826	0.848	0.74	0.799
	Inefficiency of Desirable Output	0.31	0.391	0.383	0.283	0.33	0.195
	Inefficiency of Undesirable Output	-	-	-	-	-	-
Overall Industrial Process	Input Inefficiency	0.845	0.845	0.858	0.871	0.763	0.831
	Inefficiency of Desirable Output	0.338	0.422	0.413	0.312	0.356	0.216
	Inefficiency of Undesirable Output	0.546	0.644	0.618	0.562	0.495	0.39

types of efficiencies. While counterintuitive, this *indicates that inefficiency of pollution control is the driving force for inefficient industrial productivity.*

Further Analysis

By combining Figure 21-2 with the data from Figures 21-4 and 21-5, we can better explain the differences and sources of pollution control efficiency among urban agglomerations. For example, Figure 21-2 shows that Beijing-Tianjin-Hebei and Pearl River Delta have similar industrial development and environmental investment, while SO_2 emissions in the former region are much higher than those in the latter. Intuitively, the pollution control in Beijing-Tianjin-Hebei should be less efficient compared to the latter, and this is confirmed in Figure 21-4.

The industrial pollution emissions in the Middle Reaches of Yangtze River are growing faster than those in Chengdu-Chongqing region (see Figure 21-2). While this would suggest lower pollution control efficiency in the Middle Reaches, it is higher. The reason may be that the inputs such as environmental investment during the pollution control sub-process in Middle Reaches of Yangtze River are less than those in Chengdu-Chongqing agglomeration. The desirable outputs of the Middle Reaches region are higher; both input efficiency and output efficiency are shown in Table 21-5, and hence resulted in greater pollution control efficiency.

Furthermore, by contrast with Yangtze River Delta and Pearl River Delta, Middle Reaches of Yangtze River show large amounts of SO_2 emissions and very low environmental investment (see Figure 21-2). It seems that pollution control in this agglomeration should be more efficient, which is proved to be contrary to the evidence (see Figure 21-4). This may be due to the fact that desirable outputs, such as treatment of industrial wastewater (55.0 million tons) is about half of that in Yangtze River Delta (101.4 million tons) and Pearl River Delta (112.9 million tons). Industrial soot, dust removal and green coverage are much less than the other two regions. This indicates less desirable pollution governance and thus the lower pollution control efficiency.

The empirical results indicate that the efficiency of regional pollution control cannot be accurately modeled by relying solely on either industrial pollution emissions or investments in pollution control. By combining these indicators with desirable outputs and undesirable outputs, more practical environmental decisions are possible.

CONCLUSIONS

In this chapter, the overall industrial process consists of two sub-stages—industrial production and pollution treatment. By using the network DEA model, considering undesirable outputs, and the industrial data of 2000-2014 from the TFUA, the overall industrial efficiency, industrial production efficiency and pollution governance efficiency are estimated. The evolution trends and spatial differences among urban agglomerations are also analyzed. Industrial inefficiency is decomposed and calculated using inefficiencies, desirable output inefficiencies and undesirable output inefficiencies. The primary difference of pollution governance efficiency among urban agglomerations is illustrated.

This study shows that since 2000, the industrial production efficiency of the TFUA has increased remarkably. However, the overall industrial efficiency did not improve substantially due to fluctuations in pollution governance efficiency. Among the TFUA, Beijing-Tianjin-Hebei and Yangtze River Delta were in leading positions in overall industrial efficiency and industrial production efficiency, while Pearl River Delta has the highest pollution treatment efficiency. Meanwhile, all efficiencies in Beijing-Tianjin-Hebei and Middle Reaches of Yangtze River increased steadily, which varies from the TFUA regions. Generally, there is potential for industrial efficiency to be improved in Middle Reaches of Yangtze River and Chengdu-Chongqing compared with other urban agglomerations.

It was determined by decomposition of inefficiency that input efficiency is always higher than desirable output efficiency, either in the overall process or in the pollution governance sub-process. The desirable output efficiency is lower when insufficient desirable outputs in the pollution governance sub-process offsets the desirable outputs in the industrial production sub-process. *Therefore, raising the level of pollution treatment improves the overall industrial efficiency.*

Due to the constraints of data availability, this study was limited in the selection of industrial input and output indicators. There remain questions that need to be analyzed in-depth. For example, what are the main factors that affect the efficiency of industrial pollution control? During the process of industrial transfer and optimization, how can the synergy of environmental policies within an urban agglomeration or among urban agglomerations be optimized? We plan to focus on these extended problems in our future research.

Note: If there was no special description, all of the mean values mentioned in this chapter use the geometric mean to eliminate the influence of extreme values.

Acknowledgments

This research is financially supported by the National Natural Science Foundation of China, Ministry of Education of China, the Education Department of Anhui province of China, and the Philosophy and Social Science Project of the Anhui province of China.

References

[1] Färe, R., Grosskopf, S., Lovell, C. and C. Pasurka (1989). Multilateral productivity comparisons when some outputs are undesirable: a nonparametric approach. *The Review of Economics and Statistics*, 71(1), pages 90-98.

[2] Zawaydeh, S. (2015). Energy efficiency, renewable energy targets, and CO_2 reductions expected by 2020. *Strategic Planning for Energy and the Environment*, 35(2), pages 18-47.

[3] Hoang, V. and C. Tim (2011). Measurement of agricultural total factor productivity growth incorporating environmental factors: a nutrients balance approach. *Journal of Environmental Economics and Management*, 62(3), pages 462-474.

[4] Thatte, L. and H. Chande (2014). Measurement of environmental performance index: a case study of Thane city. *IOSR Journal of Humanities and Social Science*, 19(4), pages 1-7.

[5] Piao, W., Kim, Y., Kim, H., Kim, M. and C. Kim (2016). Life cycle assessment and economic efficiency analysis of integrated management of wastewater treatment plants. *Journal of Cleaner Production*, 113, pages 325-337.

[6] Filippini, M., and L. Zhang (2016). Estimation of the energy efficiency in Chinese provinces. *Energy Efficiency*, 9, pages 1,315-1,328.

[7] Al-Refaie, A., Hammad, M. and M. Li (2016). DEA window analysis and Malmquist index to assess energy efficiency and productivity in Jordanian industrial sector. *Energy Efficiency*, 9, pages 1,299-1,313.

[8] Henriques, J. and J. Catarino (2016). Sustainable value—an energy efficiency indicator in wastewater treatment plants. *Journal of Cleaner Production*, DOI: http://dx.doi.org/10.1016/j.jclepro.2016.03.173.

[9] Charnes, A., Copper, W. and E. Rhodes (1978). Measuring the efficiency of decision making units. *European Journal of Operational Research*, 2(6), pages 429-444.

[10] Seiford, L. and J. Zhu (2002). Modeling undesirable factors in efficiency evaluation. *European Journal of Operational Research*, 142, pages 16-20.

[11] Atkinson, S. and J. Dorfman (2005). Bayesian measurement of productivity and efficiency in the presence of undesirable outputs: crediting electric utilities for reducing air pollution. *Journal of Econometrics*, 126, pages 445-468.

[12] Watanabe, M. and K. Tanaka (2007). Efficiency analysis of Chinese industry: a directional distance function approach. *Energy Policy*, 35, pages 6,323-6,331.

[13] Zhou, P., Poh, K. and B. Ang (2007). A non-radial DEA approach to measuring environmental performance. *European Journal of Operational Research*, 178(1), pages 1-9.

[14] Mahdiloo, M. and R. Saen (2012). A novel data envelopment analysis model for solving supplier selection problems with undesirable outputs and lack of inputs. *International Journal of Logistics Systems and Management*, 11(3), pages 285-305.

[15] Zhao, Z. and T. Song (2013). China's regional environmental governance efficiency and its affected factors: based on empirical analysis of four-stages DEA and bootstrap-DEA model. *Nanjing Journal of Social Sciences*, 3, pages 18-25.

[16] Shi, F. (2014). A study on regional treatment efficiency of industrial water pollution in China—based on three-stage DEA method. *East China Economic Management*, 28(8), pages 40-45.

[17] Liu, B. and B. Wang (2016). An assessment of public spending efficiency of environment protection in local China: based on three-stage bootstrapped DEA. *Journal of Zhongnan University of Economics and Law*, 1, pages 89-95.

[18] Tone, K. and M. Tsutsui (2014). Network DEA models: a basic framework. In: Cook, W. and Zhu J., Data envelopment analysis: a handbook of modeling internal structure and network. Springer: New York. Page 239.

[19] Song, M., Zhang, L., Liu, W., et al. (2013). Bootstrap-DEA analysis of BRIC's energy efficiency based on small sample data. *Applied Energy*, 112, pages 1,049-1,055.

[20] Hadi-Vencheh, A., Jablonsky, J. and A. Esmaeilzadeh (2015). The slack-based measure

model based on supporting hyper planes of production possibility set. *Expert Systems with Applications*, 42(1), pages 6,522-6,529.

[21] Lan, Q. and C. Chen (2015). Soft institution, public recognition and efficiency of air pollution abatement. *China Population, Resources and Environment*, 25(9), pages 145-152.

[22] Castellet, L. and M. Molinos-Senante (2016). Efficiency assessment of wastewater treatment plants: a data envelopment analysis approach integrating technical, economic, and environmental issues. *Journal of Environmental Management*, 167, pages 160-166.

[23] Tu, Z. and R. Shen (2013). Does environment technology efficiency measured by traditional method underestimate environment governance efficiency? From the evidence of China's industrial provincial panel data using environmental directional distance function based on the network DEA model. *Economic Review*, 5, pages 89-99.

[24] Aviles-Sacoto, S., Cook, W., Imanirad, R., et al. (2015). Two-stage network DEA: when intermediate measures can be treated as outputs from the second stage. *Journal of the Operational Research Society*, 66(11), pages 1,868-1,877.

[25] Yu, Y., Zhu, W., Shi, Q., et al. (2016). Network-like DEA approach for environmental assessment: evidence from U.S. manufacturing sectors. *Journal of Cleaner Production*, 139, pages 277-286.

[26] Lozano, S. and E. Gutiérrez (2014). A slacks-based network DEA efficiency analysis of European airlines. *Transportation Planning and Technology*, 37(7), pages 623-637.

[27] Guo, G. and Z. Zheng (2009). Assessment on efficiency of environmental pollution control in Henan province based on DEA model. *On Economic Problems*, 1, pages 48-51.

[28] Zhang, Y. and Q. Li (2014). Fiscal decentralization, public awareness and local environmental treatment efficiency. *On Economic Problems*, 3, pages 65-68.

[29] Tone, K. (2001). A slack-based measure of efficiency in data envelopment analysis. *European Journal of Operational Research,* 130(3), pages 498-509.

[30] Wang, B. and Y. Luo (2015). Empirical study on industrial process efficiency, environmental treatment efficiency and overall efficiency: using network DEA based on RAM model. *World Economic Papers*, 1, pages 99-119.

[31] Fan, C. and H. Jiang (2016). Efficiency of industrial air pollution treatment and its difference in China. *Ecological Economy*, 32(8), pages 153-157.

[32] Charnes, A. and W. Cooper (1962). Programming with linear fractional functionals. *Naval Research Logistics Quarterly*, 15, pages 333-334.

[33] Cooper, W., Seiford, L. and J. Zhu (2011). Handbook on data envelopment analysis. Springer Science + Business Media, LLC: New York. Page 164.

Chapter 22

The Guangdong Emissions Trading Scheme

Yuejun Luo, Wenjun Wang, Xueyan Li and Daiqing Zhao

A pilot carbon emissions trading scheme (ETS) has been launched for three years in the Guangdong (GD) Province in China, with the power industry contributing nearly 66% of the covered CO_2 emissions. This chapter reviews the policy design of the power sector in the GD ETS, and finds that the percentage of paid allowance is the primary factor reflected in the carbon cost for generators with an average efficiency. The ways the GD ETS influences the costs and profits of power plants are our primary focus.

The impacts of carbon cost on the overall cost of 300 MW, 600 MW and 1,000 MW plants are analyzed. The results indicate that the ratio of carbon cost to total cost is about 0.5% for the power plants in the GD ETS. This small percentage has little influence on plant operations. The impacts of the carbon cost on the cash flow of the three sizes of plants are assessed by their internal rates of return. A critical curve is developed and shows the benefit scope for the plants at a specific paid allowance and carbon price. This can be used by governments to improve policy design and by the enterprises to manage their carbon assets.

INTRODUCTION

With China's rapid economic development, energy consumption and carbon dioxide emissions have grown rapidly. To alleviate being the world's largest greenhouse gas emitter [1], a number of policies have been introduced to reduce emissions in China. Given that administrative measures are effective but not efficient, China is harnessing market forces to reduce its greenhouse gas emissions. Emission trading systems (ETS) are effective tools, and emitters have incentive to reduce emissions when they are well-designed. Designing and operating emissions trading schemes has become one of the main policies adopted to reduce CO_2 emissions globally, Examples include the European Union's ETS, and those in the U. S. including the Regional Greenhouse Gas Initiative (RGGI) and the Western Climate Initiative (WCI) [2].

In October 2011, China's National Development and Reform Commission (NDRC) approved carbon ETS pilot programs in seven regions, including two provinces (Guangdong and Hubei) and five municipalities (Beijing, Tianjin, Shanghai, Chongqing and Shenzhen) [3]. Duan provided an overview of the status of these seven ETS pilots [4]. The Hubei ETS pilot [5], Shanghai ETS pilot [6], and Shenzhen ETS pilot [7] were detailed including their coverage sectors, allowance allocation, monitoring, reporting, verification, compliance and related mechanisms.

As one of the pilot provinces designated by Chinese government for a carbon ETS, Guangdong (GD) ranks highest in total emissions among the seven pilot regions [8,9]. The emissions reduction targets of GD include lowering the energy intensity per unit of GDP by 18% and the carbon intensity by 19.5% from 2011 to 2015 [10]. After its pilot phase (2013-2015), the GD ETS added power, cement, steel and petrochemical sectors in 2016. It now covers approximately 56% of the total CO_2 emissions of Guangdong Province. The power sector is by far the biggest CO_2 emitter within this emissions scheme, accounting for 230 million metric tons of CO_2 emissions of the total 350 million metric tons [11]. The power-generating facilities, especially the conventional ones, is the sector most seriously affected by the GD ETS.

There have been no similar studies on the power plants in the GD ETS and only a few in China that considered electricity sectors in the ETS. Cong and Wei adopted the agent-based model to study the potential impact of an assumed ETS on China's power sector and found that an ETS would internalize the external environmental cost of carbon, influencing the relative costs of different power generation technologies through pricing [12]. Zhao et al. applied DEA-Malmquist Index to empirically analyze the impact of various environmental regulations on the efficiency and CO_2 emissions of power plants in China, and concluded that market-based regulations (namely a possible ETS) has an irreplaceable role in promoting green development among power plants [13]. Both papers focus on predicting possible results by building some models. Teng et al. provided an analysis of institutional barriers in China's electricity pricing and dispatching systems that may affect the performance of a presumed ETS and discussed several options to reconcile the ETS and electricity market [14].

In contrast to these previous studies, ours investigates the actual power plants covered in an operating ETS pilot and adopted a basic and classic parameter—the internal rate of return (IRR) to analyze whether the power plants can operate profitably. As the IRR is widely used in the actual investment and management of businesses, the result can be easily understood and then adopted by governments and power generators to improve ETS rule structures and manage carbon as an enterprise asset. The characteristics of policy design for the power sector in the GD ETS are also considered. Next, we compare and analyze the overall cost of coal-fired power plants in the GD ETS and determine the critical point of the impact of carbon cost on the profit of the plants. Finally, we offer conclusions and policy suggestions to improve the GD ETS.

POWER SECTOR POLICY DESIGN IN THE GUANGDONG ETS

The greenhouse gas (GHG) emissions in the power system can be divided into two parts: 1) direct emissions from the production; and 2) indirect emissions from consumption.

Greenhouse Gas Emission Sources from the Power Sector in the GD ETS

In the EU ETS, RGGI, and WCI, the installations which emit GHG directly are included in the schemes by the principle that "those who produce, take responsibility." The direct emissions or the production from the power sector are covered in the ETS, as appears in Scenario 1 (see Figure 22-1). Power plant owners may argue that the power generation process is simply an energy transition from fossil fuel to electricity. In their opinion, the facilities which consume electricity should be included in the schemes by the principle that "those who consume, take responsibility."

The indirect emissions or consumption including the plant service power are covered in the ETS, as appears in Scenario 2 (see Figure 22-1). In Scenario 1, industrial facilities, like cement and steel companies, make little effort to reduce emissions as their emissions (excluding electricity consumption) are comparatively trivial and not covered by the ETS, or their complied emissions are ignored when covered by an ETS. Owners of industrial facilities often fail to notice the increased cost passed by the power suppliers because of the Chinese government's strict control of the fixed feed-in and retail electricity prices. Industrial enterprises could be inclined to change from fossil fuels to electricity consumption in order to reduce their recorded emissions. A large amount of electricity consumed in GD is imported from other regions such as Guangxi and Yunnan Provinces.

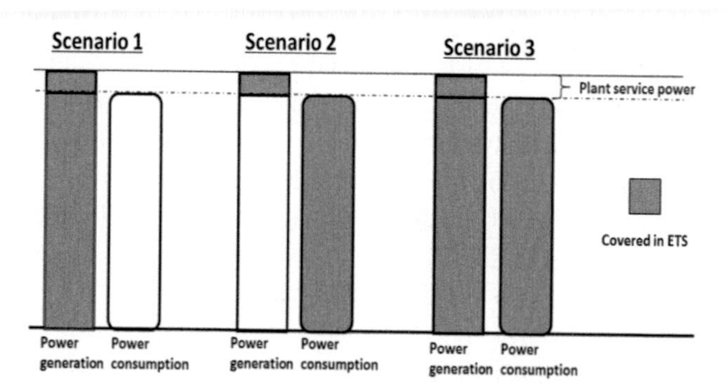

Figure 22- 1. Scenario analysis of power system covered in ETS.

In Scenario 2, since electricity used by the power plant itself (plant service power) usually accounts for about 5% of the total electricity it produces, most of the emissions from the power sector should be controlled by the thousands of electricity consumption entities. This is more difficult than regulating only the power plants. Since only a small percentage of electricity is used by the power plants, regulating points of consumption fails to motivate the power plants to reduce emissions.

For the GD ETS, the emissions from both power generation and consumption are covered in Scenario 3 (See Figure 22-1). This scenario uses the "double counting" principle. It offers regulation at both the source of production and point of consumption, alleviating pollution transfer. For electric generation, the power plants must improve generator efficiency or transfer to cleaner fuels to reduce emissions. For electric consumption, enterprises such as cement, steel or construction companies must adopt advanced technology to decrease the electricity used. With more emissions counted, more allowances are needed, which may incentivize the carbon market to a greater extent. The double accounting of emissions can be alleviated by baselining. Many enterprises are experienced with measuring and controlling electricity consumption to meet energy-saving targets. Using historical energy consumption data, the emissions accounting and allowance allocations can be calculated by the enterprises in the GD ETS, rather than by specific installations as in the EU ETS.

Allowance Allocations in the GD ETS Power Sector

The indirect emissions from consumption are counted according to the consumed electricity quantities multiplied by the emissions coefficient of the South Power Grid. These are published by the NDRC, and the allowance allocation of indirect emissions vary based on each industry's production processes. This assessment focuses on the direct emissions from power generation.

Benchmarks are used to allocate allowances for the pure-power plants in the GD ETS. Benchmarking avoids penalizing enterprises that have previously taken action to reduce emissions. This also provides for inefficient and smaller power production facilities. Many smaller generation facilities are essential for mountainous and remote regions to ensure stable working and living conditions. It may be difficult for them to improve their efficiency to the level of larger producers. Different benchmark values can be set for the various types of generation using different fuels (see Table 22-1).

Table 22-1. Benchmark values for power units in 2013 [11].

Unit type	*Coal fired generator*				*Gas fired generator*	
	1000MW	600MW	300MW	Under 300MW	390MW	Under 390MW
Benchmark value (gCO_2/kWh)	770	815	865	930	415	482

Based on the Notice of the First Allowance Allocation Plan in the Guangdong ETS [10], the allowance allocation of the pure-power unit can be calculated as the equation follows:

$$EA = \Sigma^{n}_{i=1} (HP_i \times BM_i) \times CF \times (1 - PA) \quad (22\text{-}1)$$

Emissions allowance, *EA*, is the free allowance of one power plant which may include *n* generator units. Historical production, HP_i, is the unit *i* average power generation, in years for the period from 2010 to 2012, excluding those when the unit was not operated continually for at least six months. BM_i is the benchmark value of the unit *i* shown in Table 22-1, or the average of a similar unit covered by the GD ETS. The compliance factor, *CF*, is used for coordinating the results of setting emission caps and allowance allocations (typically 1). The percentage of the paid allowance, *PA*, reflects allowance shortages compared to historical emissions when power unit efficiency is at mean output and generation is steady. The *PA* was 3% in 2013.

After operating for the compliance period, the GD ETS exposed some problems. With an economic downturn and more inflows of electricity from other provinces in 2013, the power generated by plants in Guangdong Province was far less than that in the previous three years, which led to a large surplus of allowances. To avoid productive volatility, ex-post adjustments were applied in the second compliance period of the GD ETS. The allowance allocation equation of the pure-power unit was changed to:

$$EA = \Sigma^{n}_{i=1} (CP_i \text{ X } BM_i) \text{ X } CF \text{ X } (1 - PA) \tag{22-2}$$

The *HP* in Equation (22-1) was changed to *CP*, current production, the plant's actual power generation in the compliance year. As emissions verification in 2013 was more accurate and detailed than the emissions inventory for previous years, the *BM* and unit type are updated for Table 22-2. To balance the surplus allowance from 2013 and maintain the constraints of the ETS, the *PA* was increased to 5% in 2014.

As output uncertainties are eliminated by ex-post adjustments, the percentage of paid allowance is the only variable of carbon cost for a generating unit with an average efficiency. We next focus on how the different percentag-

Table 22-2. Benchmark values for power units in 2014 and 2015 [15,16].

Unit type			*Benchmark value (gCO_2/kWh)*
Coal-fired generator	1,000MW		825
	600MW	Ultra-supercritical	850
		Supercritical	865
		Subcritical	880
	300MW	Non-circulating fluidized bed	905
		Circulating fluidized bed	927
	Under 300MW	Non-circulating fluidized bed	965
		Circulating fluidized bed	988
Gas fired generator	390MW		390
	Under 390MW		440

es of paid allowances impact the costs and profits of coal-fired thermal power plants since they will continue to be dominate in the future [17].

ANALYSIS OF COAL-FIRED THERMAL PLANTS

There are four types of coal-fired generators in Guangdong ETS categorized by their size—under 300 MW, 300 MW, 600 MW and 1,000 MW. The number of 300 MW generators is greatest in Guangdong Province which also has the largest total installed capacity of 600 MW generators. The number of generators and the total installed capacity of the 1,000 MW generators are increasing while the number of generators rated below 300 MW are decreasing.

Coal-fired power plants with generators sized 300 MW, 600 MW and 1,000 MW are chosen as reference power plants. Coal-fired plants usually have at least two identical generators to achieve the maximal efficiency, so each reference power plant is assumed to have two generators.

Cost of the Standard Reference Power Plants

According to the investigation on the coal-fired power plants in Guangdong Province and the reference cost index on the design limitations of thermal power engineering [18], the basic financial parameters for 2014 are as follows:

- Coal price 800 yuan/tce
- Limestone price is 100 yuan/t
- Equipment operates 4,500 hours for 20 years
- Loan percent of total static investment 80%
- Loan term is 15 years
- Interest rate for the loan is 6.55%
- Depreciation period is 15 years
- Residual value is 5%
- Insurance premium rate is 0.25% of total investment
- Repair costs are 2% of total investment
- Staff salaries total 50 thousand yuan annually plus 60% for welfare

The feed-in tariff is pegged at 0.502 yuan/kWh including tax [19]. Coal cost can be derived from the benchmark value. The average low calorific value of standard coal is 29,307 MJ/tce with a carbon content of 26.37 gC/MJ [20].

The parameters of coal-fired power plants with two 300 MW generators (300 MW plant) are as follows: investment of 4,394 yuan/kW, staff of 234, material fee is 6 yuan/kWh, other costs of 12 yuan/kWh, limestone consumption 8 tons/hour (2% sulphur content), discharge fees including SO_2, NO_x and fume emissions are 1,430, 1,620 and 80,000 yuan/coiler/year respectively. The total cost of a 300 MW plant is 1,094,210 thousand yuan.

The parameters of coal-fired power plants with two 600 MW generators (600 MW plant) are as follows: investment of 3,367 yuan/kW, staff of 247, material fee 5 yuan/kWh, other cost 10 yuan/kWh, limestone consumption 16 tons/hour (2% sulphur content), discharge fees including SO_2, NO_X and fume emissions are 2,600, 2,930 and 150,000 yuan/coiler/year respectively. The total cost of a 600 MW plant is 1,913,100 thousand yuan.

The parameters of coal-fired power plants with two 1,000 MW generators (1,000 MW plant) are as follows: investment of 3,334 yuan/kW, staff of 300, material fee is 4 yuan/kWh, other cost is 8 yuan/kWh, limestone consumption 8 tons/hour (0.9% sulphur content), discharge fees including SO_2, NO_x and fume emissions are 3,600, 4,100 and 240,000 yuan/coiler/year respectively. The total cost of a 1,000 MW plant is 3,032,550 thousand yuan.

The costs of the three kinds of plants are shown respectively in Figures 22-2, 22-3 and 22-4.

As shown in Figures 22-2, 22-3 and 22-4, the main costs of the plants include coal, depreciation, interest, maintenance and other costs, which account for over 95% of the total cost. The cost of coal is two thirds of the total cost. Any fluctuation in the coal cost may determine whether the

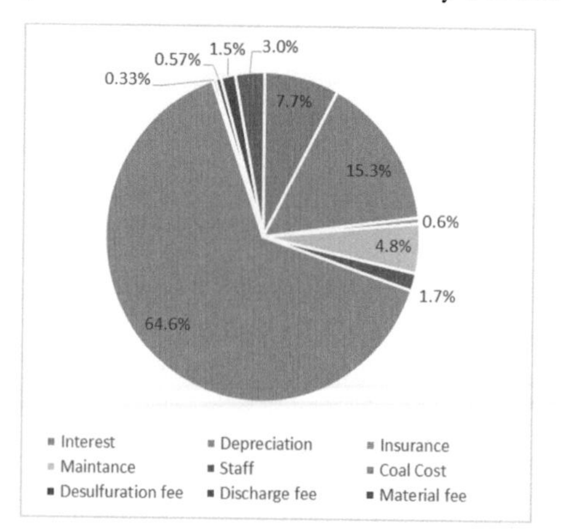

Figure 22-2. The overall costs of the 300 MW plant.

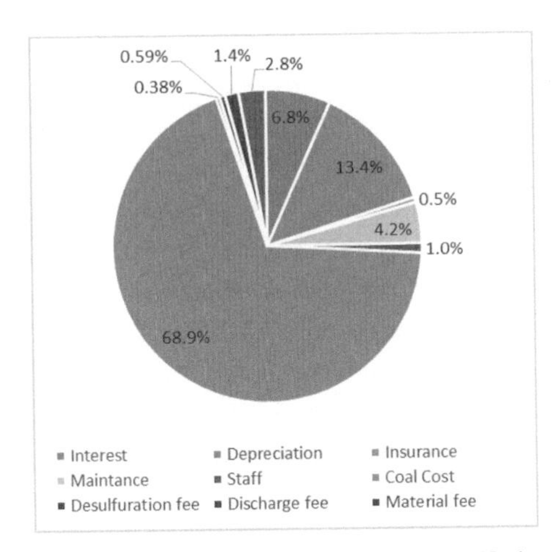

Figure 22-3. The overall costs of the 600 MW plant.

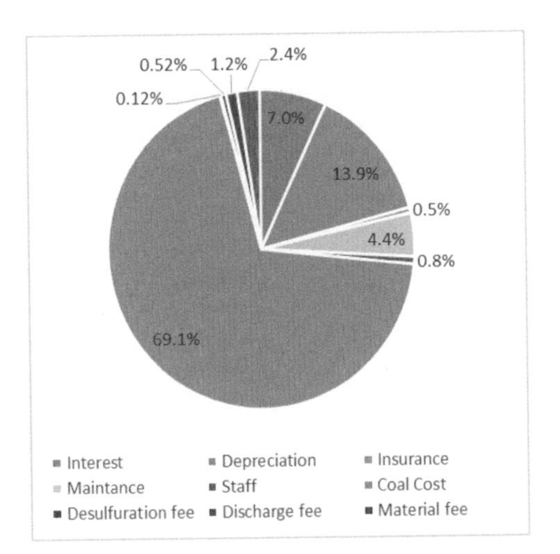

Figure 22-4. The overall costs of the 1,000 MW plant.

enterprise is profitable. With capacity increases, the proportion of human cost, material cost, and management cost decrease. Using higher quality coal with lower sulfur content, the proportional cost of desulfurization is lower for the 1,000 MW plants than for the smaller plants.

Impact of ETS on the Cost of Coal-fired Power Plants

As an external environmental management regulation, ETS influences plants covered in ETS in both the short and long terms. The plants may benefit from carbon trading, or suffer from additional expenses. For the short term, the plants need to increase the corresponding manpower and resources in carbon asset management. If the carbon emissions of the plants are more than the free quota, extra credits must be purchased for compliance. If excess allowance is validated by an emission reduction action, the plants profit by selling the surplus allowance, offsetting the cost of the emission reduction activity. The various allowance allocation schemes have different carbon costs. When the allowance is available at no cost and the benchmark value is set as the average efficiency in the industry, plants with the average efficiency are hardly influenced by the carbon cost. With more stringent carbon limits, carbon emission caps will be gradually tightened. With the total allowance in the ETS tightened, the free allowance would be reduced on the basis of the constant benchmark value. In such cases, the enterprises must purchase more allowances from the government or market, increasing their production cost.

For the different scenarios with different carbon prices and *PA*, the impacts of ETS on the coal-fired thermal power plants can be calculated and analyzed. According to the carbon allowance allocation scheme in Guangdong Province and the sampled generating plants, the hypotheses of this study are as follows:

1) The carbon cost in this study only involves the expense that the enterprise needs to purchase for allowances, regardless of the hidden costs such as the resources that the enterprise invests in carbon asset management or carbon trading.

2) The efficiency of the standard reference power plants is the average of the unit type, so the total enterprise allowances are equal to its carbon emissions when the allowance allocation benchmarks are the average of the energy consumption for specific type of generator.

3) When free quotas issued by the government are less than the carbon emissions of an enterprise, purchasing allowances in the carbon market or taking measures to reduce carbon emissions should be adopted to achieve carbon management goals. The enterprise bears the cost. Various technical mitigation actions are possible and calculating carbon reduction costs are complex. Therefore, the cost from taking carbon reduction actions is estimated to be equal to purchasing equivalent allowances.

4) The *PA* is set as 3%, 5%, 10% and 100%. According to Trial Management Measures on Carbon Emissions in Guangdong Province, "the allowance is partly free, and the free quota proportion will be gradually reduced." The free allowance proportion of the power industry is 97% in 2013 and 95% in 2014 [11,15]. The allowances of the power industry in second and third phase of the EU-ETS and in the U.S. RGGI carbon market arc almost all auctioned.

5) The carbon prices are set as 5 yuan/tCO_2, 60 yuan/tCO_2 and 120 yuan/tCO_2. The allowance auction price was set as 60 yuan/tCO_2 in Guangdong Province in 2013. The carbon price in China varies in the various carbon market pilot programs with the highest price being 120 yuan/tCO_2 in Shenzhen and the lowest of 5 yuan/tCO_2 in Shanghai.

According to the research hypothesis and the characteristics of the electric power industry, the calculation formula of carbon cost is:

Carbon cost = Installed capacity X Operational hours X
quota allocation benchmark X *PA* X Carbon price

The carbon costs of the three types of plants were calculated and analyzed based on this formula.

As shown in Figures 22-5, 22-6 and 22-7, when the carbon price is low (5 yuan/tCO_2), the ratio of carbon cost to total cost is small, and is only about 1% even when the *PA* is 100%. The carbon cost has little influence on the enterprise cost. When the carbon price is high (120 yuan/tCO_2), the proportion of carbon cost to total cost increases, accounting for 1% at the *PA* of 3% and 35% at the *PA* of 100%, which is second in magnitude to the cost of coal. In this case, the carbon cost has a significant effect on the enterprises. The impact of the ETS is closely related to the carbon price and *PA*. When carbon emissions are stable and the *PA* increases, the more allowance the plants must purchase and the carbon price becomes key to quota expenditures. As the carbon price rises, the carbon cost has increasing influence on enterprise cost. When both the carbon price and *PA* are high, the carbon cost rises quickly and may have disruptive effects on production decisions.

With the *PA* of coal-fired thermal power plants in the GD ETS equal to 3% in 2013 and 5% in 2014, and the carbon price at 60 yuan/tCO_2, the

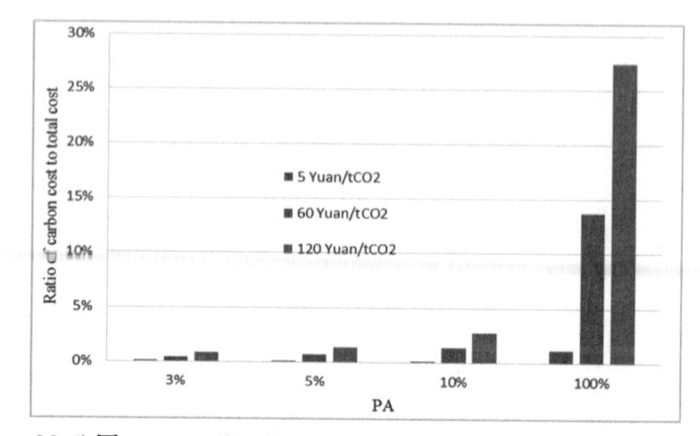

Figure 22-5. The ratio of carbon cost to total cost for the 300 MW plant.

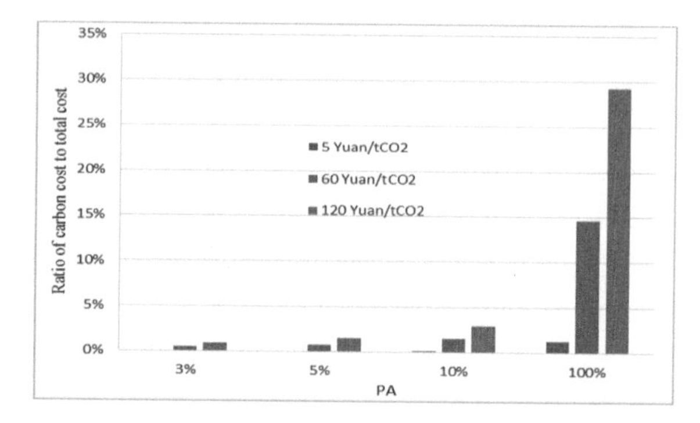

Figure 22-6. The ratio of carbon cost to total cost for the 600 MW plant.

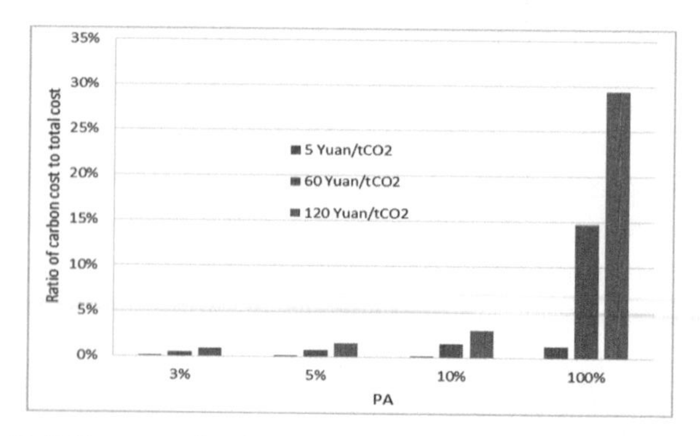

Figure 22-7. The ratio of carbon cost to total cost for the 1,000 MW plant.

ratio of carbon cost to total cost is only 0.5%. This is roughly equal to the ratio of gas pollutant discharge costs.

The power industry may be more sensitive to the carbon cost because of its large carbon emissions and higher *PA* relative to other industries in the GD ETS. Presently, the impact of the ETS is limited as carbon cost accounts for only a small portion of the total cost.

IMPACT OF THE GD ETS ON THE PROFITS OF THERMAL PLANTS

To further analyze the dynamic impacts of carbon cost on enterprise production cycles, the internal rate of return (IRR) was used to measure and compare the profitability of the three types of power plants (300 MW, 600 MW and 1,000 MW) in hypotheses 1-3. Hypotheses 4 and 5 were designed to study the effect of different carbon emission constraints. The *PA* values were set as 3%, 5%, 10%, 20%, 30%, 50% and 100%, and the carbon price was from 5 yuan to 300 yuan, with 36 increments with 5 yuan intervals below 60 yuan, and 10 yuan intervals above 60 yuan.

Impact of Carbon Cost on the Cash Flow of Thermal Power Plants

The IRR is used in capital budgeting to measure the profitability of potential investments or projects and is an indicator of the efficiency or yield of an investment. Investors should only undertake projects or investments with IRRs that exceed the cost of capital. The higher a project's IRR, the more appealing the investment becomes. The IRR of an investment or project is the "annualized effective compounded return rate." The greater the projected rate of return from a particular investment after incorporating all cash flows (both positive and negative), the greater the likelihood that investors will risk their funds. Given the (period, cash flow) pairs (n, C_n) where n is a positive integer, the total number of periods N, we find in the equation for the net present value (NPV) the IRR represented by the variable r:

$$NPV = \Sigma^{n}_{i=1} [C_n/(1 + r)^n] = 0 \qquad (22\text{-}3)$$

The IRR can be calculated using software. The results indicate that the IRR of the 300 MW plant, 600 MW plant and 1,000 MW plant are

8.7%, 15.6% and 17.7% respectively without ETS. According to the Electric Power Planning and Engineering Institute [18], the baseline IRR is 8%, meaning that the power plant can operate economically only when the IRR of the power plant is greater than 8%. Without a carbon cost, all three types of thermal power plants can retain their investment value within the assumed parameters. With the added carbon cost, whether the plants can retain their investment value depends on how much their IRR declines due to the increased cash outflows.

IRRs for the standard reference power plants were calculated for the different carbon cost scenarios using *PA* and carbon prices for the IRRs of 300 MW, 600 MW and 1,000 MW (see Figures 22-8, 22-9 and 22-10).

As shown in the Figures 22-8, 22-9 and 22-10, the IRRs decline as the carbon cost increases. When *PA* is 3% and the carbon price is below 150 yuan/t, the IRR's curve is a straight line and begins to tilt when the carbon price exceeds 150 yuan/t. When *PA* is 5% and the carbon price is below 100 yuan/t, there is a slight lean to the IRR's curve, which is more pronounced when carbon prices exceed 100 Yuan/t. When *PA* is greater than 10%, the slope of the IRR curve changes as the carbon price increases.

The influence of ETS on the IRR is dependent on the generation unit size. The results show that as the installed capacity of the generator increases, the slope of the IRR curve is less, the initial IRR (without carbon cost)

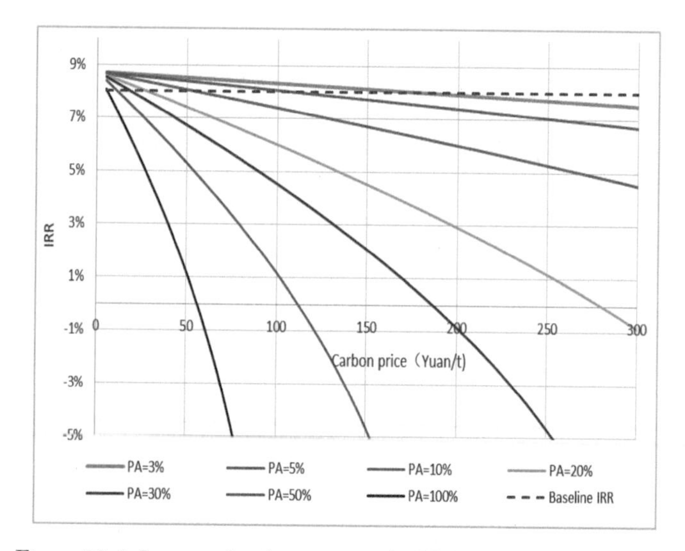

Figure 22-8. Impact of carbon cost on the IRR of the 300 MW plant.

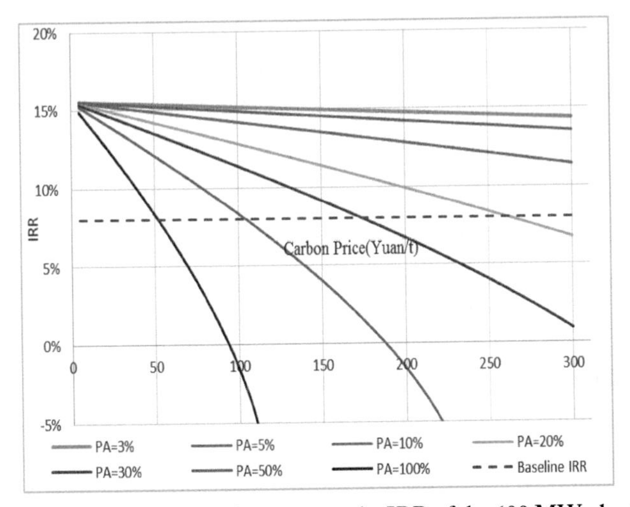

Figure 22-9. Impact of carbon cost on the IRR of the 600 MW plant.

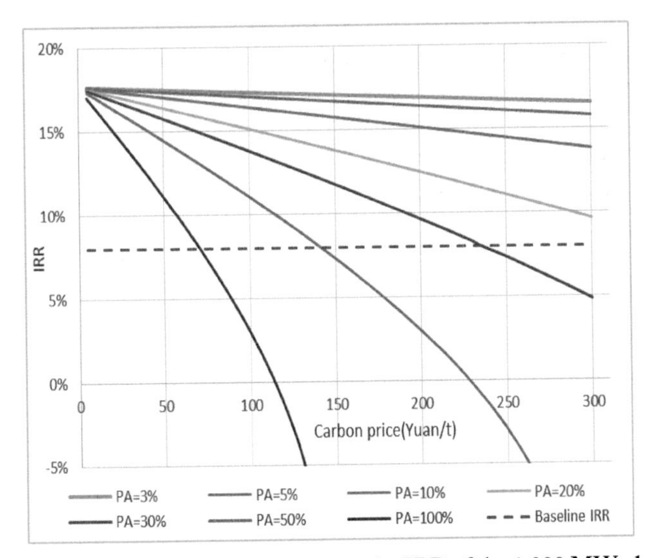

Figure 22-10. Impact of carbon cost on the IRR of the 1,000 MW plant.

is higher, and more carbon cost can be absorbed.

For the 300 MW plant, the IRRs are below 8% with *PA* =3% and the carbon price above 180 yuan/t, or when *PA* =5% and the carbon price is above 110 yuan/t, or if the *PA* =10% and the carbon price above 55 yuan/t. For these *PA* percentages, the IRRs of the 600 MW and 1,000 MW plants are over 8% if the carbon price is less than 300 yuan/t.

When *PA* is 20% and the carbon price is 263 yuan/t, the 600 MW plant can't operate profitably but the 1,000 MW plant remains feasible. If the *PA* were to reach 100%, the 300 MW, 600 MW and 1,000 MW plants are unable to retain their investment values when carbon prices are higher than 5.5 yuan/t, 52 yuan/t and 71 yuan/t respectively.

Critical Point of the Impact of Carbon Cost on the Profit of Power Plants

Given the paid allowances and carbon prices, the IRR of the three units is such that they can operate economically (IRR=8%). The critical impact points of carbon cost on the profits of the power plants are indicated in Figure 22-11.

The combinations of *PA* and carbon price where the IRR of the 300 MW plant is 8% are: 3% (80 yuan), 5% (110 yuan), 10% (55 yuan), 20% (27 yuan), 30% (18 yuan), 50% (11 yuan) and 100% (5.5 yuan). These can be used to construct the critical curve of the impact of carbon cost on the profits of the 300 MW plant. Each point in the critical curve represents that the plant can afford the maximum *PA* at some carbon price, ensuring that the IRR is equal to or greater than 8%. The points in the lower left area of the curve are the combinations of *PA* and carbon price satisfying an IRR greater than 8%, while the points in the upper right area of the curve are the

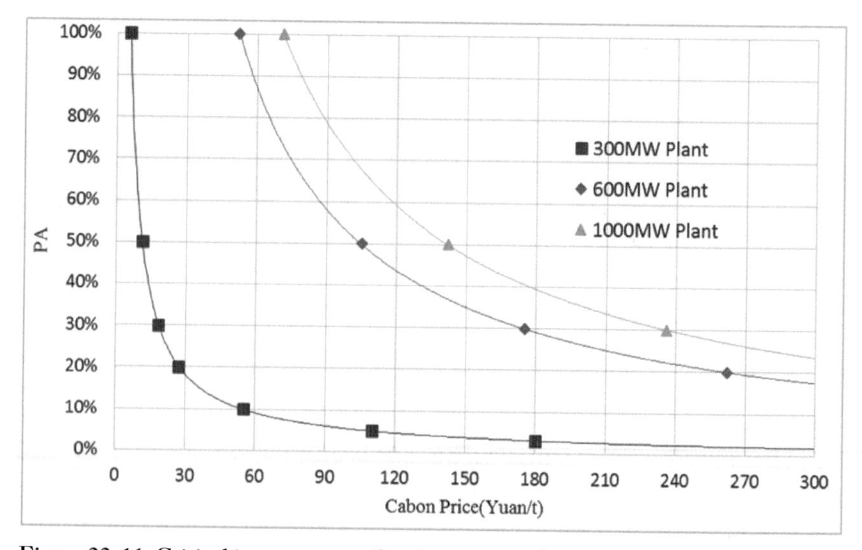

Figure 22-11. Critical impact curve of carbon cost on the profit of the three types of the plants.

combinations with the IRR less than 8%.

The combinations of *PA* and carbon price when the IRR of the 600 MW plant is 8% are: 20% (262 yuan), 30% (175 yuan), 50% (105 yuan) and 100% (52 yuan). The combinations of *PA* and carbon price when the IRR of the 1,000 MW plant is 8% are: 0% (236 yuan), 50% (142 yuan) and 100% (71 yuan). The critical curve for 600 MW and 1,000 MW plants can be constructed in a similar manner.

From Figure 22-11, the carbon cost tolerance of the various units with various combinations of *PA* and carbon prices can be compared. For example, to maintain an IRR above 8% when the carbon price is 90 yuan/tCO_2, the 300 MW plant can purchase at most 7% of its total allowance, the 600 MW plant 58% of its allowance, and the 1,000 MW plant can afford 79% of its allowance. If the power plants must purchase 30% of their total allowances to maintain their normal profits, the 300 MW, 600 MW and 1,000 MW plants can afford the carbon price of 20 yuan, 175 yuan and 238 yuan respectively.

The smaller the installed capacity of the generation unit, the lower its efficiency, and the less it's owners can pay for carbon if profit levels are to be maintained. For installed capacities from 300 MW to 1,000 MW, the generator efficiency increases, the critical curves deviate more from the axis, the profits are larger, and the ability to pay higher prices for carbon increases. If carbon prices are below 60 yuan/t, the *PA* should be less than 10% to ensure that all generating units operate profitably. As carbon prices increase, the *PA* should be reduced accordingly.

The higher the plant's operations cost, the greater the cash outflows, and the less a plant can afford to pay for carbon emissions. Since acquiring coal is the plant's largest operational cost, the price of coal impacts the cost of operation. If the coal price increases from 800 yuan/tce to 1,000 yuan/tce, the critical curve of 1,000 MW plant will move lower and to the left, as shown in the Figure 22-12.

In the U.S., RGGI distributes nearly all CO_2 allowances through quarterly, regional auctions, which is similar to distributing 100% of the *PA* for the GD ETS. In RGGI, the volume-weighted average auction clearing price of CO_2 allowance per ton was \$1.86, \$1.89, \$1.93, \$2.92, \$4.72 and \$6.10 respectively from 2010 to 2015 [21]. In the GD ETS, the ceiling carbon price is 5.5 yuan/t, 52 yuan/t and 71yuan/t for the IRR for the 300 MW, 600 MW and 1,000 MW plants above 8% when the *PA* is 100%. This means that the ceiling carbon price for the 300 MW generator unit is roughly equal to the average auction clearing price in the early phases

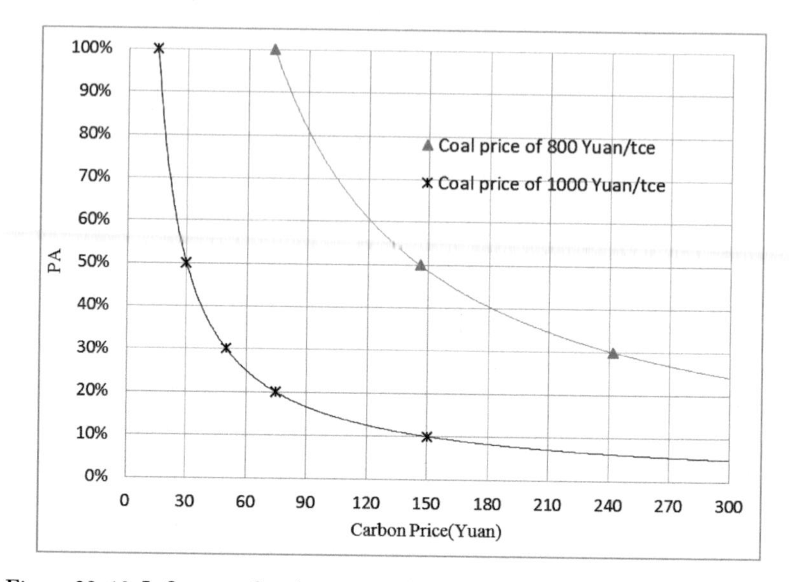

Figure 22-12. Influence of coal prices on the critical curve of the 1,000 MW plant.

of RGGI. The average carbon auction clearing price available from RGGI from 2013 to 2015 approached the ceiling price of our example 600 MW generator. As generator efficiency improves from 300 MW to 600 MW, and the energy efficiency improves, the CO_2 emissions rate for RGGI electric generation sources declines by almost 20%. The evidence indicates that the threshold values of carbon prices that thermal power plants can afford to pay appear reasonable.

CONCLUSIONS AND POLICY IMPLICATIONS

The emissions from both power generation and consumption are covered in the GD ETS, which control the emissions in two ways and alleviate pollution transfer. Based on the allowance allocation method of the power plants in the GD ETS, the paid allowance is the only carbon cost factor for generators with average efficiency, as the output uncertainty is eliminated by ex-post adjustment.

The paid allowance of 5% was adopted by the GD ETS in 2014 and 2015. The carbon prices varied from 60 yuan/t in 2013 to 20 yuan/t in 2015. The carbon cost represents a small proportion of the total cost and is acceptable for the thermal power plants in Guangdong Province.

In our study, the threshold values of carbon cost that the three types of coal-fired thermal power plants can afford in terms of *PA* and carbon price have been calculated using the IRR. We determined that the values are reasonable compared with the volume-weighted average auction clearing price of CO_2 allowances in RGGI. We also identified the conditions under which thermal power plants can economically operate.

Considering the impact of carbon cost on the profit of the power plants, the regional governments should notice that different generators can afford various combinations of allocations and carbon price. This improves the allowance allocation and justifies the allowance distribution when coal prices or on-grid tariffs fluctuate.

Several policy recommendations should be considered. First, it is possible to set different *PA*s for different units when carbon prices are too high, as different units can afford different *PA*s at specific carbon prices. All units can operate economically with the paid allowance of 5% when the carbon price is about 20 yuan/t. When the carbon price exceeds 100 yuan/t as in the Shenzhen ETS in 2013, the 300 MW units will not operate economically. A solution might be to reduce the *PA* for 300 MW units to 3% and not change the *PA* for the other units. Secondly, a linkage mechanism between *PA* and coal prices or on-grid tariffs might mitigate the cost burden. The generating plant's affordability of *PA* weakens when the coal prices increase from 800 yuan/tce to 1,000 yuan/tce for the 1,000 MW units when the on-grid tariff is invariant (see Figure 22-12). It is likely necessary to justify the *PA* according to coal prices or on-grid tariffs. To enable plant owners to budget costs, future year *PA*s should be published in advance. Finally, the complex trading strategies that include carbon derivatives should be encouraged, to allow the enterprises to hedge their carbon emission procurement strategies.

Enterprises are fully capable of managing their carbon assets and controlling carbon market risk. For this assessment, the main parameters of the three types of coal-fired thermal power plants were averaged and standardized for reference. There are variable cash flows for different operational schedules of the generating plants which are impacted by interest and depreciation. Each enterprise should apply the parameters according to their actual situation.

The power plants may facilitate other measures to help achieve their carbon management goals. These might include improving efficiency, developing renewables, fuel substitution, purchasing certified emission reduction certificates or developing infrastructure for carbon capture and storage.

Acknowledgments

This work was co-funded by the Guangdong Province Science Foundation and Guangdong Province Low Carbon Development Special Funds.

References

[1] International Energy Agency (2012). World energy outlook.

[2] Arzbaecher, C. and K. Parmenter (2014). Carbon policy impact on industrial facilities. *Strategic Planning for Energy and the Environment*, 34(1), pages 11–39.

[3] National Development and Reform Commission (2011, October 29). Notice on the work of piloting carbon emissions trading. http://www.ndrc.gov.cn/zcfb/zcfbtz/2011tz/t20120113_456506.htm.

[4] Duan, M., Pang, T. and X. Zhang (2014). Review of carbon emissions trading pilots in China. *Energy and Environment*, 25(3-4), pages 527-550.

[5] Qi, S., Wang, B. and J. Zhang (2014). Policy design of the Hubei ETS pilot in China. *Energy Policy*, 75, pages 31–38.

[6] Wu, L., Qian, H. and Li (2014). Advancing the experiment to reality: perspectives on Shanghai pilot carbon emissions trading scheme. *Energy Policy*, 75, pages 22–30.

[7] Jiang, J., Ye, B. and X. Ma (2014). The construction of Shenzhen's carbon emissions trading scheme. *Energy Policy*, 75, pages 17–21.

[8] Hart, C. and M. Zhong (2014). China's regional carbon trading experiments and the development of a national market: lessons from China's SO_2 trading program. *Energy and Environment*, 25(3-4), pages 577–592.

[9] Wang, P., Dai, H., Ren, S., Zhao, D. and T. Masui (2015) Achieving Copenhagen target through carbon emissions trading: economic impacts assessment in Guangdong Province of China. *Energy*, 79, pages 212–227.

[10] People's Government of Guangdong Province (2012, August 20). Inform on issuing the scheme of greenhouse gas emissions control for Guangdong Province during the 12th Five-Year Plan. http://zwgk.gd.gov.cn/006939748/201208/t20120828_341198.html.

[11] Guangdong Provincial Development and Reform Commission (2013, November 25). Notice of the first allowance allocation plan in Guangdong ETS. http://www.gddpc.gov.cn/xxgk/tztg/201311/t20131126_230325.htm.

[12] Cong, R. and Y. Wei (2010). Potential impact of carbon emissions trading (CET) on China's power sector: a perspective from different allowance allocation options. *Energy*, 35, pages 3,921–3,931.

[13] Zhao, X., Yin, H. and Y. Zhao (2015). Impact of environmental regulations on the efficiency and CO_2 emissions of power plants in China. *Applied Energy*, 149, pages 238–247.

[14] Teng, F., Wang, X. and L. Zhiqiang (2014). Introducing the emissions trading system to China's electricity sector: challenges and opportunities. *Energy Policy*, 75, pages 39–45.

[15] Guangdong Provincial Development and Reform Commission (2014, August 18). Notice of the allowance allocation plan in Guangdong ETS in 2014.http://210.76.72.13:9000/pub/gdsfgw2014/zwgk/zcfg/gfxwj/201503/t20150309_305043.html.

[16] Guangdong Provincial Development and Reform Commission (2015, July 10). Notice of the allowance allocation plan in Guangdong ETS in 2015.http://210.76.72.13:9000/pub/gdsfgw2014/zwgk/tzgg/zxtz/201507/t20150713_322106.html.

[17] Liu, L., Zong, H., Zhao, E., Chen, C. and J. Wang (2014). Can China realize its carbon emissions reduction goal in 2020 from the perspective of thermal power development. *Applied Energy*, 124, pages 199–212.

[18] Electric Power Planning and Engineering Institute (2014). Reference cost index on the limitation design of thermal power engineering. China electric power press.

[19] Guangdong Province Development and Reform Commission (2014, September 16). Notice of lowering feed-in tariff of the power generation enterprises. http://210.76.72.13:9000/pub/gdsfgw2014/zwgk/tzgg/jggg/201502/t20150216_304054.html.

[20] National Development and Reform Commission (2011, March). Provincial greenhouse gas inventory compilation guidelines (try out).

[21] Regional Greenhouse Gas Initiative. Market monitor reports. http://www.rggi.org/market/market_monitor, accessed 5 July 2017.

Chapter 23

Networking through International Energy Organizations

Sang Yoon Shin and Jae-Kung Kim

This study investigates differences regarding international energy cooperation among countries by energy trade type. In particular, it classifies 49 major oil and natural gas trading countries into exporting, importing, and balanced countries (i.e., remarkable countries in both export and import), and compares two network characteristics of each group within a network formed through participation in 28 international energy organizations. The analysis results confirm that both the importing and balanced groups have higher values for both degree and power centrality indices compared to the exporting group. However, there is no significant difference between the importing group and the balanced one.

The results indicate that countries having considerable imports occupy more central positions with more partners, which increases their network influence. In other words, countries that are active oil and natural gas importers, regardless of their exports, are more active participants in international energy organizations than are countries that focus only on exports. One reason for this distinction is the lower energy bargaining power of importers. This study offers the expectation that exporting countries participate in more international energy organizations to compensate for their decreasing bargaining power.

The distinction between energy importing and exporting countries is a basic method of classifying countries. In particular, several studies explain the characteristics of crude oil (i.e., a representative source in energy trade) importing and exporting countries. For oil exporting countries, the relationships among energy consumption and other factors such as economic growth and export diversification have been studied [1-3]. For oil importing countries, vulnerability to oil supply and relevant risks are the primary issues considered in the literature [4,5]. Moreover, the effects of oil price volatility on the stock markets have been widely researched for both types of countries [6,7]. Energy importing and exporting countries tend to participate in distinct international energy organizations (e.g., the International Energy Agency versus the Organization of the Petroleum Export Countries). However, relevant research about their differences in international energy cooperation has been limited.

The purpose of our study is to compare the network characteristics of energy importing and exporting countries in their networks formed through participation in international energy organizations. We suggest how the participation in these two groups evolves. The contributions of this study are:

- It expands the understanding of international energy cooperation among countries by deriving a network of major energy trade countries formed through their participation in international energy organizations.
- It demonstrates a significant difference in energy cooperation between energy importing and exporting countries by comparing the distinct network characteristics each group holds within the network.
- It explains a cause for the difference between the two groups and presents a future direction for network change.

This chapter describes our study and is organized as follows. We initially explain the meaning and influence of the country network formed through participation in international organizations. Next, we derive the network formed by major energy trade countries through their participation in international energy organizations and reconfigure the derived network by presenting significant links. Then, we classify the countries in three groups by trade patterns and compare the network characteristics of each group within the reconfigured network. Finally, we explain the results of the comparison and present our conclusions.

COUNTRY NETWORKS

The Asia Infrastructure Investment Bank (AIIB) was established in January 2016. Prior to its launch, participation by members of western countries posed a dilemma [8]. Some objected to its being the first international financial institution led by China, emphasizing that it would lead to loss of their vested interests within global governance. Others favored participation, believing that cooperation with China would offer advantages and that non-member countries would be alienated from Asia's large infrastructure construction markets. However, the confrontation between the two groups ended as the United Kingdom became the first Western nation to join the AIIB. The balance quickly shifted after France, Germany, Italy, and the Netherlands joined. This new organization listed 57 countries as founding members. This example shows that membership in an international organization has more than the symbolic meaning of participating in international meetings.

Being a member of an international organization provides another meaningful consequence—members are connected through their common affiliation to the organization [9]. Likewise, member countries of an international organization are linked by their affiliations within the organization and form a network [10]. These country networks have been extensively studied. While early studies simply explained the networks with existing theories, recent ones have addressed international relations by exploring a more complex reality [11]. Countries holding a better position within a network formed through participation in international organizations could establish discussion agendas, subsequently formulating policies favorable to them [12]. Hafner-Burton and Montgomery have shown that the characteristics of each country in a network formed through participation in international organizations significantly impact international disputes and their consequences [10]. The impact of networks formed through participation in international organizations on another network (e.g., actual trade volume network) has also been explained [13]. While energy is a critical factor in international relations, and international energy organizations have active roles, research about international energy organizations and their networks has been limited. Next, existing networks are derived and analyzed.

METHODOLGIES

This study assumes that member countries in an international energy organization are linked to one another by their common affiliation and form a network [10]. To identify international energy organizations, we referred to the "Survey of G20 countries: gaps and duplication in the existing mandates and work plans of international energy organizations" approved by the G20 in 2014 [14]. Among the international energy organizations considered in our survey, we excluded the United Nations Framework Convention on Climate Change (UNFCCC), which has most of the world's countries as members, and the International Atomic Energy Agency (IAEA) with about 170 members. Also excluded were organizations whose membership is not clearly identified, such as the International Confederation of Energy Regulators (ICER) and the World Energy Forum (WEF).

Network Identification

Our study addresses 28 organizations (see Table 23-1). However, since the number of countries reached 142, it was difficult to draw meaningful implications about energy trading or cooperation among them. Therefore, only major countries engaged in energy trading were included in the analysis. Rather than addressing all energy sources, we focused on oil and natural gas, which account for about 75% of the world energy trade. We chose the top 20 countries in both exports and imports of oil and natural gas for analysis (see Table 23-2). After excluding redundancies, 49 countries were included in this study.

The configuration of the network of our 49 major countries, formed through their memberships in the 28 international energy organizations indicates that all countries actively participate in the organizations. Each country is connected to most of the other 48 countries. The average number of other actors with which any one actor is connected, called the average number of degrees is 42.4. The density, measured as the number of actual connections between two actors over the number of possible connections is 0.94, or almost 1, meaning nearly perfect connectivity within the network.

Centralization, a measure of the degree to which a network is concentrated in its center, decreases when the centrality of each actor becomes similar and increases when the variance of actors' centrality increases. The centralization value for our network is 0.06, or approximately 0, defining a completely dispersed network. Freeman's study refers to measurement, as shown in Equation 23-1 [15]. $Cx(pi)$ is the centrality value of the i actor and $Cx(p^*)$ is the value of the most central actor in the network. The numerator of this index is the sum of differences between the centrality value of the most central actor and the centrality values of the other actors in the

Table 23-1. International energy organizations.

	International Energy Organizations
1	Organization of the Petroleum Exporting Countries (OPEC)
2	Gas Exporting Countries Forum (GECF)
3	EU Energy Initiative Partnership Dialogue Facility (EUEI PDF)
4	OPEC Fund for International Development
5	ASEAN Center for Energy
6	Latin American Energy Organization
7	European Association for the Promotion of Cogeneration (EAPC)
8	Energy and Climate Partnership of the Americas (ECPA)
9	International Energy Agency (IEA)
10	International Energy Forum (IEF)
11	International Gas Union (IGU)
12	International Partnership on Energy Efficiency Cooperation (IPEEC)
13	International Renewable Energy Agency (IRENA)
14	Clean Energy Ministerial (CEM)
15	Carbon Sequestration Leadership Forum (CSLF)
16	World Energy Council (WEC)
17	Energy Working Group, Asia-Pacific Economic Cooperation (APEC)
18	Asia Pacific Energy Research Centre (APERC)
19	Energy Cooperation Task Force, East Asia Summit (EAS)
20	Economic Research Institute for ASEAN (ERIA) and East Asia
21	Major Economies Forum on Energy and Climate (MEFEC)
22	World Petroleum Council (WPC)
23	Renewable Energy and Energy Efficiency Partnership (REEEP)
24	Global Carbon Capture and Storage Institute
25	Global Bioenergy Partnership (GBEP)
26	Renewables Club
27	Energy Charter Treaty (ECT)
28	Energy Regulators Regional Association (ERRA)

network. The denominator is the maximum value of the sum which a network of the same number of actors might have.

$$Cx = [\Sigma Ni=1Cx(px) - Cx(pi)] \div [max\Sigma Ni=1\ Cx(px) - Cx(pi)] \quad (23\text{-}1)$$

With these data from the network analysis, no meaningful implications or conclusions were possible due to the large number connections. To resolve this, we excluded statistically insignificant connections and reconfigured the network using only with the major connections.

Table 23-2. Major oil and natural gas trading countries.

Top 20 Oil Exporting Countries	Top 20 Oil Importing Countries	Top 20 Gas Exporting Countries	Top 20 Gas Importing Countries
Saudi Arabia	USA	Russia	Germany
Russia	China	Canada	Japan
Iran	India	Norway	Italy
Iraq	Japan	Algeria	United Kingdom
Nigeria	Republic of Korea	Netherlands	Republic of Korea
United Arab Emirates	Germany	Turkmenistan	France
Angola	Italy	Qatar	USA
Venezuela	France	Indonesia	Russia
Norway	Netherlands	Malaysia	Turkey
Canada	Singapore	USA	Spain
Mexico	Spain	Nigeria	China
Kazakhstan	United Kingdom	Australia	Ukraine
Kuwait	Thailand	Trinidad & Tobago	Netherlands
Qatar	Canada	Egypt	Canada
Libya	Belgium	Uzbekistan	Belgium
Algeria	Poland	Oman	Belarus
Azerbaijan	Australia	Germany	United Arab Emirates
Colombia	Greece	Bolivia	Mexico
Oman	Sweden	United Kingdom	Brazil
United Kingdom	Indonesia	Myanmar	India

Network Reconfiguration

The methodology of reconfiguring a network by simplification has been widely used in the natural sciences, where complex networks are often addressed. The methodology of Serrano et al. considers the number of connections between two actors as the weight of the connection and interprets each normalized weight as a random variable [16]. Connections with weights that deviate from the uniform distribution are excluded. In this case, the excluded connections vary by significance level. As the significance level decreases, the number of excluded connections increases.

We compared the reconfigured networks by altering the significance level to 1%, 5% and 10%. However, it was difficult to derive meaningful conclusions and implications since too few connections remained after the reconfiguration. Therefore, we adjusted the significance level to 20% and a network composed of the significant links at this level was derived and analyzed. Figure 23-1 presents the reconfigured network, which is composed of 84 links with a total weight of 1,858 (reduced from the original weight of 14,932) by reconfiguration. A link repeated more than twice is displayed as bold, meaning that the two countries tied by a bold link are more firmly connected through common membership in more international energy organizations. For visual convenience, the bolder links show the strongest connections.

In the reconfigured network, the average number of links each for country is 6.4, meaning each country connects with 6.4 countries on average. In terms of the centralization index, the value increased from 0.06 to 0.32, which means that the network became more centralized after being reconfigured. The density index decreased sharply from the previous value of 0.94 to 0.13 as only significant links remained.

The reconfigured network includes 12 countries in the center, including France, Germany, the United Kingdom, Indonesia, Brazil, the United States, Canada, Australia, Japan, Korea, China, and India, while the other 37 countries are in the periphery. Among them, Angola, Kuwait, Bolivia, Egypt, Trinidad and Tobago, Belarus, Ukraine, Uzbekistan, Myanmar, and Turkmenistan have no significant connection with other countries. The overall composition of the network shows that countries form two major groups. Saudi Arabia, Libya, Algeria, the United Arab Emirates, Venezuela, Iraq, Iran, Qatar, and Nigeria form a group of exporters, while the other countries form a larger group of both exporters and importers. In addition, the two groups are linked by Oman and Kazakhstan, which are connected to each group through Russia, Nigeria and Azerbaijan.

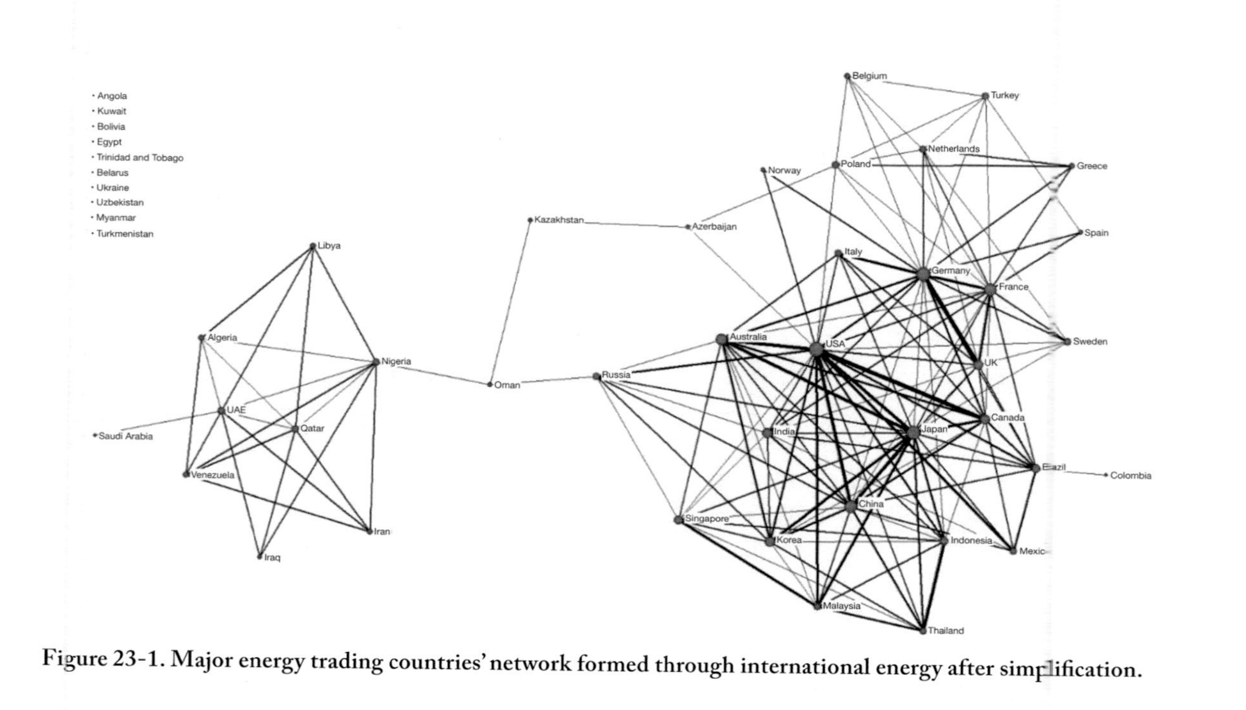

Figure 23-1. Major energy trading countries' network formed through international energy after simplification.

Comparison Analysis

This study compares network characteristics of three groups of trade type. While the initial sample is composed of two groups (i.e., exporting and importing countries), ten countries are included in both groups. These countries were classified as balanced countries, namely Australia, Canada, Germany, Indonesia, Mexico, the Netherlands, Russia, the United Arab Emirates, the United Kingdom and the U.S. The importing country group of 17 countries includes Belarus, Belgium, Brazil, China, France, Greece, India, Italy, Japan, Poland, Korea, Singapore, Spain, Sweden, Thailand, Turkey and Ukraine. The exporting country group consists of 22 countries including Angola, Algeria, Azerbaijan, Bolivia, Colombia, Egypt, Iran, Iraq, Kazakhstan, Kuwait, Libya, Malaysia, Myanmar, Nigeria, Norway, Oman, Qatar, Saudi Arabia, Trinidad and Tobago, Turkmenistan, Uzbekistan and Venezuela.

To compare these three groups, we focus on two widely used characteristics in social network analysis. First, the degree index is derived by normalizing the number of other countries with which a focal country connects. The higher index is associated with the greater number of countries with which the focal country directly cooperates. This enables the country to cooperate closely with more countries. The next characteristic is about the influence of a focal country within the network. Bonacich's power centrality [17] was used to measure the influence, as shown in Equation 23-2:

$$c(\alpha,\beta) = \alpha \sum_{k=0}^{\infty} \beta^k R^{k+1} 1, \tag{23-2}$$

In this equation, α is a scaling factor, β is a weighting factor, R is a matrix of relationships, and 1 is a column vector of 1s. In the matrix, all main diagonal elements are set to 0. And each element rij and rji in the matrix R takes the value of 1 if a tie occurs or 0 otherwise. Additionally, the designation of β follows the example of previous research, which sets it equal to three-quarters of the reciprocal of the largest eigenvalue [18]. According to this measure, a country's power is a positive function of the number of its links and the power of other countries with which the focal country forms links.

Table 23-3 shows degree and power centrality indices of countries belonging to the three groups. As the sample numbers of the three groups are 18, 22 and 9, which appear to not follow a normal distribution, a nonparametric test is used for comparing the two indices. Moreover, instead of a simultaneous comparison among the three groups, each pair was compared by using Wilcoxon's rank sum test three times [19,20].

Analysis Results

The comparison analysis found that the exporting country group showed a significant difference from the other two groups. Between the importing and exporting groups, the former had significantly higher values in terms of both degree and power centrality ($p < 0.001$). With regard to the comparison between the balanced and exporting country groups, the exporting country group had significantly lower degree and power centrality ($p < 0.001$) as the first comparison. Categorized as a balanced country, the U.S. had both the highest degree (0.438) and power centrality (2.628) of

Table 23-3. Network characteristics of countries: degree and power centrality.

Import Country	*Degree*	*Power Centrality*
Belarus	0	0
Belgium	0.104	0.282
Brazil	0.208	1.210
China	0.333	2.070
France	0.354	1.536
Greece	0.083	0.303
India	0.250	1.380
Italy	0.146	1.095
Japan	0.375	2.425
Poland	0.167	0.472
Korea	0.229	1.493
Singapore	0.208	1.203
Spain	0.063	0.259
Sweden	0.125	0.652
Thailand	0.146	1.079
Turkey	0.125	0.304
Ukraine	0	0

Balanced Country	*Degree*	*Power Centrality*
Australia	0.292	2.018
Canada	0.250	1.858
Germany	0.396	1.864
Indonesia	0.208	1.383
Mexico	0.125	0.969
Netherlands	0.167	0.461
Russia	0.167	0.982
UAE	0.167	0.007
UK	0.250	1.552
USA	0.438	2.628

Export Country	*Degree*	*Power Centrality*
Algeria	0.104	0.005
Angola	0	0
Azerbaijan	0.063	0.185
Bolivia	0	0
Colombia	0.021	0.072
Egypt	0	0
Iran	0.083	0.004
Iraq	0.063	0.003
Kazakhstan	0.042	0.015
Kuwait	0	0
Libya	0.083	0.004
Malaysia	0.167	1.260
Myanmar	0	0
Nigeria	0.167	0.010
Norway	0.042	0.335
Oman	0.063	0.061
Qatar	0.146	0.006
Saudi Arabia	0.021	0.001
Trinidad and Tobago	0	0
Turkmenistan	0	0
Uzbekistan	0	0
Venezuela	0.104	0.005

all countries considered. Comparisons between the importing and balanced country groups did not show any significant difference in either indices. Table 23-4 presents the analysis results. Both the importing and balanced countries occupy more central positions with more partners in the network than the exporting countries.

CONCLUSIONS

This study classified 49 major oil and natural gas trading countries into exporting, importing, and balanced country groups. It then compared two network characteristics of each group within a network formed through participation in 28 international energy organizations. Both importing and

Table 23-4a. Wilcoxon's rank sum test: degree comparison between import and export groups.

Degree	*Observation*	*Actual rank sum*	*Expected rank sum*
Import country group	17	460.5	340
Export country group	22	319.5	440
Combined group	39	780	780

Unadjusted variance	1,246.67
Adjusted variance	1,222.06
z	3.447
Prob > |z|	0.0006

Table 23-4b. Wilcoxon's rank sum test: power centrality comparison between balanced and export groups.

Degree	*Observation*	*Actual rank sum*	*Expected rank sum*
Balanced country group	10	269	165
Export country group	22	259	363
Combined group	32	528	528

Unadjusted variance	605
Adjusted variance	592.47
z	4.273
Prob > |z|	0.0000

Table 23-4c. Wilcoxon's rank sum test: degree comparison between import and balanced groups.

Degree	*Observation*	*Actual rank sum*	*Expected rank sum*
Balanced country group	10	175.5	140
Import country group	17	202.5	238
Combined group	27	378	378
Unadjusted variance	396.67		
Adjusted variance	393.76		
z	1.789		
Prob > |z|	0.0736		

Table 23-4d. Wilcoxon's rank sum test: power centrality comparison between import and export groups.

Degree	*Observation*	*Actual rank sum*	*Expected rank sum*
Import country group	17	477	340
Export country group	22	303	440
Combined group	39	780	780
Unadjusted variance	1246.67		
Adjusted variance	1225.59		
z	3.913		
Prob > |z|	0.0001		

Table 23-4e: Wilcoxon's rank sum test: power centrality comparison between balanced and export groups.

Degree	*Observation*	*Actual rank sum*	*Expected rank sum*
Balanced country group	10	265	165
Export country group	22	263	363
Combined group	32	528	528
Unadjusted variance	605		
Adjusted variance	595.46		
z	4.098		
Prob > |z|	0.0000		

Table 23-4f. Wilcoxon's rank sum test: power centrality comparison between import and balanced groups.

Degree	*Observation*	*Actual rank sum*	*Expected rank sum*
Balanced country group	10	169	140
Import country group	17	209	238
Combined group	27	378	378
Unadjusted variance	396.67		
Adjusted variance	396.55		
z	1.456		
Prob > |z|	0.1453		

balanced countries showed higher values for degree and power centrality indices than the exporting countries. However, there was no significant difference between the importing and balanced countries. The result indicates that countries having considerable imports occupy more central positions with more partners, which increases their influences within the network. In other words, countries that are active in oil and natural gas imports, regardless of their exports, are more active participants in international energy organizations than countries focusing only on exports. Therefore, oil and natural gas imports seem to be a more critical factor in determining international energy organization participation than exports. This conclusion is reinforced by the case of wealthy countries, such as Kuwait, Norway, and Qatar, included in the exporting country group. This implies that the differences are not simply due to national wealth.

One of the reasons that led to this difference might be the efforts of importing countries to strengthen their energy security. In particular, importing countries suffered from the oil shocks in 1970s and their experiences reinforced the importance of energy security. Their efforts to reduce this risk through collective actions included creating organizations such as the International Energy Agency [21]. Their collective efforts extended to more active international energy cooperation and memberships in international energy organizations. The higher participation of importing countries continued until recently, as the oil and gas markets have become more supplier-oriented. With shale oil development and weaker growth in energy demand, the difference in participation between exporting countries and others is expected to disappear.

As responses to climate change increase worldwide, the bargaining power of fossil fuel-exporting countries will decline. The more their markets become consumer-oriented, the more effort exporting countries will make not to lose their current market shares. Exporting countries who in the past focused on bilateral oil and natural gas trading are likely to join multilateral cooperation mechanisms, such as international energy organizations. Oil and natural gas exporting countries tend to participate in international energy organizations due to the reduction of their bargaining power and their greater involvement in climate change responses. Through various international energy organizations, oil and gas exporting countries are attempting to either affect the international energy cooperation agenda for their own interests or utilize policy references from other countries.

Countries that have been passive in international energy cooperation are realizing the need to increase their participation in international energy organizations. With greater concerns regarding climate change, interest in renewable energy and energy efficiency is increasing. International organizations such as the International Renewable Energy Agency and the International Partnership for Energy Efficiency Cooperation are expanding their operations. With increasing energy market uncertainty and risks, international cooperation among governments will increase along with private cooperation. All these changes will be reflected in the international energy cooperation network of countries. The network investigated in this study is expected to have denser linkages in the future.

References

[1] Mehrara, M. (2007). Energy consumption and economic growth: the case of oil exporting countries. *Energy Policy*, 35(5), pages 2,939-2,945.

[2] Mohammadi, H. and S. Parvaresh (2014). Energy consumption and output: evidence from a panel of 14 oil-exporting countries. *Energy Economics*, 41, pages 41-46.

[3] Omgba, L. (2014). Institutional foundations of export diversification patterns in oil-producing countries. *Journal of Comparative Economics*, 42(4), pages 1,052-1,064.

[4] Gupta, E. (2008). Oil vulnerability index of oil-importing countries. *Energy Policy*, 36(3), pages 1,195-1,211.

[5] Wu, G., Liu, L. and Wei, Y. (2009). Comparison of China's oil import risk: results based on portfolio theory and a diversification index approach. *Energy Policy*, 37(9), pages 3,557-3,565.

[6] Filis, G., Degiannakis, S. and C. Floros (2011). Dynamic correlation between stock market and oil prices: the case of oil-importing and oil-exporting countries. *International Review of Financial Analysis*, 20(3), pages 152-164.

[7] Guesmi, K. and S. Fattoum. (2014). Return and volatility transmission between oil prices and oil-exporting and oil-importing countries. *Economic Modelling*, 38, pages 305-310.

[8] Wright, T. (2015). A special argument: the U.S., U.K., and the AIIB, Brooking Institution.

https://www.brookings.edu/blog/order-from-chaos/2015/03/13/a-special-argument-the-u-s-u-k-and-the-aiib, accessed 7 February 2017.
[9] Wasserman, S. and K. Faust (1994). Social network analysis: methods and applications, volume 8. Cambridge University Press.
[10] Hafner-Burton, E. and A. Montgomery (2006). Power positions in international organizations, social networks and conflict. *Journal of Conflict Resolution*, 50(1), pages 3-27.
[11] Skjelsbaek, K. (1972). Peace and the structure of the international organization network. *Journal of Peace Research*, 9(4), pages 315-330.
[12] Beckfield (2003). Inequality in the world polity: the structure of international organization. *American Sociological Review*, 68(3), pages 401-424.
[13] Ingram, P., Robinson, J. and M. Busch (2005). The intergovernmental network of world trade: IGO connectedness, governance and embeddedness. *American Journal of Sociology*, 111(3), pages 824-858.
[14] G20 (2014). Survey of G20 countries: gaps and duplication in the existing mandates and work plans of international energy organizations.
[15] Freeman, L. (1979). Centrality in social networks conceptual clarification. *Social Networks*, 1(3), pages 215-239.
[16] Serrano, M., Boguna, M. and A. Vespignani (2009), Extracting the multiscale backbone of complex weighted networks. *PNAS*, 106(16), pages 6,483-6,488.
[17] Bonacich, P. (1987). Power and centrality: a family of measures. *American Journal of Sociology*, 92, pages 1,170-1,182.
[18] Podolny, J. (1993). A status-based model of market competition. *American Journal of Sociology*, 98(4), pages 829-872.
[19] Peto, R. and J. Peto. (1972). Asymptotically efficient rank invariant test procedures. *Journal of the Royal Statistical Society*. Series A (General), pages 185-207.
[20] Powell, T. and I. Reinhardt. (2010). Rank friction: an ordinal approach to persistent profitability. *Strategic Management Journal*, 31(11), pages 1,244-1,255.
[21] Miller, R. (2011). Future oil supply: the changing stance of the International Energy Agency. *Energy Policy*, 39(3), pages 1,569-1,574.

Index

E